NUMERICAL CALCULATION OF LUBRICATION

NUMERICAL CALCULATION OF LUBRICATION

METHODS AND PROGRAMS

Ping Huang
South China University of Technology, Guangzhou, China

WILEY

TSINGHUA UNIVERSITY PRESS

This edition first published 2013

Published by John Wiley & Sons Singapore Pte. Ltd., 1 Fusionopolis Walk, #07-01 Solaris South Tower, Singapore 138628, under exclusive license by Tsinghua University Press in all media throughout the world excluding Mainland China and excluding Simplified and Traditional Chinese languages.

For details of our global editorial offices, for customer services and for information about how to apply for permission to reuse the copyright material in this book please see our website at www.wiley.com.

Library of Congress Cataloging-in-Publication Data

Huang, Ping, 1957-
Numerical calculation of lubrication : methods and programs / Huang Ping.
pages cm
Includes bibliographical references and index.
ISBN 978-1-118-45119-9 (cloth)
1. Lubrication and lubricants—Mathematical models. I. Title.
TJ1077.H92 2013
621.8'9—dc23

2013014001

Set in 11/13 pt Times by Thomson Digital, Noida, India

1/2013

Contents

Preface

Lubrication calculation is the most successful area of tribology, using mathematical methods to obtain numerical solutions. Due to the development of computer science in the recent half a century, it has made remarkable achievements.

However, most books on tribology mainly introduce theories or calculation methods of lubrication. They rarely give and discuss numerical calculation programs. Not only is this inconvenient for research or the production of practical lubrication, but also many similar programming tasks have to be carried out repeatedly. Furthermore, because of the limitations of lubrication theory, some numerical calculation programs may give mistaken solutions, leading to wrong conclusions.

This book is different from previous theoretical books or monographs on lubrication; it mainly introduces the numerical calculation programs of lubrication. This is the main feature of the book. Some of the programs have been used for many years in lubrication calculations and research by the author.

Due to the complexity of lubrication problems, the book mainly focuses on how to numerically solve the Reynolds equation, energy equation, elastic deformation equation and their combinations. The analyzed lubrication problems include line, surface and point contacts, which correspond to thrust bearings, journal bearings and rolling contact bearings. Furthermore, the working conditions include incompressible, compressible, nonthermal, thermal, isoviscosity, variable viscosity, Newtonian fluid, non-Newtonian fluid (only grease), rigid and elastic deformation situations.

The book is divided into four parts, covering 24 chapters:

The first part (Chapters 1–9) is about the solution of the Reynolds equation, which is the basic technique for the numerical analysis of lubrication. The contents include the boundary and connection conditions of the Reynolds equation, discretizing the Reynolds equation, numerical methods and programs of slider lubrication, numerical methods and programs of journal bearing lubrication, numerical methods and programs of dynamic bearing lubrication, numerical methods and programs of gas lubrication (especial magnetic hard disk/head) and so on. In this part, we also discuss the rheology of lubricants. As an example, grease lubrication is discussed, which is a kind of non-Newtonian fluid.

The second part of the book (Chapters 10 and 11) is on temperature calculation. First, we give a discrete form of the energy equation and the temperature–viscosity equation. Then, combining the Reynolds equation and the energy equation, we give numerical methods and programs of thermal hydrodynamic lubrication.

Elastohydrodynamic lubrication (EHL) is a difficult topic in lubrication calculation, because of its poor convergence. In the third part (Chapters 12–20), calculations of elastic deformation equations for line and point contacts are given first. Then, combined with the Reynolds equation, the pressure–viscosity equation and the elastic deformation equation, calculation programs of EHL are introduced in detail. Furthermore, combined with the energy equation, numerical methods and programs of thermal EHL are given. We also give numerical methods and programs of EHL and thermal EHL for grease in this part, and we consider the rheological effect.

Finally, in the last part of the book (Chapters 21–(24), we introduce some programs developed for practical lubrication design. These programs include a lubrication calculation package and its optimized design package for the herringbone groove bearing of a micro motor and a calculation program and balancing attitude program of ultra thin gas lubrication for magnetic hard disk/head design. Because these packages and programs have some special requirements, pre-treatment and post-treatment have been added for easy usage in engineering. Although the basic theories of these contents are introduced at the front of the book, more details about the function and usages of the packages and programs can be found on the Wiley Companion Website: www.wiley.com/go/huang/lubrication.

The reason why we provide all source codes and an attached source code disc for all the programs is that most users need not repeat programming tasks even if they have well mastered the principles of lubrication. Especially, those who are not familiar with lubrication analysis can directly use the programs to carry out lubrication calculation. If some users have enough lubrication knowledge, they can use the programs or need only rewrite the pre-assignment or data sentences to input the different parameters to solve their own lubrication problems more easily. This will bring great convenience for researchers and technical staff in this field.

The book is mainly aimed at teachers, post-graduate students and doctoral students at colleges and universities. It can also be used as a reference book for technical personnel and research staff in engineering.

I would like to thank all of my post-graduate students who participated in the program writing and debugging and the book writing. Among them, I thank Li Ping for Chapters 2 and 3, Sun Zhonghua for Chapters 5 and 6, Niu Rongjun for Chapter 8, Wang Qiliang, Glenn and Liu Ping for Chapters 10 and 11, Wang Yazhen for Chapters 12–14, Yu Mei for Chapters 15 and 19, Lai Tianmao for Chapters 16 and 20, Yao Huaping for Chapters 21 and 22, and Wang Hongzhi for Chapters 23 and 24.

Ping Huang
South China University of Technology
31 August, 2012

Part One

Numerical Method for Reynolds Equation

1

Reynolds Equation and its Discrete Form

1.1 General Reynolds Equation and Its Boundary Conditions

1.1.1 Reynolds Equation

The general form of the Reynolds equation is

$$\frac{\partial}{\partial x}\left(\frac{\rho h^3}{\eta}\cdot\frac{\partial p}{\partial x}\right)+\frac{\partial}{\partial y}\left(\frac{\rho h^3}{\eta}\cdot\frac{\partial p}{\partial y}\right)=6\left[\frac{\partial}{\partial x}(U\rho h)+\frac{\partial}{\partial y}(V\rho h)+2\frac{\partial \rho h}{\partial t}\right] \quad (1.1)$$

where $U = U_0 – U_h$; $V = V_0 – V_h$. If we assume that the fluid density does not change with time, we have $\frac{\partial \rho h}{\partial t} = \rho(w_h - w_0)$.

1.1.2 Definite Condition

The definite conditions of the Reynolds equation usually include the boundary conditions, the initial conditions and the connection conditions.

1.1.2.1 Boundary Condition

In order to solve the Reynolds equation, the pressure boundary conditions should be used to determine the integration constants. There are commonly two forms of pressure boundary conditions, namely

Coercive boundary condition $p|_s = 0$

Natural boundary condition $\left.\frac{\partial p}{\partial n}\right|_s = 0$

Numerical Calculation of Lubrication: Methods and Programs, First Edition. Ping Huang.

where s is the boundary of the solution domain; and n is the normal direction of the border.

Usually, the inlet and outlet pressure boundaries for an oil film can be easily determined according to its geometry and the situation of the oil supply. However, such as a journal bearing which has both a convergence clearance and a divergence clearance, the position of the outlet cannot be determined in advance. Therefore, it can be assumed that both pressure and pressure derivative are equal to zero at the same time to determine the location of the outlet. Such a boundary condition is known as the Reynolds boundary condition, which is in this form

$$p|_s = 0 \quad \text{and} \quad \left.\frac{\partial p}{\partial n}\right|_s = 0$$

Here are two examples of boundary conditions.

One-dimensional boundary conditions in the region of $0 \leq x \leq l$

If the boundaries are known, we have $p|_{x=0} = 0$ and $p|_{x=l} = 0$.

If the outlet is unknown, we have $p|_{x=0} = 0$, $p|_{x=x'} = 0$ and $\left.\frac{\partial p}{\partial x}\right|_{x=x'} = 0$, where x' is the outlet boundary to be determined.

Two-dimensional boundary conditions in the rectangular area of $(0 \leq x \leq l, -b/2 \leq y \leq b/2)$

If the boundaries are known, we have $p|_{x=0} = 0$, $p|_{x=l} = 0$ and $p|_{y=\pm b/2} = 0$.

If the outlet is unknown, we have $p|_{x=0} = 0$, $p|_{x=x'} = 0$, $\left.\frac{\partial p}{\partial x}\right|_{x=x'} = 0$ and $p|_{y=\pm b/2} = 0$.

1.1.2.2 Initial Condition

For the nonsteady-state lubrication problem in which the velocity and/or the load change with time, such as the fluid hydrodynamic lubrication of a crankshaft bearing in the internal combustion engine, the Reynolds equation contains the squeeze term at the right end of Equation 1.1. The lubrication film thickness changes with time, so we need to give some initial conditions for solving the Reynolds equation. The general forms of the initial condition are as follows.

Initial film thickness: $h|_{t=0} = h_0(x, y)$
Initial pressure: $p|_{t=0} = p_0(x, y)$

If the lubricant viscosity and density also vary with time, their initial conditions should also be given.

1.1.2.3 Connection Condition

If the film thickness varies abruptly in several parts, like a ladder, the lubrication region also needs to divide into several subregions to solve the problem because the film thickness derivatives at the right end of Equation 1.1 do not exist at the abruptly changing positions. Therefore, the connection conditions should be given. The commonly used connection conditions are the continuity conditions of pressure and flow. If a film thickness changes abruptly at x', its connection conditions will be as follows.

Continuous pressure condition: $p|_{x=x'-0} = p|_{x=x'+0}$
Continuous flow condition: $Q|_{x=x'-0} = Q|_{x=x'+0}$

1.1.3 Computation of Lubrication Performances

After obtaining the pressure distribution from the Reynolds equation, we can calculate the static performances of lubrication, including the load carrying capacity, the friction and the flow.

1.1.3.1 Load Carrying Capacity w

The load carrying capacity of the lubricating film can be obtained by integrating the pressure $p(x,y)$ in the entire lubrication domain, that is

$$w = \iint p dx dy \tag{1.2}$$

1.1.3.2 Frictional Force f

The frictional forces of the lubricating film on a solid surface can be obtained by integrating the shear stress over the whole boundary of the lubricating film. The fluid shear stress is equal to

$$\tau = \eta \frac{\partial u}{\partial z} = \frac{1}{2} \cdot \frac{\partial p}{\partial x}(2z - h) + (U_h - U_0)\frac{\eta}{h} \tag{1.3}$$

For surfaces of $z = 0$ and $z = h$, to integrate the shear stress on both surfaces we have

$$f_0 = \iint \tau|_{z=0} dx dy \tag{1.4}$$

$$f_h = \iint \tau|_{z=h} dx dy$$

where f_0 and f_h are the frictional forces respectively on surfaces $z = 0$ and $z = h$.

After the frictional forces have been obtained, we then can determine the friction coefficient $\mu = f/w$ as well as the frictional power loss and the heat due to the friction.

1.1.3.3 Lubricant Flow Q

The side leaking flows of lubricant can be obtained by integrating the flow rates through the lubricating film boundary.

$$\begin{aligned} Q_x &= \int q_x dy \\ Q_y &= \int q_y dx \end{aligned} \tag{1.5}$$

By summing all leaking flows over all boundaries we can obtain the total flow, which gives us the amount of lubricant needed to fill the clearance. At the same time, the total leaking flow will influence the extent of convection so that we can calculate the balanced thermal temperature according to leaking flow and friction power loss.

1.2 Reynolds Equations for Some Special Working Conditions

In Section 1.1, we have given the general form of the Reynolds equation. However, for some specific engineering problems, the general Reynolds equation can be simplified, which may make solving much easier. In the following, some forms of the Reynolds equation for different conditions are given.

1.2.1 Slider and Thrust Bearing

A wedge slider is the simplest problem of lubrication design. If the geometry of the slider is not very complicated, we can obtain an analytical solution. In addition, through analysis of the slider problem, it will not only help us to understand the basic characteristics of lubrication, but will also be useful for the thrust bearing lubrication design.

Because the side leakage of lubricant need not be considered for solving an infinitely wide slider, its Reynolds equation then can be simplified into a one-dimensional ordinary differential equation:

$$\frac{d}{dx}\left(h^3 \frac{dp}{dx}\right) = 6U\eta \frac{dh}{dx} \tag{1.6}$$

The common two-pressure boundary conditions of a slider are as follows.

$p|_{x=0} = 0$; $p|_{x=x'} = 0$ (x' is the outlet boundary, $x' = b$; and b is the slider width).

$p|_{x=0} = 0$; $p|_{x=x'} = 0$ and $\left.\dfrac{\partial p}{\partial x}\right|_{x=x'} = 0$ (x' is the outlet boundary to be determined, $x' \leq b$).

If the film thickness or its derivative is discontinuous, we should divide the lubrication region into two parts at the discontinuous line so that the number of the integral constants will correspondingly increase. Therefore, the connection conditions must be used at the discontinuous line. If the discontinuous line is at x^*, the connection conditions will be:

$$\text{Pressure continuous condition } p|_{x=x^*-0} = p|_{x=x^*+0} \tag{1.7}$$

Flow continuous condition

$$\left[-\frac{h^3}{12\eta}\frac{\partial p}{\partial x} + (U_1 + U_2)\frac{h}{2}\right]_{x=x^*-0} = \left[-\frac{h^3}{12\eta}\frac{\partial p}{\partial x} + (U_1 + U_2)\frac{h}{2}\right]_{x=x^*+0} \tag{1.8}$$

1.2.2 *Journal Bearing*

By spreading the journal bearing along the circumferential direction, we can transform x into $R\theta$ so that the general form of the Reynolds equation is

$$\frac{\partial}{R^2\partial\theta}\left(\frac{\rho h^3}{\eta}\cdot\frac{\partial p}{\partial\theta}\right) + \frac{\partial}{\partial y}\left(\frac{\rho h^3}{\eta}\cdot\frac{\partial p}{\partial y}\right) = 6\left[\frac{\partial}{R\partial\theta}(U\rho h) + \frac{\partial}{\partial y}(V\rho h) + 2\frac{\partial\rho h}{\partial t}\right]. \tag{1.9}$$

The corresponding shape of the clearance can be expressed as:

$$h = e\cos\theta + c = c(1 + \varepsilon\cos\theta) \tag{1.10}$$

where e is the eccentricity, c is the clearance of the radii of the bearing and the journal, $\varepsilon = e/c$ is the eccentricity ratio and θ is the circumferential coordinate starting from the maximum film thickness position.

1.2.2.1 Infinitely Narrow Bearing

If the axial width of a bearing along the y direction is much less than the circumferential length along the x direction, we have $\frac{\partial p}{\partial y} \gg \frac{\partial p}{R\partial\theta}$ so that we can set $\frac{\partial p}{R\partial\theta} = 0$. Because the film thickness h is only related to θ, but independent of y, the Reynolds equation becomes

$$\frac{d}{dy}\left(h^3\cdot\frac{dp}{dy}\right) = 6U\eta\frac{dh}{Rd\theta} \tag{1.11}$$

The above Reynolds equation has only side boundary conditions. They are $p=0$ at $y=\pm\frac{b}{2}$ and $\frac{dp}{dy}=0$ at $y=0$ due to symmetry.

1.2.2.2 Infinitely Wide Bearing

We can approximately choose $\frac{dp}{dy}=0$ for an infinitely wide bearing because the side leakage can be ignored. Therefore, the Reynolds equation changes into an ordinary differential equation.

$$\frac{d}{Rd\theta}\left(h^3\frac{dp}{Rd\theta}\right)=6U\eta\frac{dh}{Rd\theta} \tag{1.12}$$

Its boundary conditions usually are $p|_{\theta=0}=0$, $p|_{\theta=\theta_2}=0$ and $\left.\frac{\partial p}{\partial\theta}\right|_{\theta=\theta_2}=0$ (where θ_2 is the outlet boundary to be determined, $\theta_2\leq 2\pi$).

1.2.3 Hydrostatic Lubrication

The oil film for hydrostatic lubrication is formed by a fluid forced in under pressure from the outside. Therefore, even if two lubricating surfaces have no relative motion, a thick enough lubricating film can be achieved. The advantages of hydrostatic lubrication are: (1) its load carrying capacity and the oil film thickness have no relationship with the sliding velocity; (2) the film stiffness is so strong that it has a very high accuracy; (3) its friction coefficient is so low that we can ignore the influence of the static friction. The main disadvantages of hydrostatic lubrication are: its structure is complex and a pressure oil supply system must be required which often affects the working life and reliability of hydrostatic lubrication.

Substituting the condition of no relative sliding velocity into the Reynolds Equation 1.1, we have the Reynolds equation for hydrostatic lubrication as follows

$$\frac{\partial}{\partial x}\left(\frac{\rho h^3}{\eta}\frac{\partial p}{\partial x}\right)+\frac{\partial}{\partial y}\left(\frac{\rho h^3}{\eta}\frac{\partial p}{\partial y}\right)=0 \tag{1.13}$$

For a rectangular region, the outer pressure boundary conditions are usually $p|_{x=0}=0$; $p|_{x=l}=0$; $p|_{y=\pm b/2}=0$; and the boundary pressure condition in the oil chamber is: $p=p_s$, where p_s is the supplied oil pressure.

For a journal hydrostatic bearing, Reynolds Equation 1.13, the film thickness equation and the boundary conditions can be solved easily in the form of cylindrical coordinates. For solving the above equations, we can determine the variation relationship between the load and the film thickness. Furthermore, if we consider the working conditions, such as equal film thickness, incompressibility or isoviscosity, the Reynolds Equation 1.13 can be further simplified.

1.2.4 Squeeze Bearing

The relative sliding between two bearing surfaces is usually assumed to be zero when analyzing squeeze lubrication, so that the Reynolds Equation 1.1 can be written as follows

$$\frac{\partial}{\partial x}\left(\rho h^3 \frac{\partial p}{\partial x}\right) + \frac{\partial}{\partial y}\left(\rho h^3 \frac{\partial p}{\partial y}\right) = 12\eta \frac{\partial(\rho h)}{\partial t} \tag{1.14}$$

Usually, for a rectangular region, the boundary conditions are $p|_{x=0} = 0; p|_{x=l} = 0;$ and $p|_{y=\pm b/2} = 0$. To solve the above equation we can determine the variation relationship between the load and the film thickness.

1.2.5 Dynamic Bearing

Most actual bearings withstand a varying load whose direction, rotational speed or other parameters change with time. Such bearings are collectively referred to as dynamic bearing or nonstable load bearing. Obviously, the axis or the thrust plate of a dynamic bearing moves along a certain trajectory. If the working parameters are periodic functions of time, the trajectory of the axis is a complex and closed curve.

The working principles of the dynamic bearing can be divided into two types. The first is where the journal does not rotate around its central axial, that is, there is no relative sliding. Therefore, the journal axial moves along a certain trajectory under the load. In this case, the journal and the bearing surfaces move mainly in the direction of the film thickness so that the film pressure is generated by the squeeze effect. The other type is where the journal rotates around its own center and the journal center also moves. Therefore, the film pressure originates from the squeeze effect of these two movements, that is, the journal rotation and the axis movement.

The general Reynolds equation for incompressible and dynamic lubrication is the basic equation to analyze dynamic bearings. It can be written as follows

$$\frac{\partial}{\partial x}\left(\frac{h^3}{\eta}\frac{\partial p}{\partial x}\right) + \frac{\partial}{\partial y}\left(\frac{h^3}{\eta}\frac{\partial p}{\partial y}\right) = 6U\frac{\partial h}{\partial x} + 12W \tag{1.15}$$

In Equation 1.15 the first term on the right is the hydrodynamic effect; the second term represents the squeeze effect; and when the Reynolds Equation 1.15 is applied to a stable bearing, the term of the squeeze effect can be omitted, that is,

$$W = w_h - w_0 = \frac{\partial h}{\partial t} = 0.$$

The problem of calculating the axis trajectory of a dynamic bearing by Equation 1.15 belongs to an initial value problem. The stepping method is usually used to determine the axis of the trajectory according to the given initial position of the axis.

1.2.6 Gas Bearing

The main feature of gas lubrication is that a gas is compressible. Therefore, the density of the gas must be treated as a variable, that is, by using the Reynolds equation for a variable density

$$\frac{\partial}{\partial x}\left(\frac{\rho h^3}{\eta}\frac{\partial p}{\partial x}\right)+\frac{\partial}{\partial y}\left(\frac{\rho h^3}{\eta}\frac{\partial p}{\partial y}\right)=6\left[U\frac{\partial(\rho h)}{\partial x}+2\frac{\partial(\rho h)}{\partial t}\right] \tag{1.16}$$

Because the gas density varies with temperature and pressure, the ideal gas equation can be expressed as follows

$$\frac{p}{\rho}=RT \tag{1.17}$$

where T is the absolute temperature, and R is the gas constant which does not change for a certain gas.

For a usual gas lubrication problem, gas lubrication can be regarded as an isothermal process and this assumption has an error less than a few percent. For such a problem, Equation 1.15 becomes

$$p=k\rho \tag{1.18}$$

where k is a proportional constant.

In addition, if a gas lubrication process is so fast that the heat cannot be conducted in time, the process can be thought to be adiabatic. The gas state equation of the adiabatic process is as follows

$$p=k\rho^n \tag{1.19}$$

where n is the gas specific heat ratio relate to the atomic number of the gas molecules. For air, $n=1.4$.

For an isothermal process, the Reynolds equation becomes

$$\frac{\partial}{\partial x}\left(h^3p\frac{\partial p}{\partial x}\right)+\frac{\partial}{\partial y}\left(h^3p\frac{\partial p}{\partial y}\right)=6\eta\left[U\frac{\partial}{\partial x}(ph)+2\frac{\partial}{\partial t}(ph)\right] \tag{1.20}$$

Equation 1.20 is the basic equation for solving gas lubrication problems.

1.3 Finite Difference Method of Reynolds Equation

If the boundary conditions are given for solving a differential equation, this is known as a boundary value problem. In hydrodynamic lubrication calculations, the finite difference method is commonly used to numerically solve the Reynolds equation. The major steps of finite difference method are as follows.

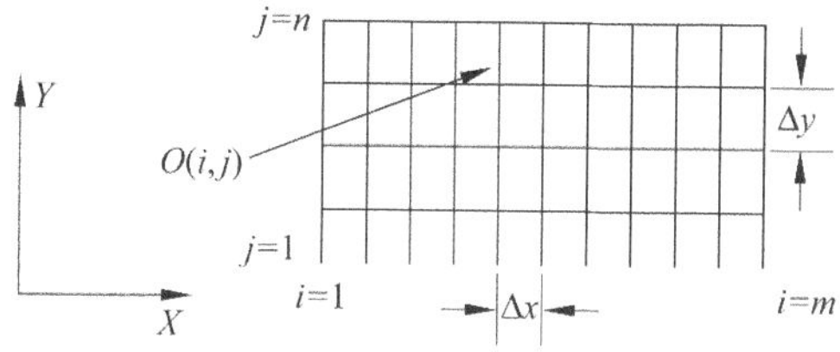

Figure 1.1 Uniform mesh.

1.3.1 Discretization of Equation

First, change the partial differential equations into dimensionless forms. This is accomplished by expressing the variables in a universal form.

Then divide the solution region into a mesh with uniform or nonuniform grids. Figure 1.1 shows a uniform mesh, with nodes m in the x direction and with nodes n in the y direction so that the total nodes are equal to $m \times n$. The division of a mesh is determined by calculation accuracy. For a common hydrodynamic lubrication problem, using $m = 12$–25 and $n = 8$–10 will usually meet the requirement of accuracy. Sometimes, in order to improve the accuracy, where the unknown variables have a rapid change in the region, the grid needs refining by using two or more different subgrids.

Take the pressure p as an example. The distribution of p in the whole region can be expressed by each node p_{ij}. According to the differential regularities, the partial derivatives at the node $O(i, j)$ can be represented by the surrounding node variables.

As shown in Figure 1.2, the expression of the partial derivatives of the intermediate difference at the node $O(i, j)$ have the following forms

$$\begin{aligned}\left(\frac{\partial p}{\partial x}\right)_{i,j} &= \frac{p_{i+1,j} - p_{i-1,j}}{2\Delta x}\\ \left(\frac{\partial p}{\partial y}\right)_{i,j} &= \frac{p_{i,j+1} - p_{i,j-1}}{2\Delta y}\end{aligned} \tag{1.21}$$

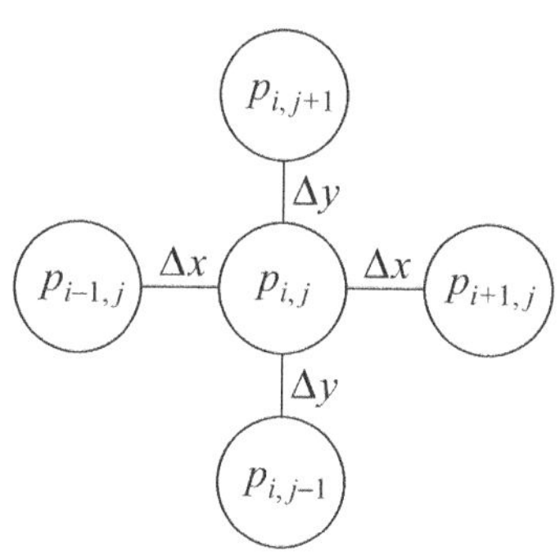

Figure 1.2 Relationship of difference.

The second-order partial derivatives of the intermediate difference are as follows

$$\begin{aligned}\left(\frac{\partial^2 p}{\partial x^2}\right)_{i,j} &= \frac{p_{i+1,j} + p_{i-1,j} - 2p_{i,j}}{(\Delta x)^2}\\ \left(\frac{\partial^2 p}{\partial y^2}\right)_{i,j} &= \frac{p_{i,j+1} + p_{i,j-1} - 2p_{i,j}}{(\Delta y)^2}\end{aligned} \tag{1.22}$$

In order to obtain the unknown variables near the border, forward or backward difference formulas are used as follows

$$\begin{aligned}\left(\frac{\partial p}{\partial x}\right)_{i,j} &= \frac{p_{i+1,j} - p_{i,j}}{\Delta x}\\ \left(\frac{\partial p}{\partial y}\right)_{i,j} &= \frac{p_{i,j+1} - p_{i,j}}{\Delta y}\end{aligned} \tag{1.23}$$

$$\begin{aligned}\left(\frac{\partial p}{\partial x}\right)_{i,j} &= \frac{p_{i,j} - p_{i-1,j}}{\Delta x}\\ \left(\frac{\partial p}{\partial y}\right)_{i,j} &= \frac{p_{i,j} - p_{i,j-1}}{\Delta y}\end{aligned} \tag{1.24}$$

Usually, the accuracy of the intermediate difference is high. The following intermediate difference formulas can also be used in calculation

$$\left(\frac{\partial p}{\partial x}\right)_{i,j} = \frac{p_{i+1/2,j} - p_{i-1/2,j}}{\Delta x} \tag{1.25}$$

1.3.2 Difference Form of Reynolds Equation

According to the above formulas, the two-dimensional Reynolds equation can be written in a standard form of the second-order partial differential equation

$$A\frac{\partial^2 p}{\partial x^2} + B\frac{\partial^2 p}{\partial y^2} + C\frac{\partial p}{\partial x} + D\frac{\partial p}{\partial y} = E \tag{1.26}$$

where A, B, C, D and E are known parameters.

Equation 1.26 can be applied to each node. According to Equations 1.21 and 1.22, the relationship of pressure $p_{i,j}$ at node $O(i,j)$ with the adjacent pressures can be written as follows

$$\tilde{p}^k_{i,j} = C_N p^k_{i,j+1} + C_S p^{k+1}_{i,j-1} + C_E p^k_{i+1,j} + C_W p^{k+1}_{i-1,j} + G \tag{1.27}$$

where, p with superscript k is the original pressure, and with superscript $k+1$ is the iterated one; and

$$C_N = \left[\frac{B}{\Delta y^2} + \frac{D}{2\Delta y}\right] \Big/ K; C_S = \left[\frac{B}{\Delta y^2} - \frac{D}{2\Delta y}\right] \Big/ K; C_E = \left[\frac{A}{\Delta x^2} + \frac{C}{2\Delta x}\right] \Big/ K;$$

$$C_W = \left[\frac{A}{\Delta x^2} - \frac{C}{2\Delta x}\right] \Big/ K; G = -\frac{E}{K}; K = 2\left[\frac{A}{\Delta x^2} + \frac{B}{\Delta y^2}\right].$$

By using Equation 1.27, we can write out the calculation equations of the finite difference method at each node. And, in the border nodes, variables should satisfy the boundary conditions because their values are known. In this way, a set of linear algebraic equations can be obtained. The number of equations obtained is equal to the number of unknown variables, so it is a definite problem to be solved. By using an elimination method or iterative method, we can solve these algebraic equations. If the convergent result meets with the given precision, the solution of each node has been found.

The following describes how to solve hydrodynamic lubrication problems with the finite difference method.

1.3.3 Iteration of Differential Equation

Equation 1.27 is the finite difference calculation equation. For each node, we can write an equation in the domain. On the boundary, the node variables should satisfy the boundary conditions so that their values are known. In this way, a set of linear algebraic equations can be obtained whose number is the same as the number of unknown variables so that the set can be solved by using the elimination method or the iteration method. When the accuracy meets the convergence requirement, the calculation ultimately stops and the variables for each node in the entire solution domain can be obtained.

During iteration, in order to guarantee convergence the relaxation or super relaxation iteration method is often used, that is, weighted to add the old pressure to the iterated pressure as a new pressure, as in Equation 1.28

$$p_{ij}^{k+1} = (1-\alpha)p_{ij}^k + \alpha\tilde{p}_{ij}^k \tag{1.28}$$

where p_{ij}^k is the old pressure, p_{ij}^{k+1} is the new pressure, $\tilde{p}_{ij}^k$ is the iterated pressure by Equation 1.27 and α is a positive number, larger than 0 and usually less than 1.

Then, we use the new pressure to check whether convergent. If not, we carry out the next round of iteration.

1.3.4 Iteration Convergence Condition

Using the differential Equation 1.27, the next iterative pressure is obtained to combine the new and old pressures of a node; and the correction of all nodes in the lubrication area will be completed only when a whole iteration is finished.

Equation 1.27 is usually not satisfied. However, through iteration, the pressure often converges to the true solution.

There are two general methods to be used to make sure whether or not the iterative process meets the accuracy requirement. They are the absolute and relative accuracy criteria.

1.3.4.1 Absolute Accuracy Criterion

When a new iteration finishes, Equation 1.27 can be rewritten as follows

$$r_{ij}^{k+1} = p_{i,j}^{k+1} - C_N p_{i,j+1}^{k+1} - C_S p_{i,j-1}^{k+1} - C_E p_{i+1,j}^{k+1} - C_W p_{i-1,j}^{k+1} - G \tag{1.29}$$

where r_{ij}^{k+1} is the residual of the differential equation at the node i, j.

According to the accuracy criterion, the residual of all nodes should be less than a very small positive number ε_1, namely

$$\left| r_{ij}^{k+1} \right| \le \varepsilon_1 \tag{1.30}$$

Sometimes, in order to facilitate the use of two iterations the pressure difference can be used to determine the convergence, that is

$$\left| p_{ij}^{k+1} - p_{ij}^{k} \right| \le \varepsilon_2 \tag{1.31}$$

where ε_2 is a very small positive number.

1.3.4.2 Relative Accuracy Criterion

For most lubrication problems, because the residuals of the equation may vary, to accurately give the absolutely accuracies ε_1 or ε_2 is quite difficult. Therefore, relative accuracy criteria are often used in practice.

There are two commonly used criteria for relative accuracy. The stricter one is to have the pressure on each node met the relative accuracy, that is

$$\left| \frac{p_{i,j}^{k+1} - p_{i,j}^{k}}{p_{i,j}^{k+1}} \right| \le \varepsilon_3 \tag{1.32}$$

The looser criterion of relative accuracy is to have the relative accuracy of the two iteration loads met the following requirement

$$\frac{\sum \sum \left| p_{i,j}^{k+1} - p_{i,j}^{k} \right|}{\sum \sum p_{i,j}^{k+1}} \le \varepsilon_4 \tag{1.33}$$

In the above two equations, ε_3 and ε_4 are the relative accuracies. According to the convergence difficulty of the problem, they are often chosen between $0.01 \sim 10^{-6}$, that is, the relative precision of convergence is from 1% to one per million.

Next, a hydrostatic lubricated bearing is taken as an example to show how a hydrodynamic lubrication problem is solved with finite difference method.

In stable operating conditions, the oil film thickness h of hydrostatic lubrication is a constant. If we do not consider relative sliding and thermal effects, the viscosity η is also a constant. Then, its Reynolds equation can be simplified to the Laplace equation

$$\nabla^2 p = \frac{\partial^2 p}{\partial x^2} + \frac{\partial^2 p}{\partial y^2} = 0 \tag{1.34}$$

Set the dimensionless parameters $X = x/l$, $Y = /b$ and $P = p/p_r$, where l is the length of the bearing, b is the width and p_r is the pressure of the oil chamfer; and $\alpha = l^2/b^2$. Then, the dimensionless Reynolds equation will be

$$\frac{\partial^2 P}{\partial X^2} + \alpha \frac{\partial^2 P}{\partial Y^2} = 0 \tag{1.35}$$

The boundary conditions of Equation 1.35 are

In the oil chamber $P = 1$.
At the outside borders of the bearing $P = 0$.

Substituting Equation 1.22 into Equation 1.35, we have

$$\frac{P_{i+1,j} + P_{i-1,j} - 2P_{i,j}}{\Delta X^2} + \alpha \frac{P_{i,j+1} + P_{i,j-1} - 2P_{i,j}}{\Delta Y^2} = 0 \tag{1.36}$$

or

$$P_{i,j} = \frac{P_{i+1,j} + P_{i-1,j} + \alpha(\Delta X/\Delta Y)^2 P_{i,j+1} + P_{i,j-1}}{2[1 + \alpha(\Delta X/\Delta Y)^2]} \tag{1.37}$$

Substituting the boundary conditions into Equation 1.36 or 1.37, the numerical solution of the oil film pressure distribution can be easily obtained with an elimination or iteration method.

2

Numerical Method and Program for Incompressible and Steady Lubrication of One-dimensional Slider

2.1 Basic Equations

In all lubrication calculations, one-dimensional slider lubrication is the simplest. The analytical solution can be obtained if the geometry of the slider is not very complicated. In addition, analysis of the slider not only helps us to know the basic characteristics of the slider lubrication, but it also is the basis for lubrication calculation of the thrust bearing. A typical one-dimensional slider lubrication model is shown in Figure 2.1.

If the width of the slider is much wider, it can be considered to be infinite and the side leakage can be neglected. Then, the Reynolds equation can be simplified into a one-dimensional ordinary differential equation. When the film thickness equation is known, the general pressure solution can often be obtained. Combined with the boundary conditions and continuity conditions, we can get the pressure distribution analytically. The lubrication characteristics such as load, frictional force and flow can be further calculated by using the obtained pressure.

2.1.1 Reynolds Equation

The Reynolds equation to solve the one-dimensional slider lubrication problems is

$$\frac{d}{dx}\left(\frac{h^3}{\eta}\frac{dp}{dx}\right) = 6U\frac{dh}{dx} \tag{2.1}$$

Numerical Calculation of Lubrication: Methods and Programs, First Edition. Ping Huang.

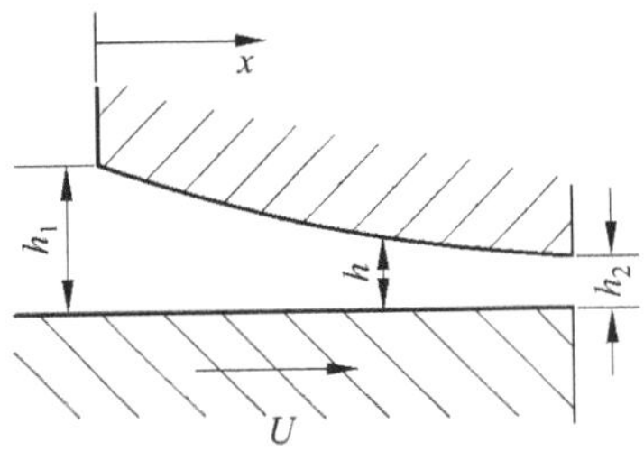

Figure 2.1 Model of one-dimensional slider lubrication.

2.1.2 Boundary Conditions

In order to solve the Reynolds equation, two integral constants must be determined based on two pressure boundary conditions. Generally the pressure boundary conditions are

$$\begin{aligned} p|_{x=0} &= 0 \\ p|_{x=l} &= 0 \end{aligned} \tag{2.2}$$

2.1.3 Continuity Equation

If the thickness function or its derivative is not continuous, the connecting condition should be used at the discontinuous point.

2.1.3.1 Pressure Continuity Condition

$$p|_{x=x'-0} = p|_{x=x'+0} \tag{2.3}$$

where x' is the coordinate of the discontinuous point.

2.1.3.2 Fluid Continuity Condition

$$\left[-\frac{h^3}{12\eta}\frac{\partial p}{\partial x}+\frac{Uh}{2}\right]_{x=x'-0} = \left[-\frac{h^3}{12\eta}\frac{\partial p}{\partial x}+\frac{Uh}{2}\right]_{x=x'+0} \tag{2.4}$$

2.2 Numerical Method for Incompressible and Steady Lubrication of One-dimensional Slider

In addition to linear sliders, other types of one-dimensional sliders are curve sliders, composite sliders and stepped sliders. Their characteristics mainly depend on the film thickness h.

2.2.1 Discrete Reynolds Equation

As for the incompressible and steady lubrication of a slider, if its density and viscosity are thought to be constants, the dimensionless Reynolds equation is as follows

$$\frac{d}{dX}\left(H^3\frac{dP}{dX}\right)=\frac{dH}{dX}. \tag{2.5}$$

where X is the dimensionless coordinate, $X=x/l$; l is the length of the slider; P is the dimensionless pressure, $P=p\frac{h_2^2}{6\eta Ul}$; and H is the dimensionless film thickness, $H=h/h_2$.

The dimensionless boundary conditions are

$$\begin{aligned} P|_{X=0}&=0\\ P|_{X=1}&=0 \end{aligned} \tag{2.6}$$

By using the equidistance differential formula, the discrete Reynolds Equation 2.5 can be written as follows

$$\frac{H_{i+1/2}^3P_{i+1}-(H_{i+1/2}^3+H_{i-1/2}^3)P_i+H_{i-1/2}^3P_{i-1}}{(\Delta X)^2}=-\frac{H_{i+1}-H_{i-1}}{2\Delta X} \tag{2.7}$$

where $H_{i+1/2}=\frac{H_{i+1}+H_i}{2}$ and $H_{i-1/2}=\frac{H_i+H_{i-1}}{2}$.

or

$$P_i=\frac{\Delta X(H_{i+1}-H_{i-1})/2+H_{i+1/2}^3P_{i+1}+H_{i-1/2}^3P_{i-1}}{H_{i+1/2}^3+H_{i-1/2}^3} \tag{2.8}$$

The discrete boundary conditions become $P_1=0$ and $P_N=0$.

After obtaining the pressure of all nodes, we can use the numerical method to calculate the lubrication properties, such as load, center of pressure, frictional force and flow. The specific formulas are given as follows.

2.2.1.1 Load

The load per length is

$$w=\int_0^l pdx=\sum_{i=1}^{N}p_i\Delta x \tag{2.9}$$

2.2.1.2 Pressure Center

The pressure center can be obtained by calculating the moment to the origin. The distance from the pressure center to the origin x_0 can be given as follows

$$x_0 = \frac{1}{w}\int_0^l pxdx = \frac{\sum_{i=1}^{N} p_i x_i \Delta x}{\sum_{i=1}^{N} p_i \Delta x} \tag{2.10}$$

2.2.1.3 Frictional Force

The shear stresses on the up and down surfaces are

$$\tau|_{z=h,0} = \eta\frac{\partial u}{\partial z} = \pm\frac{\partial p}{\partial x}\frac{h}{2} + \frac{\eta}{h}U = \pm\frac{p_{i+1} - p_i}{\Delta x}\frac{h_i}{2} + \eta U/h_i \tag{2.11}$$

Therefore, the frictional force is

$$f_{h,0} = \int_0^l \tau_{h,0}dx = \int_0^l \left(\pm\frac{\partial p}{\partial x}\frac{h}{2} + \eta\frac{U}{h}\right)dx = \sum_{i=1}^{N}\left(\pm\frac{p_{i+1} - p_i}{\Delta x}\frac{h_i}{2} + \eta U/h_i\right)\Delta x \tag{2.12}$$

where τ_h, f_h and τ_0, f_0 are the shear stresses and frictional forces on the up and down surfaces $z = h$ and $z = 0$, respectively.

2.2.1.4 Flux

Because there is no side leakage for a one-dimensional slider, the flux is

$$Q_x = \int_0^b q_x dy = \int_0^b \left(-\frac{h^3}{12\eta}\cdot\frac{dp}{dx} + \frac{Uh}{2}\right)dy \approx \left(-\frac{h_i^3}{12\eta}\cdot\frac{p_{i+1} - p_i}{\Delta x} + \frac{Uh_i}{2}\right)b \tag{2.13}$$

where b is the width of the slider.

2.3 Calculation Program for Incompressible and Steady Lubrication of One-dimensional Slider

2.3.1 Introduction

The calculation program for the incompressible and steady lubrication of a one-dimensional slider consists of a main program and three subprograms, which are the output subprogram OUTPUT, the film thickness calculation subprogram SUBH and the pressure computation subprogram SUBP.

The functions of the main program are:

2.3.1.1 Pre-assignment Parameters

Node number N = 121, speed U = 1.0, dimensionless start coordinate X1 = 0.0, dimensionless end coordinate X2 = 1.0, dimensionless maximum film thickness H1 = 1.0, dimensionless minimum film thickness H2 = 0.5, lubricant viscosity EDA = 0.02 Pa s and slider length AL = 0.01 m.

Pre-assignment parameters can be modified according to the user's working conditions, but the program needs to be recompiled and relinked before executing.

2.3.1.2 Input Parameters

The two given examples in the program are the linear slider and the curve slider, corresponding to the value of the control parameter KG. When KG = 1 (as well as KG $\neq$ 2), the program gives linear slider calculation results and the corresponding analytical solution; and when KG = 2, it gives the results for a symmetric parabolic slider. Because there is a divergence zone, the pressure in this region is set at zero.

2.3.1.3 Output Parameters

After the calculation is finished, the program will output node number N, load ALOAD and load center X0.

Dimensionless coordinates X(I), dimensionless film thickness H(I) and dimensionless pressure P(I) are given into the output file SLIDER.DAT by columns.

If the slider is a linear slider, another group of data given in the end column is dimensionless analytic pressure P0(I).

2.3.2 Calculation Diagram

The calculation diagram for the incompressible and steady lubrication of a one-dimensional slider is shown in Figure 2.2.

2.3.3 Calculation Program

```
PROGRAM SLIDER
DIMENSION X(121),H(121),P(121)
   COMMON /COM1/X1,X2,H1,H2,U,EDA,AL,ALOAD,DX
   DATA N,U,X1,X2,H1,H2,EDA,AL/ 121,1.0,0.0,1.0,1.0,0.5,0.02,0.01/
   OPEN(7,FILE='SLIDER.DAT',STATUS='UNKNOWN')
   WRITE(*,*)'If KG=1: Straight slider; KG=2: Curve slider; Input
   KG='
```

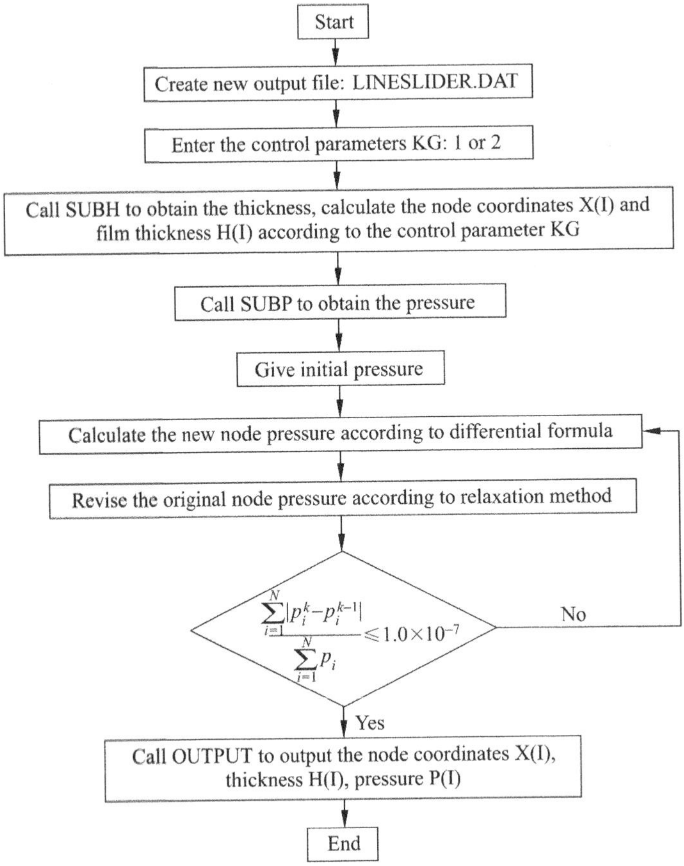

Figure 2.2 Calculation diagram for incompressible and steady lubrication of a one-dimensional slider.

```
READ(*,*)KG
IF(KG.EQ.2)THEN
X1=-1.0
X2=1.0
ELSE
KG=1
ENDIF
CALL SUBH(KG,N,X,H)
CALL SUBP(N,X,H,P)
CALL OUTPUT(KG,N,X,H,P)
STOP
END
SUBROUTINE OUTPUT(KG,N,X,H,P)
DIMENSION X(N),H(N),P(N)
```

```
      COMMON /COM1/X1,X2,H1,H2,U,EDA,AL,ALOAD,DX
      X0=0.0
      DO I=1,N
      X0=X0+P(I)*X(I)
      ENDDO
      X0=X0*AL
      ALOAD=ALOAD*DX*AL*6.0*U*EDA*AL/H2**2
      WRITE(*,*)N,ALOAD,X0
      DO I=1,N
      IF(KG.EQ.1)THEN
      P0=-(-1.0/(H(I)*H2)+H1*H2/(H1+H2)/(H2*H(I))**2+1.0/(H1+H2))/
     (H1/H2-1.0)*H2
      WRITE(7,40) X(I),H(I),P(I),P0
      ELSE
      WRITE(7,40) X(I),H(I),P(I)
      ENDIF
      END DO
40    FORMAT(1X,4(E12.6,1X))
      RETURN
      END
      SUBROUTINE SUBH(KG,N,X,H)
      DIMENSION X(N),H(N)
      COMMON /COM1/X1,X2,H1,H2,U,EDA,AL,ALOAD,DX
      DX=1./(N-1.0)
      DO I=1,N
      IF(KG.EQ.1) THEN
      X(I)=X1-(I-1)*DX*(X1-X2)
      H(I)=H1/H2+(H1/H2-1.0)*X(I)
      ELSE
      X(I)=X1-(I-1)*DX*(X1-X2)
      H(I)=1.0+(H1/H2-1.0)*X(I)*X(I)
      ENDIF
      ENDDO
      RETURN
      END
      SUBROUTINE SUBP(N,X,H,P)
      DIMENSION X(N),H(N),P(N)
      COMMON /COM1/X1,X2,H1,H2,U,EDA,AL,ALOAD,DX
      DO I=2,N-1
      P(I)=0.5
      ENDDO
      P(1)=0.0
      P(N)=0.0
      IK=0
```

```
10 C1=0.0
   ALOAD=0.0
   DO I=2,N-1
   A1=(0.5*(H(I+1)+H(I)))**3
   A2=(0.5*(H(I)+H(I-1)))**3
   PD=P(I)
   P(I)=(-0.5*DX*(H(I+1)-H(I-1))+A1*P(I+1)+A2*P(I-1))/(A1+A2)
   P(I)=0.3*PD+0.7*P(I)
   IF(P(I).LT.0.0)P(I)=0.0
   C1=C1+ABS(P(I)-PD)
   ALOAD=ALOAD+P(I)
   ENDDO
   ERO=C1/ALOAD
   IK=IK+1
   IF(ERO.GT.1.E-7)GOTO 10
   RETURN
   END
```

2.3.4 *Calculation Results*

In Figure 2.3, dimensionless pressure distributions with corresponding film thicknesses are given according to the given conditions above for: (a) a linear slider (KG = 1) and (b) a curve slider (KG = 2).

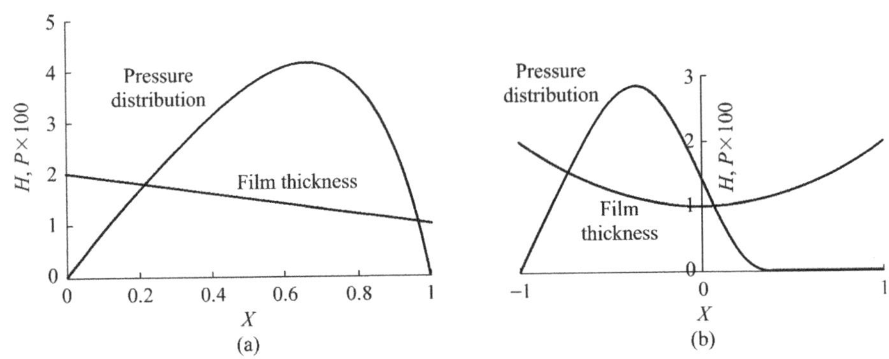

Figure 2.3 Calculation results for: (a) a linear slider, (b) a curve slider.

3

Numerical Method and Program for Incompressible and Steady Lubrication of Two-dimensional Slider

3.1 Basic Equations

The Reynolds equation for two-dimensional slider lubrication is

$$\frac{\partial}{\partial x}\left(\frac{\rho h^3}{6\eta}\frac{\partial p}{\partial x}\right)+\frac{\partial}{\partial y}\left(\frac{\rho h^3}{6\eta}\frac{\partial p}{\partial y}\right)=-U\frac{\partial(\rho h)}{\partial x} \tag{3.1}$$

If l is the slider length and b is the width, the pressure boundary conditions are

$$\begin{aligned} p|_{x=0} &= 0 \\ p|_{x=l} &= 0 \\ p|_{y=\pm b/2} &= 0 \end{aligned} \tag{3.2}$$

If we further set the density ρ and viscosity η to be constants, the dimensionless Reynolds equation becomes

$$\frac{\partial}{\partial X}\left(H^3\frac{\partial P}{\partial X}\right)+\alpha\frac{\partial}{\partial Y}\left(H^3\frac{\partial P}{\partial Y}\right)=-\frac{\partial H}{\partial X} \tag{3.3}$$

Numerical Calculation of Lubrication: Methods and Programs, First Edition. Ping Huang.

The corresponding dimensionless boundary conditions are

$$\begin{aligned} P|_{X=0} &= 0 \\ P|_{X=1} &= 0 \\ P|_{Y=\pm\frac{1}{2}} &= 0 \end{aligned} \tag{3.4}$$

3.2 Discrete Reynolds Equation

The discrete form of the Reynolds equation based on equidistance differential formulas is

$$\begin{aligned} &\frac{H^3_{i+1/2,j}P_{i+1,j} - (H^3_{i+1/2,j} + H^3_{i-1/2,j})P_{i,j} + H^3_{i-1/2,j}P_{i-1,j}}{\Delta X^2} \\ &+\alpha\frac{H^3_{i,j+1/2}P_{i,j+1} - (H^3_{i,j+1/2} + H^3_{i,j-1/2})P_{i,j} + H^3_{i,j-1/2}P_{i,j-1}}{\Delta Y^2} \\ &= -\frac{H_{i+1,j} - H_{i-1,j}}{2\Delta X} \end{aligned} \tag{3.5}$$

Or it can be rewritten as

$$P_{i,j} = \frac{\Delta X(H_{i+1,j} - H_{i-1,j})/2 + H^3_{i+1/2,j}P_{i+1,j} + H^3_{i-1/2,j}P_{i-1,j} + \alpha(\Delta X/\Delta Y)^2(H^3_{i,j+1/2}P_{i,j+1} + H^3_{i,j-1/2}P_{i,j-1})}{H^3_{i+1/2,j} + H^3_{i-1/2,j} + \alpha(\Delta X/\Delta Y)^2(H^3_{i,j+1/2} + H^3_{i,j-1/2})} \tag{3.6}$$

The boundary conditions after discretization are as follows.

In the length direction, we set the initial pressure as $P_{1,j}=0$. The terminal pressure should be determined by the conditions of $P_{N,j}=0$ or $P_{i,j}=0$ and $\left(\frac{\partial P}{\partial X}\right)_{ij}=0$. In iteration, if the terminal pressures are variational, we set $P_{i,j}$ to be equal to 0 if $P_{i,j}<0$. By this method, we can ultimately determine the outlet position. In the width direction, we set $P_{i,1}=0$ and $P_{i,M}=0$. According to the symmetrical condition, we may set $\left(\frac{\partial P}{\partial Y}\right)_{i,M/2+1}=0$. In this case, we only need to solve half of the whole region.

3.3 Calculation Program for Incompressible and Steady Lubrication of Two-dimensional Slider

3.3.1 Introduction

The calculation program for the incompressible and steady lubrication of a two-dimensional slider consists of a main program and three subprograms, which are output subprogram OUTPUT, film thickness calculation subprogram SUBH and pressure computation subprogram SUBP.

The functions of the main program are:

3.3.1.1 Pre-assignment Parameters

Node number $N \times M = 121 \times 121$, speed $U = 1.0$, dimensionless start coordinate $X1 = 0.0$, dimensionless end coordinate $X2 = 1.0$, dimensionless maximum film thickness $H1 = 0.1$, dimensionless minimum film thickness $H2 = 0.05$, lubricant viscosity $EDA = 0.02$ Pa·s, slider length $ALX = 0.03$ m and slider width $ALY = 0.024$ m.

Pre-assignment parameters can be modified according to the user's working conditions, but the program needs to be recompiled and relinked before executing.

3.3.1.2 Input Parameters

The given two example programs are the plane slider and the surface slider, corresponding to the value of the control parameter KG. When $KG = 1$ (and $KG \neq 2$), it gives the plane slider calculation results and the corresponding analytical solution; and when $KG = 2$, it gives the results of the symmetric parabolic surface slider. Because there is a divergence zone, the pressure in this region is set at zero.

3.3.1.3 Output Parameters

After calculation is finished, the dimensionless pressures P(I,J) are output into PRESSURE. DAT by columns of X(I) and by rows of Y(J); and the dimensionless film thicknesses H(I,J) are output into SLIDER.DAT by columns of X(I) and by rows of Y(J).

3.3.2 Calculation Diagram

The calculation diagram for the incompressible and steady lubrication of a two-dimensional slider is shown in Figure 3.1.

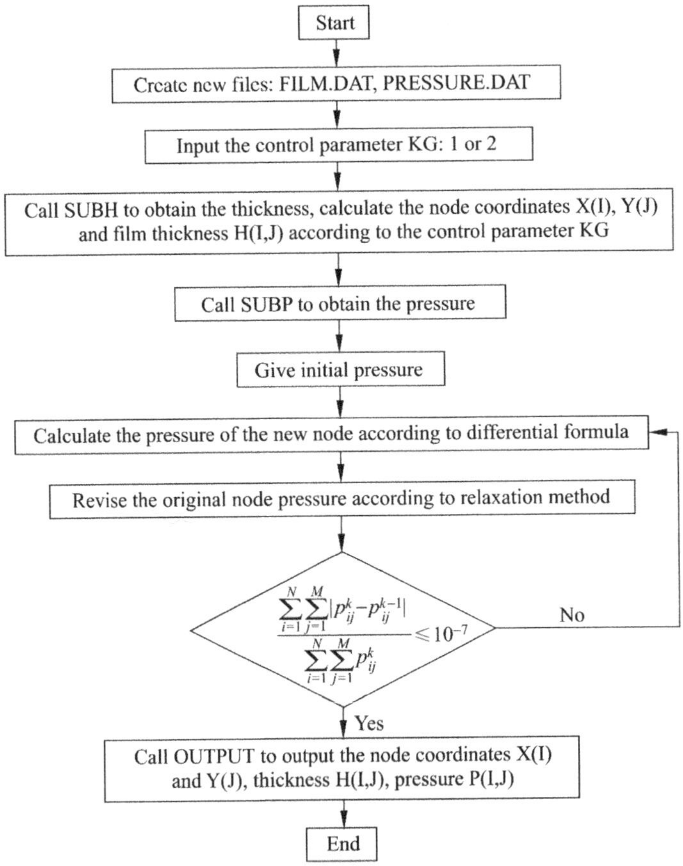

Figure 3.1 Calculation diagram for the incompressible and steady lubrication of a two-dimensional slider.

3.3.3 Calculation Program

```
      PROGRAM SURFACESLIDER
      DIMENSION X(121),Y(121),H(121,121),P(121,121)
      COMMON
/COM1/DX,DY,ALFA,X1,X2,Y1,Y2,H1,H2,ALOAD,U,EDA,ALX,ALY,ALENDA
      DATA
N,M,EDA,ALX,ALY,U,H1,H2,X1,X2,Y1,Y2/121,121,0.02,0.03,0.024,
1.0,0.1,0.05,0.0,1.0,-0.5,0.5/
      OPEN(8,FILE='PRESSURE.DAT',STATUS='UNKNOWN')
      OPEN(9,FILE='FILM.DAT',STATUS='UNKNOWN')
```

```
      WRITE(*,*)'If KG=1: Plane surface; KG=2: Curve surface; Input
KG='
      READ(*,*)KG
      CALL SUBH(KG,N,M,X,Y,H)
      CALL SUBP(N,M,X,Y,H,P)
      CALL OUTPUT(N,M,X,Y,H,P)
      STOP
      END
      SUBROUTINE SUBH(KG,N,M,X,Y,H)
      DIMENSION X(N),Y(N),H(N,M)
      COMMON
/COM1/DX,DY,ALFA,X1,X2,Y1,Y2,H1,H2,ALOAD,U,EDA,ALX,ALY,ALENDA
      IF(KG.EQ.2)THEN
      X1=-1.0
      X2=1.0
      Y1=-1.0
      Y2=1.0
      ELSE
      KG=1
      ENDIF
      DX=1.0/(N-1.0)
      DY=1.0/(M-1.0)
      ALFA=(ALX/ALY)**2
      DO I=1,N
      X(I)=X1-(I-1)*DX*(X1-X2)
      ENDDO
      DO J=1,M
      IF(KG.EQ.1)Y(J)=-0.5-(J-1)*DY*(Y1-Y2)
      IF(KG.EQ.2)Y(J)=-1.0-(J-1)*DY*(Y1-Y2)
      ENDDO
      DO I=1,N
      DO J=1,M
      IF(KG.EQ.1)H(I,J)=H1/H2-X(I)*(H1/H2-1.0)
      IF(KG.EQ.2)H(I,J)=1.0+(X(I)*X(I)+Y(J)*Y(J))*(H1/H2-1.0)
      ENDDO
      ENDDO
      RETURN
      END
      SUBROUTINE SUBP(N,M,X,Y,H,P)
      DIMENSION X(N),Y(N),H(N,M),P(N,M)
      COMMON
/COM1/DX,DY,ALFA,X1,X2,Y1,Y2,H1,H2,ALOAD,U,EDA,ALX,ALY,ALENDA
      DO I=1,N
      P(I,1)=0.0
      P(I,N)=0.0
```

```
      ENDDO
      DO J=1,M
      P(1,J)=0.0
      P(N,J)=0.0
      ENDDO
      DO I=2,N-1
      DO J=2,M-1
      P(I,J)=0.05
      ENDDO
      ENDDO
      IK=0
10    C1=0.0
      ALOAD=0.0
      DO I=2,N-1
      I1=I-1
      I2=I+1
      DO J=2,M-1
      J1=J-1
      J2=J+1
      PD=P(I,J)
      A1=(0.5*(H(I2,J)+H(I,J)))**3
      A2=(0.5*(H(I,J)+H(I1,J)))**3
      A3=ALFA*(0.5*(H(I,J2)+H(I,J)))**3
      A4=ALFA*(0.5*(H(I,J)+H(I,J1)))**3
      P(I,J)=(-0.5*DX*(H(I2,J)-
H(I1,J))+A1*P(I2,J)+A2*P(I1,J)+A3*P(I,J2)+A4*P(I,J1))/(A1+A2+A3+A4)
      P(I,J)=0.5*PD+0.5*P(I,J)
      IF(P(I,J).LT.0.0)P(I,J)=0.0
      C1=C1+ABS(P(I,J)-PD)
      ALOAD=ALOAD+P(I,J)
      ENDDO
      ENDDO
      IK=IK+1
      C1=C1/ALOAD
      WRITE(*,*)IK,C1,ALOAD
      IF(C1.GT.1.E-7)GOTO 10
      RETURN
      END
      SUBROUTINE OUTPUT(N,M,X,Y,H,P)
      DIMENSION X(N),Y(M),H(N,M),P(N,M)
      COMMON
 /COM1/DX,DY,ALFA,X1,X2,Y1,Y2,H1,H2,ALOAD,U,EDA,ALX,ALY,ALENDA
      ALENDA=6.0*U*EDA*ALX/H2**2
      ALOAD=ALOAD*ALENDA*DX*DY*ALX*ALY/(N-1.0)/(M-1.0)
      WRITE(8,40)Y(1),(Y(J),J=1,M)
```

```
      DO I=1,N
      WRITE(8,40)X(I),(P(I,J),J=1,M)
      ENDDO
      WRITE(9,40)Y(1),(Y(J),J=1,M)
      DO I=1,N
      WRITE(9,40)X(I),(H(I,J),J=1,M)
      ENDDO
40    FORMAT(122(E12.6,1X))
      STOP
      END
```

3.3.4 Calculation Results

Figures 3.2 and 3.3 give the results of dimensionless film thicknesses and dimensionless pressure distributions, according to the conditions given above for a plane slider (KG = 1) or a curve surface slider (KG = 2).

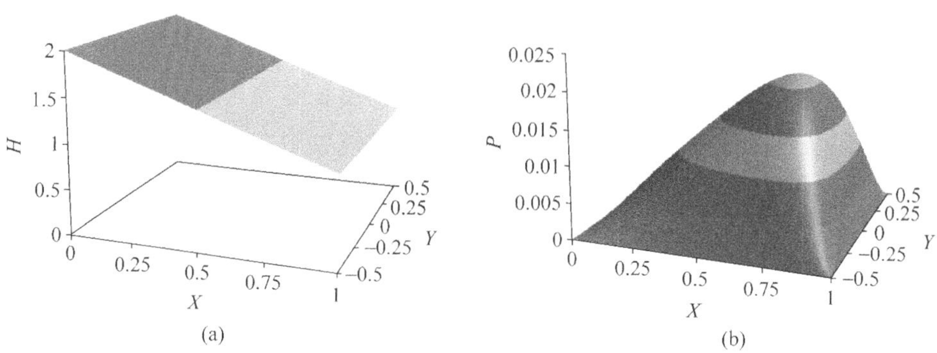

Figure 3.2 Results for a plane slider: (a) dimensionless film thickness, (b) dimensionless pressure distribution.

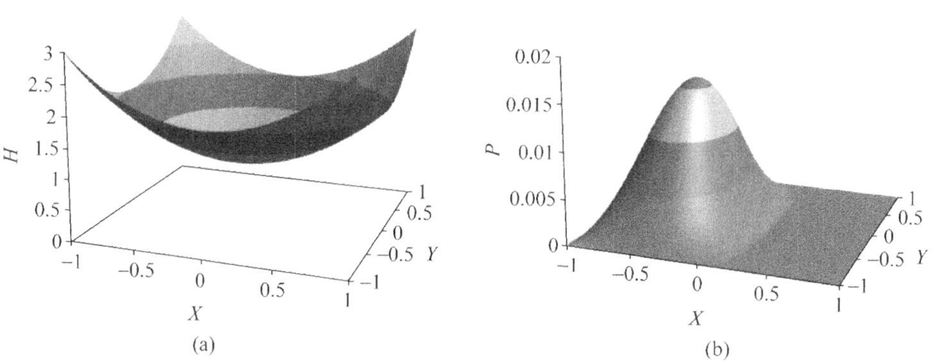

Figure 3.3 Results for a curve surface slider: (a) dimensionless film thickness. (b) dimensionless pressure distribution.

4

Numerical Method and Program for Incompressible and Steady Lubrication of Journal Bearing

4.1 Basic Equations

4.1.1 Axis Position and Clearance Shape

As the journal rotates to bring lubricant into the convergent gap to produce hydrodynamic pressure, its pressure resultant balances the load. The equilibrium position of the journal inside the bearing is as shown in Figure 4.1.

The journal center O_2 at the equilibrium position can be determined by two parameters: deviation angle ψ and eccentricity e. The deviation angle ψ is the angle between the line of load W and the connection line O_1O_2 of the bearing and the journal centers. The dimensionless eccentricity ε is often used in calculation, where $\varepsilon = e/c$. Here, c is the clearance of the radii of the bearing and the journal, that is, $c = R_1 - R_2$.

We can see from Figure 4.2 that film thickness h is a function of θ, which is an angle counted from the deviation angle ψ. In $\Delta\ O_1O_2P$, using the sine law we have

$$\frac{e}{\sin\alpha} = \frac{R_1}{\sin\theta}, \quad \text{that is,} \quad \sin\alpha = \frac{e}{R_1}\sin\theta \tag{4.1}$$

and

$$\cos\alpha = (1 - \sin^2\alpha)^{1/2} = \left(1 - \frac{e^2}{R_1^2}\sin^2\theta\right)^2 = 1 - \frac{e^2}{2R_1^2}\sin^2\theta + \cdots \tag{4.2}$$

Numerical Calculation of Lubrication: Methods and Programs, First Edition. Ping Huang.

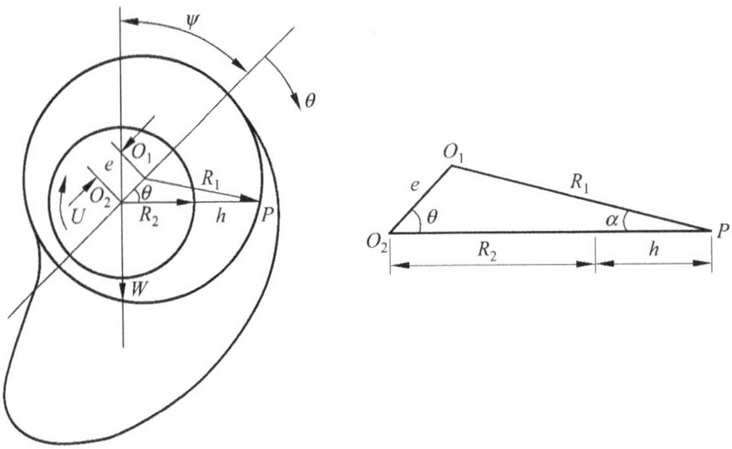

Figure 4.1 Axis position.

Because $e/R_1 << 1$, it can be omitted. Therefore, we have the geometric relationship

$$h + R_2 = e\cos\theta + R_1\cos\alpha = e\cos\theta + R_1$$

or

$$h = e\cos\theta + c = c(1 + \varepsilon\cos\theta) \tag{4.3}$$

The shape of the bearing film given by Equation 4.3 is a cosine function, which can be expanded as Figure 4.2. The expression error is only about 0.1%.

4.1.2 *Reynolds Equation*

The Reynolds equation for isoviscosity lubrication is

$$\frac{\partial}{\partial x}\left(\frac{h^3}{6\eta}\frac{\partial p}{\partial x}\right) + \frac{\partial}{\partial y}\left(\frac{h^3}{6\eta}\frac{\partial p}{\partial y}\right) = U\frac{dh}{dx} \tag{4.4}$$

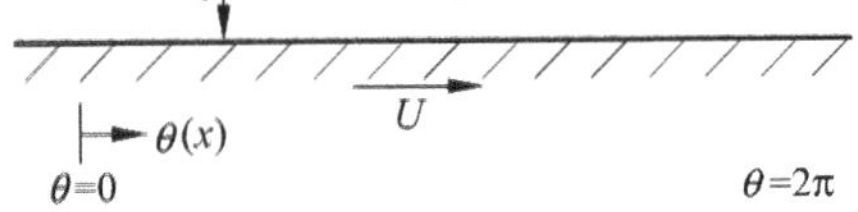

Figure 4.2 Journal bearing expand.

Film thickness h is a function of x. If we set $x = R\theta$ and $U = R\omega$, the Reynolds equation becomes

$$\frac{\partial}{R^2 \partial\theta}\left(\frac{h^3}{6\eta}\frac{\partial p}{\partial\theta}\right) + \frac{h^3}{6\eta}\frac{\partial^2 p}{\partial y^2} = \omega\frac{dh}{d\theta} \tag{4.5}$$

For a journal bearing, the ranges of the two variables y and θ in Equation 4.5 are: y from 0 to $\pm b/2$ (i.e., from the axial middle section to the edges) and θ from 0 to 2π. Therefore, the boundary conditions are as follows.

4.1.2.1 Axial Direction

At the edges $y = \pm b/2$, $p = 0$; and in the middle section $y = 0$, $\frac{\partial p}{\partial y} = 0$.

4.1.2.2 Circumferential Direction

At the film starting point where $\theta = 0$, $p = 0$. The end of the film (outlet) is in the divergent interval. According to the Reynolds boundary conditions, we have $p = 0$ and $\frac{\partial p}{\partial \theta} = 0$. Its position must be determined during the solving process so that it is a floating boundary condition. While we use the iteration method to solve the algebraic equations, we must set the node pressures to zero if the pressures on these nodes are less than zero during every iteration. Therefore, we can finally determine the film end position.

4.2 Numerical Method for Incompressible and Steady Lubrication of Journal Bearing

4.2.1 Dimensionless Reynolds Equation

The dimensionless Reynolds equation is

$$\frac{\partial}{\partial\theta}\left(H^3\frac{\partial P}{\partial\theta}\right) + \alpha H^3\frac{\partial^2 P}{\partial Y^2} = -\varepsilon\sin\theta \tag{4.6}$$

where R is the journal radius, b is the bearing length, $Y = y/b$, $\alpha = (R/b)^2$, $P = pc^2/6U\eta R$ and $H = h/c = (1 + \varepsilon\cos\theta)$.

The corresponding dimensionless boundary conditions are:

Axial direction

$$P|_{Y=\pm\frac{1}{2}} = 0 \tag{4.7}$$

Circumferential direction

$$
\begin{aligned}
&P|_{\theta=0}=0\\
&P|_{\theta=\theta_2}=0\\
&\frac{\partial P}{\partial \theta}|_{\theta=\theta_2}=0
\end{aligned}
\tag{4.8}
$$

In order to obtain the outlet of the film, an iterative method can be used. We set the pressure to zero at the node where $P<0$. Then, the first zero pressure node along the θ direction is the outlet of the oil film.

4.2.2 Discrete Form of Reynolds Equation

Substituting the equidistance discrete differential formula into Equation 4.6, we have the numerical calculation equation, as follows

$$
\frac{H^3_{i+1/2}(P_{i+1,j}-P_{i,j})-H^3_{i-1/2}(P_{i,j}-P_{i-1,j})}{\Delta\theta^2}+\alpha H^3_i\frac{(P_{i,j+1}-2P_{i,j}+P_{i,j-1})}{\Delta Y^2}=-\varepsilon\sin\theta_i
\tag{4.9}
$$

where $H_{i+1/2}=\left[1+\varepsilon\cos\left(\frac{\theta_i+\theta_{i+1}}{2}\right)\right]\approx\frac{H_{i+1}+H_i}{2}$

and

$$
H_{i-1/2}=\left[1-\varepsilon\cos\left(\frac{\theta_i+\theta_{i-1}}{2}\right)\right]\approx\frac{H_i+H_{i-1}}{2}.
$$

The discrete boundary conditions are as follows.

In the axial direction: $P_{i,1}=0$; using the symmetric condition we have $\left(\frac{\partial P}{\partial Y}\right)_{i,j=\frac{M}{2}+1}=0$ (M is the node total number in the y direction). This condition means that we only need to solve half of the area.

In the circumferential direction: initial pressure boundary conditions $P_{1,j}=0$; the end pressure boundary should also meet the conditions of $P_{i,j}=0$ and $\left(\frac{\partial P}{\partial \theta}\right)_{i,j}=0$. Because the end pressure boundary varies during iteration, we set the pressure to be zero for all nodes if $P_{i,j}<0$. Then, we can ultimately determine the end position of the lubricating film.

4.3 Calculation Program for Incompressible and Steady Lubrication of Journal Bearing

4.3.1 Calculation Diagram

The calculation program for incompressible steady-state journal bearing lubrication includes a main program and three subroutines, which are the subroutine of film thickness calculation SUBH, the subroutine of pressure calculation SUBP and the output subroutine OUTPUT.

In the main program the main function is pre-parameter assignment. In the present program, bearing width B = 0.06 m, bearing radius R = 0.025 m, clearance of radius $C0 = 5.0 \times 10^{-5}$ m, rotational speed AN = 60 000 r min^{-1}, viscosity of lubricating oil EDA = 0.05 Pa s and eccentricity EPSON = 0.7. The calculation diagram of a hydrodynamic lubrication calculation for a journal bearing is shown in Figure 4.3.

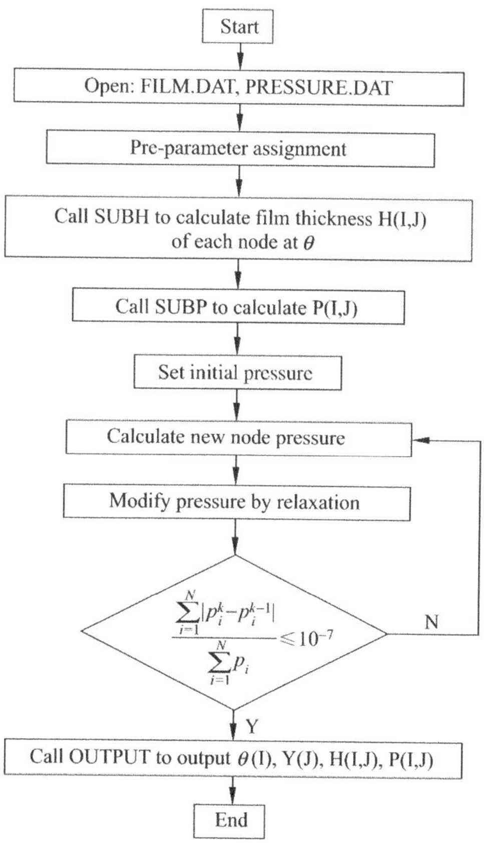

Figure 4.3 Calculation diagram of the hydrodynamic lubrication calculation of a journal bearing.

4.3.2 Calculation Program

```
PROGRAM JOURNAL
DIMENSION H(61,21),P(61,21)
DATA B,R,C0,AN,EDA,EPSON/60.0E-3,25.0E-3,5.0E-5,6.0E4,0.05,0.7/
OPEN(9,FILE='PRESSURE.DAT',STATUS='UNKNOWN')
OPEN(8,FILE='FILM.DAT',STATUS='UNKNOWN')
PI=3.1415926
N=61
M=21
DX=2.0*PI/FLOAT(N-1)
DY=1./FLOAT(M-1)
OMEGA=AN*2.0*PI/60.0
U=OMEGA*R
ALFA=(R/B*DX/DY)**2
CALL SUBH(N,M,DX,EPSON,H)
CALL SUBP(N,M,DX,EPSON,ALFA,H,P)
CALL OUTPUT(N,M,DX,DY,H,P)
STOP
END
SUBROUTINE SUBH(N,M,DX,EPSON,H)
DIMENSION H(N,M)
DO I=1,N
SETA=(I-1.0)*DX
DO J=1,M
H(I,J)=1.0+EPSON*COS(SETA)
ENDDO
ENDDO
RETURN
END
SUBROUTINE SUBP(N,M,DX,EPSON,ALFA,H,P)
DIMENSION H(N,M),P(N,M)
DO I=1,N
DO J=2,M-1
P(I,J)=0.5
ENDDO
ENDDO
DO J=1,M
P(1,J)=0.0
P(N,J)=0.0
ENDDO
DO I=1,N
P(I,1)=0.0
P(I,M)=0.0
ENDDO
```

```
   IK=0
10 C1=0.0
   ALOAD=0.0
   DO I=2,N-1
   I1=I-1
   I2=I+1
   DO J=2,M-1
   PD=P(I,J)
   J1=J-1
   J2=J+1
   A1=(0.5*(H(I2,J)+H(I,J)))**3
   A2=(0.5*(H(I,J)+H(I1,J)))**3
   A3=ALFA*(0.5*(H(I,J2)+H(I,J)))**3
   A4=ALFA*(0.5*(H(I,J)+H(I,J1)))**3
   P(I,J)=(-DX*(H(I2,J)-H(I1,J))+A1*P(I2,J)+A2*P(I1,J)+A3*P
(I,J2)+A4*P(I,J1))/(A1+A2+A3+A4)
   P(I,J)=0.7*PD+0.3*P(I,J)
   IF(P(I,J).LT.0.0)P(I,J)=0.0
   C1=C1+ABS(P(I,J)-PD)
   ALOAD=ALOAD+P(I,J)
20 CONTINUE
   ENDDO
   ENDDO
   IK=IK+1
   C1=C1/ALOAD
   WRITE(*,*)IK,C1,ALOAD
   IF(C1.GT.1.E-7)GOTO 10
   RETURN
   END
   SUBROUTINE OUTPUT(N,M,DX,DY,H,P)
   DIMENSION Y(21),H(N,M),P(N,M)
   DO J=1,M
   Y(J)=(J-1.)*DY-0.5
   ENDDO
   WRITE(8,40)Y(1),(Y(J),J=1,M)
   WRITE(9,40)Y(1),(Y(J),J=1,M)
   DO I=1,N
   AX=(I-1.0)*360.0/(N-1.0)
   WRITE(8,40)AX,(H(I,J),J=1,M)
   WRITE(9,40)AX,(P(I,J),J=1,M)
   ENDDO
40 FORMAT(22(E12.6,1X))
   STOP
   END
```

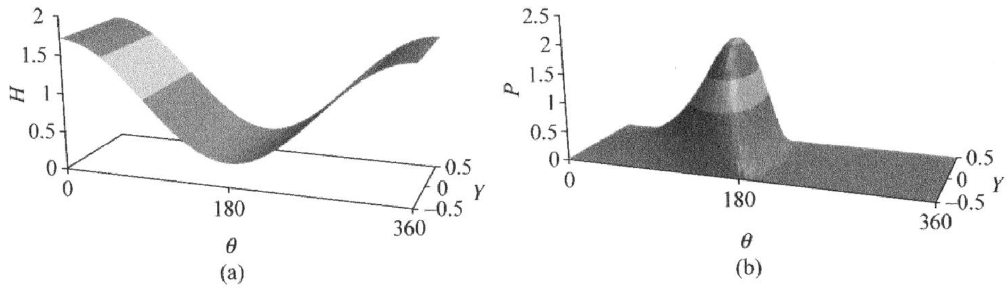

Figure 4.4 Film thickness (a) and pressure distribution (b) for the numerical calculation of a journal bearing.

4.3.3 Calculation Results

With the given program and working conditions, the film thickness and pressure distribution can be obtained. The drawings of the film thickness and pressure distribution are shown in Figure 4.4.

5

Numerical Method and Program for Incompressible Squeeze Lubrication

5.1 Basic Equation

In order to analyze squeeze lubrication, we assume that there is no relative sliding between the two bearing surfaces and the lubricant viscosity is constant. Therefore, the Reynolds equation becomes

$$\frac{\partial}{\partial x}\left(h^3\frac{\partial p}{\partial x}\right)+\frac{\partial}{\partial y}\left(h^3\frac{\partial p}{\partial y}\right)=6\eta U\frac{\partial h}{\partial x}+12W \tag{5.1}$$

where U is the tangential velocity and W is the normal velocity. The first term on the right of Equation 5.1 comes from the dynamic pressure effect and the second from the squeeze effect.

When the Reynolds Equation 5.1 is used in a steady load bearing, the squeeze effect does not exist, that is, $W=0$.

If the carrying load changes in the thickness direction, the lubricant will be squeezed. Because the lubricant between the bearing surfaces is usually not squeezed out of the bearing immediately, the film can produce pressure to balance the load. For instance, in an aero engine piston pin bearing, the squeezed pressure is larger than 35 MPa and, in a shearing machine or punch crank bearing, the pressure may reach 55 MPa.

Numerical Calculation of Lubrication: Methods and Programs, First Edition. Ping Huang.

5.2 Numerical Method and Program for Rectangular Plane Squeeze

5.2.1 Basic Equations

As shown in Figure 5.1, there are two plates, subjected to load W. The film thickness between the two plates is h, filled with some viscous lubricant.

Because there is no relative sliding between the two surfaces, the Reynolds Equation 5.1 can be written as

$$\frac{\partial}{\partial x}\left(h^3\frac{\partial p}{\partial x}\right)+\frac{\partial}{\partial y}\left(h^3\frac{\partial p}{\partial y}\right)=12\eta\frac{\partial h}{\partial t} \tag{5.2}$$

The pressure boundary conditions are $p|_{x=\pm\frac{l}{2}}=0$ and $p|_{y=\pm\frac{b}{2}}=0$.

The dimensionless form of Equation 5.2 is

$$\frac{\partial}{\partial X}\left(\frac{\partial P}{\partial X}\right)+\alpha\frac{\partial}{\partial Y}\left(\frac{\partial P}{\partial Y}\right)=\frac{1}{H^3}\frac{\partial H}{\partial t} \tag{5.3}$$

where l is the length of the rectangular plane, b is its width, $X=x/l$, $Y=y/b$, $H=h/l$, $P=p/12\eta$ and $\alpha=(l/b)^2$.

The corresponding dimensionless pressure boundary conditions are

$$P|_{X=\pm\frac{1}{2}}=0$$
$$P|_{Y=\pm\frac{1}{2}}=0$$

5.2.2 Numerical Method

The discrete form of the Reynolds Equation 5.3 can be written as

$$\frac{P_{i+1,j}+P_{i-1,j}-2P_{i,j}}{\Delta X^2}+\alpha\frac{P_{i,j+1}+P_{i,j-1}-2P_{i,j}}{\Delta Y^2}=\frac{1}{H^3}\frac{dH}{dt} \tag{5.4}$$

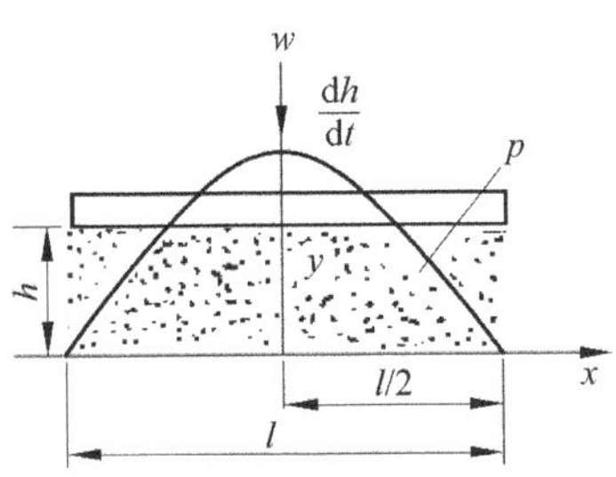

Figure 5.1 Rectangular plane squeeze film bearing.

The process solves the rectangular plane squeeze problem in the following two steps:

1. Calculate the instantaneous pressure distribution with the given film thickness and the load squeeze velocity, and then, the load bearing capacity.
2. With the given initial film thickness and the time-dependent load, calculate the film thickness at each period.

5.2.3 Calculation Diagram

The calculation diagram for rectangular plane squeeze is shown in Figure 5.2.

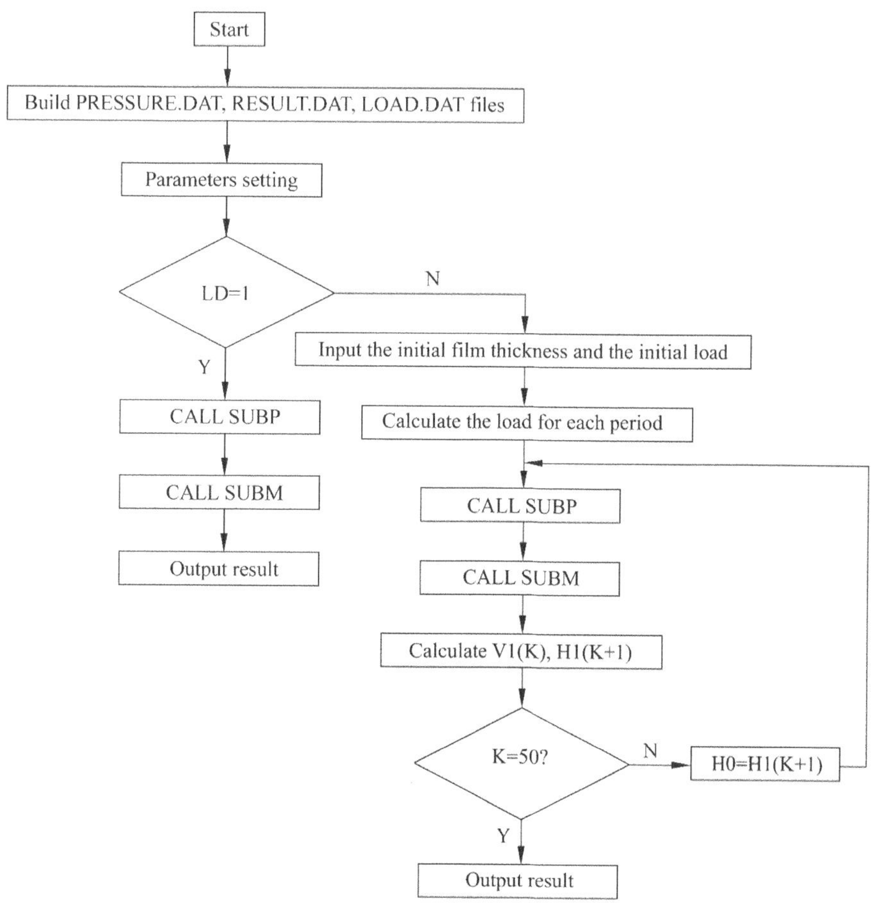

Figure 5.2 Calculation diagram for rectangular plane squeeze.

5.2.4 Calculation Program

5.2.4.1 Parameter Pre-assignment

In the present program, we set: lubricant viscosity EDA = 0.02 Pa s; length of the rectangular plane L = 30 mm; width of the rectangular plane B = 24 mm. The parameter preassignment can be modified according to the user's need, but the program needs to be recompiled and relinked before running.

5.2.4.2 Parameter Output

```
Pressure distribution  P(I,J)     in PRESSURE.DAT
Bearing capacity       SUM1       in LOAD.DAT
Load W, film thickness H, velocity of each time V  in RESULT.DAT
```

5.2.4.3 Program

```
      PROGRAM SQUARESQUEEZE
      DIMENSION X(121),Y(121),P(121,121),PD(121,121),H1(51),T(51),
V1(51),W(51)
      DATA N,N1,N2,N3/121,120,51,50/
      DATA AL,B,EDA,PI/30.0,24.0,0.02,3.14159265/
      OPEN(8,FILE='PRESSURE.DAT',STATUS='UNKNOWN')
      OPEN(9,FILE='RESULT.DAT',STATUS='UNKNOWN')
      OPEN(10,FILE='LOAD.DAT',STATUS='UNKNOWN')
      WRITE(*,*)'LD=0 squeeze process calculation; LD=1 instantaneous
squeeze pressure distribution and load calculation?'
      READ(*,*)LD
      IF(LD.EQ.1)THEN
      H0=0.641032E-02
      V=-0.194504E-01
      ELSE
      SUM0=1000.0
      H0=0.05
      ENDIF
      DX=1.0/N1
      DY=1.0/N1
      PI2=2.0*PI
      RATIO=AL/B
      ALF=RATIO**2
      V=V/AL
      H0=H0/AL
      S=AL*B/1000000.0
      DO I=1,N
      X(I)=-0.5+(I-1)*DX
      Y(I)=-0.5+(I-1)*DY
```

```
      ENDDO
      DO I=1,N
      DO J=1,N
      P(I,J)=0.0
      ENDDO
      ENDDO
      IF(LD.EQ.1)THEN
      SUM0=0.0
      GOTO 70
      ELSE
      GOTO 50
      ENDIF
   50 U=PI2/50.0
      DO I=1,N2
      W(I)=SUM0*(1.0+SIN(U*(I-1)))
      ENDDO
      H1(1)=H0
      DO K=1,N3
      V=-1.0
      V=V/AL
      DT=0.001
      CALL SUBP
      CALL SUBM
      ALENDA=12.0*EDA
      SUM1=ALENDA*S*SUM1
      V1(K)=W(K)*V/SUM1
      H1(K+1)=H1(K)+V1(K)*DT
      WRITE(*,*)SUM1,V1(K),H1(K+1)
      H0=H1(K+1)
      ENDDO
      H0=H1(N2)
      V=-1.0
      V=V/AL
      CALL SUBP
      CALL SUBM
      ALENDA=12.0*EDA
      SUM1=ALENDA*S*SUM1
      V1(N2)=W(N2)*V/SUM1
      WRITE(9,"('time T, load W, film thickness H, squeeze velocity V')")
      DO I=1,N2
      WRITE(9,40)DT*I,W(I),H1(I)*AL,V1(I)*AL
      ENDDO
      STOP
   40 FORMAT(122(E12.6,1X))
   70 CALL SUBP
```

```
    CALL SUBM
    ALENDA=12.0*EDA
    SUM1=ALENDA*S*SUM1
    WRITE(8,40)Y(1),(Y(I),I=1,N)
    DO I=1,N
    WRITE(8,40)X(I),(P(I,J),J=1,N)
    ENDDO
    WRITE(10,*)'bearing capacity    W'
    WRITE(10,*)SUM1
    CONTAINS
    SUBROUTINE SUBP
    DO I=1,N
    DO J=1,N
    PD(I,J)=P(I,J)
    ENDDO
    ENDDO
    IK=0
    TEMP=DX**2
    H=H0**3
20  C1=0.0
    DO I=2,N1
    I1=I-1
    I2=I+1
    DO J=2,N1
    P(I,J)=((P(I2,J)+P(I1,J)+ALF*P(I,J+1)+ALF*P(I,J-1)-TEMP*V/
H))/(2+2*ALF)
    C1=C1+ABS(P(I,J)-PD(I,J))
    PD(I,J)=P(I,J)
    ENDDO
    ENDDO
    IK=IK+1
    IF(C1.GT.1.E-20.AND.IK.LE.20000)GOTO 20
    RETURN
    END SUBROUTINE SUBP
    SUBROUTINE SUBM
    SUM2=0.0
    DO I=1,N
    DO J=1,N
    SUM2=SUM2+P(I,J)*DX*DY
    ENDDO
    ENDDO
    SUM1=SUM2
    RETURN
    END SUBROUTINE SUBM
    END
```

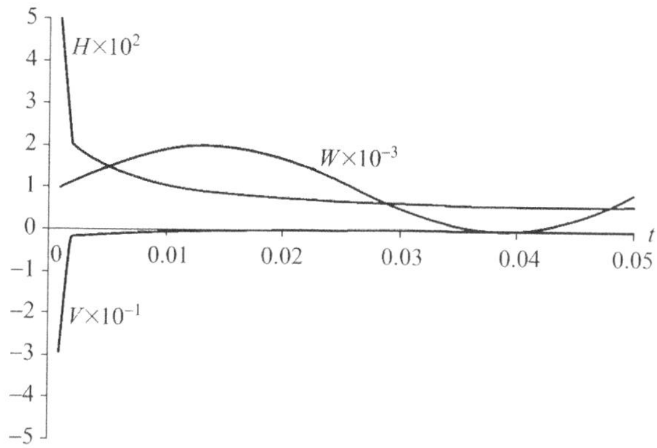

Figure 5.3 Calculation results of rectangular plane squeeze.

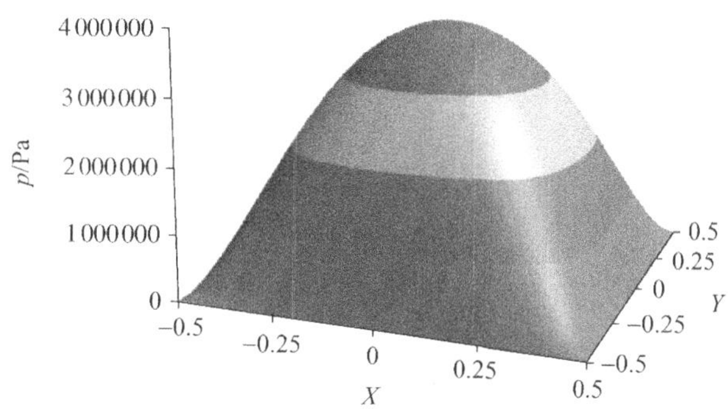

Figure 5.4 Instantaneous pressure distribution of rectangular plane squeeze.

5.2.5 Calculation Results

According to the parameters of the working conditions and the dynamic load given in the present program, the variation curve of the squeeze film with time t from 0 to 0.05 s is shown in Figure 5.3. In Figure 5.4, a pressure distribution is given for the instantaneous squeeze velocity and the corresponding film thickness.

5.3 Numerical Method and Program for Disc Squeeze

5.3.1 Basic Equations

In order to analyze disc squeeze lubrication with a radius a, the Reynolds equation in the polar form is used as follows

$$\frac{\partial}{\partial r}\left(rh^3\frac{\partial p}{\partial r}\right)+\frac{\partial}{r\partial\theta}\left(h^3\frac{\partial p}{\partial\theta}\right)=12r\eta\frac{\partial h}{\partial t} \tag{5.5}$$

If the problem is axisymmetric, that is, $\frac{\partial p}{\partial \theta} = 0$, and h is independent to r, we have

$$\frac{d}{dr}\left(r\frac{dp}{dr}\right) = \frac{12\eta r}{h^3}\frac{dh}{dt} \tag{5.6}$$

The boundary conditions of the above equation are

$$\frac{dp}{dr}|_{r=0} = 0 \text{ and } p|_{r=a} = 0$$

The dimensionless form of Equation 5.6 is

$$\frac{d}{dR}\left(R\frac{dP}{dR}\right) = \frac{R}{H^3}\frac{dH}{dt} \tag{5.7}$$

where $R = r/a$, $H = h/a$, $P = p/12\eta$ and a is the disc radius.

The corresponding dimensionless forms of the pressure boundary conditions are

$$P|_{R=1} = 0$$

$$\frac{dP}{dR}|_{R=0} = 0$$

5.3.2 Numerical Method

The discrete form of Equation 5.7 can be written as

$$\frac{P_{i+1} - P_i}{\Delta X} + X_i\frac{P_{i+1} + P_{i-1} - 2P_i}{\Delta X^2} = X_i\frac{1}{H^3}\frac{dH}{dt} \tag{5.8}$$

The corresponding discrete boundary conditions are

$$P_N = 0$$

$$P_2 - P_1 = 0$$

This process solves the disc squeeze problem in the following two steps:

1. Calculate the instantaneous pressure distribution with the given film thickness and the load squeeze velocity, and then the load bearing capacity.
2. With the given initial film thickness and the time dependent load, calculate the film thickness at each period.

5.3.3 Calculation Diagram

The calculation diagram for disc squeeze is shown in Figure 5.5.

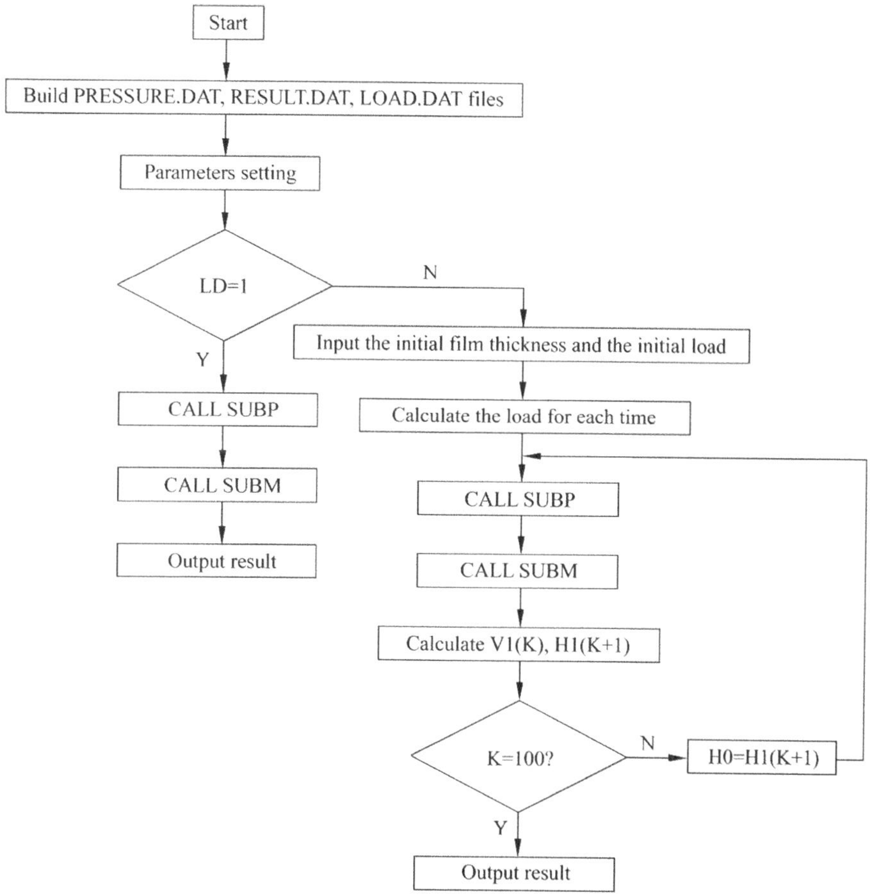

Figure 5.5 Calculation diagram for disc squeeze.

5.3.4 Calculation Program

5.3.4.1 Parameter Pre-assignment

In the program, we set lubricant viscosity EDA = 0.02 Pa s and disc radius $a = 15$ mm. The parameter preassignment can be modified according to the user's requirement, but it needs to be recompiled and relinked before running.

5.3.4.2 Parameter Output

```
Pressure distribution  P(I)   in PRESSURE.DAT
Bearing capacity       SUM1      in LOAD.DAT
Load W, film thickness H, velocity of each time V in RESULT.DAT
```

5.3.4.3 Program

```
 PROGRAM CIRCULARSQUEEZE
   DIMENSION X(121),P(121),PD(121),W(101),V1(101),T(101),H1(101)
   DATA
   N,N1,N2,N3,EDA,H0,PI,R,SUM0,V/121,120,101,100,0.02,0.05,
3.14159265,15.0,1000.0,-0.804146E-01/
   OPEN(8,FILE='RESULT.DAT',STATUS='UNKNOWN')
   OPEN(9,FILE='LOAD.DAT',STATUS='UNKNOWN')
   WRITE(*,*)'LD=0 squeeze process calculation; LD=1 instantaneous
   squeeze pressure distribution and load calculation?'
   READ(*,*)LD
   DX=1.0/N1
   PI2=2.0*PI
   S=PI*R**2/1000000.0
   V=V/R
   H0=H0/R
   DO I=1,N
   X(I)=(I-1)*DX
   ENDDO
   DO I=1,N
   P(I)=0.0
   ENDDO
   IF(LD.EQ.1)THEN
   SUM0=0.0
   GOTO 70
   ELSE
   GOTO 50
   ENDIF
50 U=PI2/100.0
   DO I=1,N2
   W(I)=SUM0*(1.0+SIN(U*(I-1)))
   ENDDO
   H1(1)=H0
   DO K=1,N3
   V=-1.0
   V=V/R
   DT=0.001
   CALL SUBP
   CALL SUBM
   ALENDA=12.0*EDA
   SUM1=ALENDA*S*SUM1
   V1(K)=W(K)*V/SUM1
   H1(K+1)=H1(K)+V1(K)*DT
   H0=H1(K+1)
```

```
      ENDDO
      H0=H1(N2)
      V=-1.0
      V=V/R
      CALL SUBP
      CALL SUBM
      ALENDA=12.0*EDA
      SUM1=ALENDA*S*SUM1
      V1(N2)=W(N2)*V/SUM1
      WRITE(8,"(' time T, load W, film thickness H, squeeze velocity V')")
      DO I=1,N2
      WRITE(8,40)DT*I,W(I),H1(I)*R,V1(I)*R
      ENDDO
      STOP
40    FORMAT(4(E12.6,1X))
70    CALL SUBP
      CALL SUBM
      ALENDA=12.0*EDA
      SUM1=ALENDA*S*SUM1
      WRITE(8,"(' Coordinates X pressure P')")
      DO I=1,N
      WRITE(8,40)X(I),P(I)
      ENDDO
      WRITE(9,*)'bearing capacity W'
      WRITE(9,*)SUM1
      CONTAINS
      SUBROUTINE SUBP
      DO I=1,N
      PD(I)=P(I)
      ENDDO
      IK=0
      TEMP=DX*DX
      H=H0*H0*H0
20    C1=0.0
      DO I=2,N1
      TEMP1=(X(I)+DX)/(DX+2.0*X(I))
      TEMP0=X(I)/(DX+2.0*X(I))
      P(I)=TEMP1*P(I+1)+P(I-1)*TEMP0-(TEMP0*V*TEMP)/H
      C1=C1+ABS(P(I)-PD(I))
      PD(I)=P(I)
      ENDDO
      P(1)=P(2)
      P(N)=0.0
      PD(1)=PD(2)
      PD(N)=0.0
```

```
IK=IK+1
IF(C1.GT.1.E-20.AND.IK.LE.20000)GOTO 20
RETURN
END SUBROUTINE SUBP
SUBROUTINE SUBM
SUM1=0.0
DO I=1,N
SUM1=SUM1+P(I)*X(I)*DX*PI2
ENDDO
RETURN
END SUBROUTINE SUBM
END
```

5.3.5 Calculation Results

According to the parameters of the working conditions and the dynamic load given in the present program, the variation curve of the squeeze film with time t from 0 to 0.01 s is shown in Figure 5.6. In Figure 5.7, a pressure distribution is given for the instantaneous squeeze velocity and the corresponding film thickness.

5.4 Numerical Method and Program for Journal Bearing Squeeze

5.4.1 Basic Equations

As shown in Figure 5.8, a journal bearing is under load W. If the shaft center has a downward speed as follows, it has squeeze lubrication.

$$\frac{de}{dt} = c\frac{d\varepsilon}{dt} \tag{5.9}$$

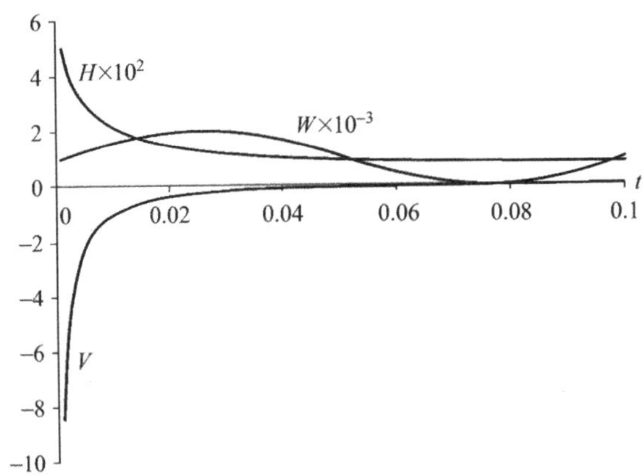

Figure 5.6 Calculation results of disc squeeze.

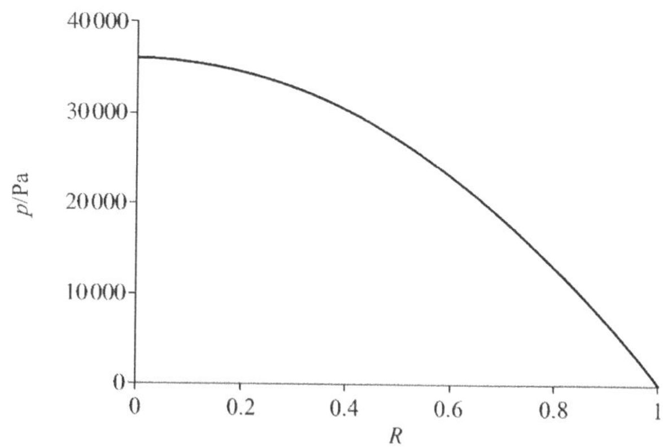

Figure 5.7 Instantaneous pressure distribution of disc squeeze.

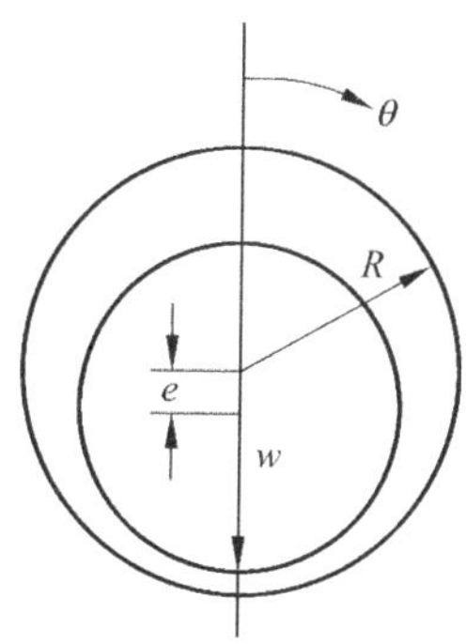

Figure 5.8 Journal bearing squeeze.

The film thickness variation is

$$\frac{dh}{dt} = c\cos\theta\frac{d\varepsilon}{dt} \tag{5.10}$$

The Reynolds equation of journal bearing squeeze can be written as

$$\frac{\partial}{\partial x}\left(h^3\frac{\partial p}{\partial x}\right) + \frac{\partial}{\partial y}\left(h^3\frac{\partial p}{\partial y}\right) = 12\eta\frac{\partial h}{\partial t} \tag{5.11}$$

By using the polar coordinates, $\theta = x/R$ so that $d\theta = dx/R$, Equation 5.11 becomes

$$\frac{\partial}{\partial \theta}\left(h^3\frac{\partial p}{\partial \theta}\right) + R^2\frac{\partial}{\partial y}\left(h^3\frac{\partial p}{\partial y}\right) = 12\eta R^2\frac{\partial h}{\partial t} \tag{5.12}$$

where $h = c(1 + \varepsilon\cos\theta)$.

The boundary conditions are

$$p|_{\theta=0} = 0$$
$$\frac{\partial p}{\partial \theta}|_{\theta=\pi} = 0$$

The dimensionless form of Equation 5.12 is

$$\frac{\partial}{\partial \theta}\left(H^3 \frac{\partial P}{\partial \theta}\right) + \alpha \frac{\partial}{\partial Y}\left(H^3 \frac{\partial P}{\partial Y}\right) = \frac{\partial H}{\partial t} = \dot{\varepsilon} \cos \theta \tag{5.13}$$

where $Y = y/b, h = cH, H = 1 + \varepsilon\cos\theta, P = pc^2/12\eta R^2, \alpha = (R/b)^2$, R is the radius of the bearing and b is the width of the bearing.

5.4.2 Numerical Method

The discrete form of the Reynolds Equation 5.13 can be written as

$$\begin{aligned} &\frac{H^3_{i+1/2,j}(P_{i+1,j} - P_{i,j}) - H^3_{i-1/2,j}(P_{i,j} - P_{i-1,j})}{\Delta\theta^2} \\ &+\alpha \frac{H^3_{i,j+1/2}(P_{i,j+1} - P_{i,j}) - H^3_{i,j-1/2}(P_{i,j} - P_{i,j-1})}{\Delta Y^2} = \dot{\varepsilon} \cos[(i-1)\Delta\theta] \end{aligned} \tag{5.14}$$

The process solves the journal bearing squeeze problem in the following two steps:

1. Calculate the pressure distribution with the given eccentricity e and the given axis moving speed de/dt.
2. With the given initial eccentricity e and the given dynamic load, calculate the eccentricity, film thickness, axis moving speed and squeeze velocity of each period.

5.4.3 Calculation Diagram

The calculation diagram for journal bearing squeeze is shown in Figure 5.9.

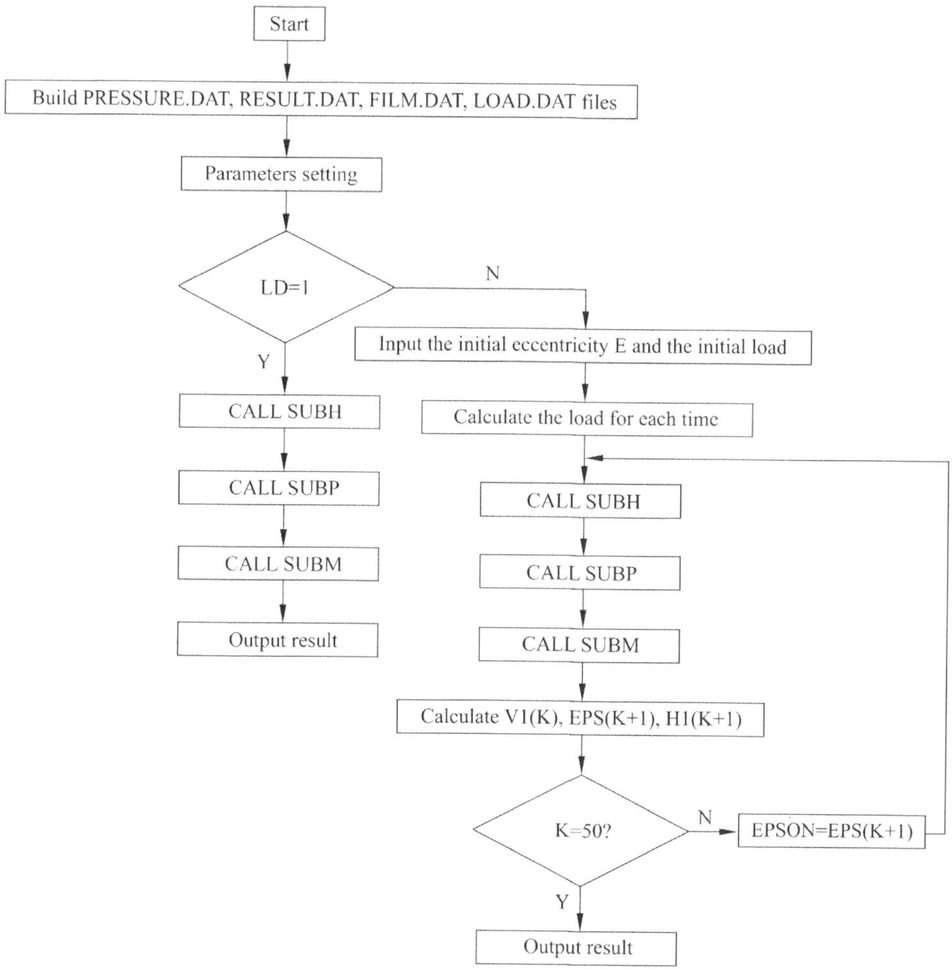

Figure 5.9 Calculation diagram for journal bearing squeeze.

5.4.4 Calculation Program

5.4.4.1 Parameter Pre-assignment

In the program, we set: lubricant viscosity EDA = 0.02 Pa s; diameter of the axle DD = 0.02992 m; diameter of the bearing D = 0.03 m; width of the bearing B = 0.024 m. The parameter preassignment can be modified according to the user's requirement, but it needs to be recompiled and relinked before running.

5.4.4.2 Output Parameter

```
Pressure distribution   P(I)    in PRESSURE.DAT
Bearing capacity        SUM1    in LOAD.DAT
W, e, de/dt, dh/dt of each period in RESULT.DAT
```

5.4.4.3 Program

```
   PROGRAM JOURNALSQUEEZE
   DIMENSION X(121),Y(121),H(121,121),P(121,121),PD(121,121),
HH(121)
   DIMENSION HXY(121,121),HXF(121,121),HXB(121,121),HYF(121,121),
HYB(121,121)
   DIMENSION H1(51),T(51),V1(51),W(51),EPS(51)
   DATA N,N1,N2,N3/121,120,51,50/
   DATA B,D,DD,E,EDA,PI,V/0.024,0.030,0.02992,0.328999E-
04,0.02,3.14159265,0.342583E-03/
   OPEN(8,FILE='PRESSURE.DAT',STATUS='UNKNOWN')
   OPEN(9,FILE='FILM.DAT',STATUS='UNKNOWN')
   OPEN(13,FILE='RESULT.DAT',STATUS='UNKNOWN')
   OPEN(14,FILE='LOAD.DAT',STATUS='UNKNOWN')
   PI2=2.0*PI
   R=D/2
   C=(D-DD)/2
   DX=PI2/N1
   DY=1.0/N1
   RATIO=(PI2*R)/B
   ALFA=RATIO**2
   DXDY=PI2/(N1*N1)
   V=V/C
   EPSON=E/C
   S=D*B
   DO I=1,N
   X(I)=(I-1)*DX
   Y(I)=-0.5+(I-1)*DY
   DO J=1,N
   P(I,J)=0.0
   ENDDO
   ENDDO
   WRITE(*,*)'LD=0 squeeze process calculation; LD=1
 instantaneous
   Squeeze pressure distribution and load calculation?'
   READ(*,*)LD
   IF(LD.EQ.1)THEN
   SUM0=0.0
```

```
      GO TO 70
      ELSE
      SUM0=1000.0
      E=0.00001
      GO TO 50
      ENDIF
 50   U=PI2/50.0
      DO I=1,N2
      W(I)=SUM0*(1.0+SIN(U*(I-1)))
      ENDDO
      EPS(1)=EPSON
      H1(1)=1-EPS(1)
      DO K=1,N3
      V=1.0
      V=V/C
      DT=0.001
      WRITE(*,*)K,V,EPSON
      CALL SUBH
      CALL SUBP
      CALL SUBM
      ALENDA=(12.0*EDA*R**2)/C**2
      SUM1=ALENDA*S*SUM1
      WRITE(*,*)SUM1
      V1(K)=W(K)*V/SUM1
      WRITE(*,*)V1(K)
      EPS(K+1)=EPS(K)+V1(K)*DT
      H1(K+1)=1-EPS(K+1)
      EPSON=EPS(K+1)
      WRITE(*,*)EPSON
      ENDDO
      EPSON=EPS(N2)
      V=1.0
      V=V/C
      CALL SUBH
      CALL SUBP
      CALL SUBM
      ALENDA=(12.0*EDA*R**2)/C**2
      SUM1=ALENDA*S*SUM1
      V1(N2)=W(N2)*V/SUM1
      WRITE(13,"('time T, load W, eccentricity E, axis moving speed
 de/dt, squeeze velocity dh/dt')")
      DO I=1,N2
      WRITE(13,40)DT*I,W(I),EPS(I),V1(I)*C,-V1(I)*C
      ENDDO
      STOP
```

```
70 CALL SUBH
   CALL SUBP
   CALL SUBM
   CALL SUBMAX
   ALENDA=(12.0*EDA*R**2)/C**2
   SUM1=ALENDA*S*SUM1
   WRITE(8,40)Y(1),(Y(I),I=1,N)
   DO I=1,N
   WRITE(8,40)X(I)*180.0/PI,(P(I,J),J=1,N)
   ENDDO
   WRITE(9,40)Y(1),(Y(I),I=1,N)
   DO I=1,N
   WRITE(9,40)X(I)*180.0/PI,(H(I,J),J=1,N)
   ENDDO
   WRITE(14,*)' bearing capacity W'
   WRITE(14,*)SUM1
   WRITE(*,*)AIMAX,J_MAX,I_MAX
40 FORMAT(122(E12.6,1X))
   CONTAINS
   SUBROUTINE SUBH
   DO I=1,N
   DO J=1,N
   H(I,J)=1.0+EPSON*COS((I-1)*DX)
   ENDDO
   ENDDO
   DO I=1,N
   I1=I-1
   I2=I+1
   IF(I.EQ.1)I1=N1
   IF(I.EQ.N)I2=2
   DO J=2,N1
   HXF(I,J)=(0.5*(H(I2,J)+H(I,J)))**3
   HXB(I,J)=(0.5*(H(I1,J)+H(I,J)))**3
   HYF(I,J)=ALFA*(0.5*(H(I,J+1)+H(I,J)))**3
   HYB(I,J)=ALFA*(0.5*(H(I,J-1)+H(I,J)))**3
   HXY(I,J)=1.0/(HXF(I,J)+HXB(I,J)+HYF(I,J)+HYB(I,J))
   ENDDO
   ENDDO
   RETURN
   END SUBROUTINE SUBH
   SUBROUTINE SUBP
   DO I=1,N
   DO J=1,N
   PD(I,J)=P(I,J)
   ENDDO
```

```
   ENDDO
   IK=0
   TEMP0=DX**2*V
10 C1=0.0
   DO I=1,N1
   I1=I-1
   I2=I+1
   IF(I1.EQ.0)I1=N1
   IF(I2.EQ.N)I2=1
   DO J=2,N1
   P(I,J)=(HXF(I,J)*P(I2,J)+HXB(I,J)*P(I1,J)+HYF(I,J)*P(I,J
+1)+HYB(I,J)*P(I,J-1)-TEMP0*COS((I-1)*DX))*HXY(I,J)
   IF(P(I,J).LE.0.0)P(I,J)=0.0
   C1=C1+ABS(P(I,J)-PD(I,J))
   PD(I,J)=P(I,J)
   ENDDO
   ENDDO
   DO J=2,N1
   P(N,J)=P(1,J)
   PD(N,J)=PD(1,J)
   ENDDO
   IK=IK+1
   IF(C1.GT.1.E-20.AND.IK.LE.20000)GOTO 10
   RETURN
   END SUBROUTINE SUBP
   SUBROUTINE SUBM
   PX=0.0
   PY=0.0
   TEMP=PI/60.0
   DO I=1,N1
   AI=(I-1)*TEMP
   DO J=1,N
   PX=PX-P(I,J)*COS(AI)*DXDY
   PY=PY+P(I,J)*SIN(AI)*DXDY
   ENDDO
   ENDDO
   SUM1=SQRT(PX*PX+PY*PY)
   RETURN
   END SUBROUTINE SUBM
   SUBROUTINE SUBMAX
   TEMP0=PI2/N1
   PMAX=P(2,2)
   DO I=1,N
   DO J=1,N
   IF(P(I,J).GE.PMAX)THEN
```

```
PMAX=P(I,J)
I_MAX=I
J_MAX=J
ENDIF
ENDDO
ENDDO
AIMAX=(I_MAX-1)*TEMP0*180/PI
RETURN
END SUBROUTINE SUBMAX
END
```

5.4.5 Calculation Results

According to the parameters of working conditions and dynamic load given in the present program, the variation curve of the squeeze film with time t from 0 to 0.05s is shown in Figure 5.10. In Figure 5.11, a pressure distribution is given for the instantaneous squeeze velocity and the corresponding film thickness.

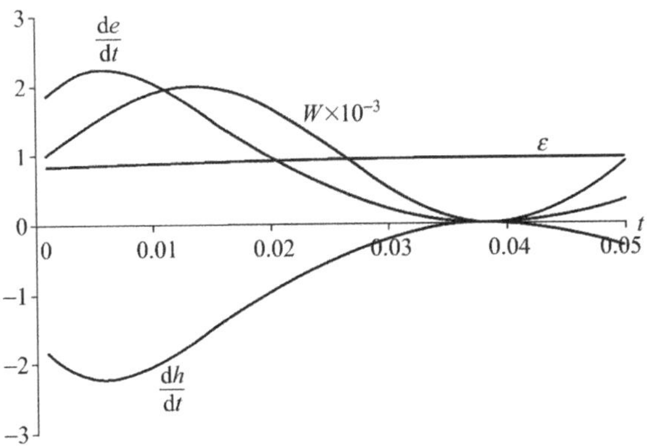

Figure 5.10 Calculation results of journal bearing squeeze.

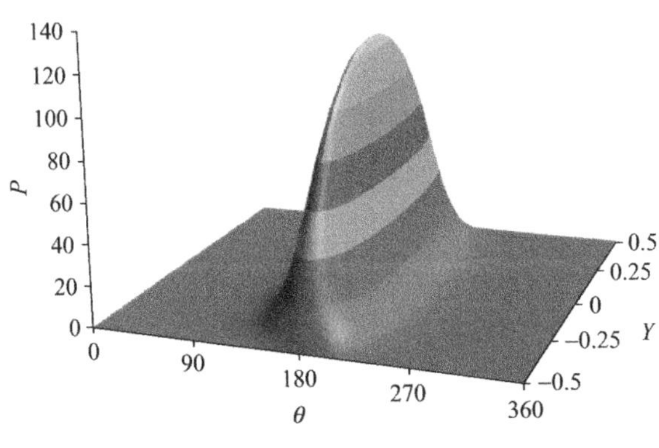

Figure 5.11 Instantaneous pressure distribution of journal bearing squeeze.

6

Numerical Method and Program for Dynamic Bearing

6.1 Basic Equations

A dynamic bearing is shown in Figure 6.1. The Reynolds equation for a dynamic bearing is

$$\frac{\partial}{\partial x}\left(h^3\frac{\partial p}{\partial x}\right)+\frac{\partial}{\partial y}\left(h^3\frac{\partial p}{\partial y}\right)=6\eta(U_j+U_b)\frac{\partial h}{\partial x}+12\eta\frac{\partial h}{\partial t} \tag{6.1}$$

where $h=c(1+\varepsilon\cos\theta)$, c is the radius clearance between the bearing and the journal, η is the lubricating oil viscosity, U_j is the journal velocity ($U_j=R_j\omega_j\approx R_b\omega_j$), U_b is the bearing velocity ($U_b=R_b\omega_b$), R is the radius, ω is the angular velocity, the subscript j represents the journal and the subscript b represents the bearing.

If we introduce the dimensionless variables $\theta=x/R_b$, $H=h/c=1+\varepsilon\cos\theta$, $Y=y/b$, $\alpha=(R_b/b)^2$ and $P=pc^2/6\omega R_b^2\eta$, where b is the width of the bearing, $\omega=\omega_j+\omega_b$, the dimensionless form of the Reynolds Equation 6.1 will be

$$\frac{\partial}{\partial\theta}\left(H^3\frac{\partial P}{\partial\theta}\right)+\alpha\frac{\partial}{\partial Y}\left(H^3\frac{\partial P}{\partial Y}\right)=-\varepsilon\sin\theta+\frac{2\varepsilon}{\omega}\sin\theta\frac{d\delta}{dt}+\frac{2}{\omega}\cos\theta\frac{d\varepsilon}{dt} \tag{6.2}$$

On the right side of Equation 6.2, the three terms express three kinds of lubrication effects: $-\varepsilon\sin\theta$ is the spin effect produced by the angular velocity ω, $\frac{2\varepsilon}{\omega}\sin\theta\frac{d\delta}{dt}$ is another spin effect produced by the journal center angular velocity $\frac{d\delta}{dt}$ in the vertical

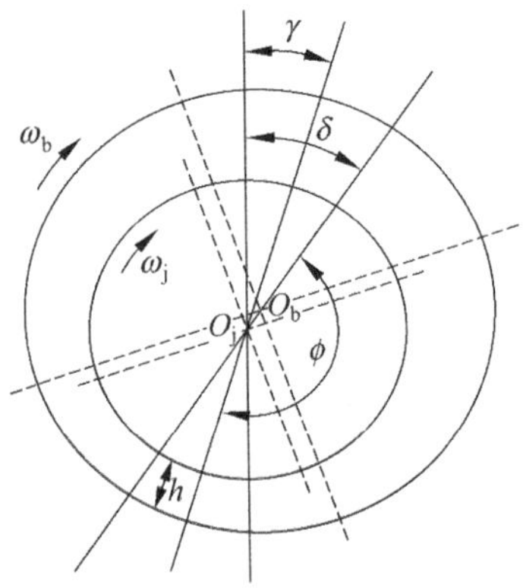

Figure 6.1 Dynamic bearing.

direction of the offsetting line O_jO_b and $\frac{2}{\omega}\cos\theta\frac{d\varepsilon}{dt}$ is the squeezing effect produced by the journal center angular velocity $\frac{d\varepsilon}{dt}$ in the direction along the offsetting line O_jO_b.

On the right side of Equation 6.2, classifying the first and the second expression, we have

$$-\varepsilon\sin\theta+\frac{2\varepsilon}{\omega}\sin\theta\frac{d\delta}{dt}=-\frac{\varepsilon}{\omega}\sin\theta\left(\omega-2\frac{d\delta}{dt}\right)=-\frac{\varepsilon}{\omega}\sin\theta\left(\omega_j+\omega_b-2\frac{d\delta}{dt}\right) \tag{6.3}$$

Then

$$\omega^*=\omega_j+\omega_b-2\frac{d\delta}{dt}=\omega-2\frac{d\delta}{dt} \tag{6.4}$$

where ω^* is called as the equivalent angular velocity.

Hence, the first and the second terms on the right side of Equation 6.2 are the spin effect produced by the equivalent angular velocity ω^*.

Now, we introduce other dimensionless parameters

$$\begin{aligned} q &= \frac{2}{\omega^*}\frac{d\varepsilon}{dt} \\ P &= c^2p/6\omega^*R_b^2\eta \end{aligned} \tag{6.5}$$

where q is the kinetic parameter.

According to Equations 6.3 and 6.5, the dimensionless Reynolds Equation 6.2 can be written as

$$\frac{\partial}{\partial\theta}\left(H^3\frac{\partial P}{\partial\theta}\right)+\alpha\frac{\partial}{\partial Y}\left(H^3\frac{\partial P}{\partial Y}\right)=-\varepsilon\sin\theta+q\cos\theta \tag{6.6}$$

The pressure boundary conditions are as follows.

In the axial direction

$$p|_{y=\pm b/2} = 0; \frac{\partial p}{\partial y}|_{y=0} = 0.$$

In the circumferential direction: According to the Reynolds boundary conditions, the pressure is equal to zero at the starting point of the oil film, that is, $p|_{\theta=0}=0$; at the terminal points of the oil film, the pressure should meet the conditions $p=0$ and $\frac{\partial p}{\partial \theta}=0$. Because the terminal points of the oil film must be determined according to the boundary conditions in the solving process, there is a floating boundary. When we use iteration to solve the problem, and if some points are $p<0$ during the iteration process, we set the pressure $p=0$. Eventually, all the terminal points can be determined in this way.

Equation 6.6 is a linear second partial differential equation.

According to the superposition principle, the solution of the equation for P can be divided into two individual results, P_1 and P_2, as follows

$$P = P_1 + qP_2 \tag{6.7}$$

Therefore, the following two Equations 6.8 and 6.9 can be individually solved.

$$\frac{\partial}{\partial \theta}\left(H^3 \frac{\partial P_1}{\partial \theta}\right) + \alpha \frac{\partial}{\partial Y}\left(H^3 \frac{\partial P_1}{\partial Y}\right) = -\varepsilon \sin\theta \tag{6.8}$$

$$\frac{\partial}{\partial \theta}\left(H^3 \frac{\partial P_2}{\partial \theta}\right) + \alpha \frac{\partial}{\partial Y}\left(H^3 \frac{\partial P_2}{\partial Y}\right) = \cos\theta \tag{6.9}$$

where $q = \frac{2}{\omega^*}\frac{d\varepsilon}{dt}$.

When $q=0$, we have a pure spin situation. When q tends to infinity, the angle of the reaction force of the oil film is relative to the offsetting position of the axis and tends to be either 0° or 180°. It is a pure squeezing state in either the forward direction or the reverse direction.

Here we use the difference method to solve the two differential Equations 6.8 and 6.9. The discrete form of Equation 6.8 can be written as

$$\begin{aligned}&\frac{H^3_{i+1/2,j}(P_{i+1,j}-P_{i,j}) - H^3_{i-1/2,j}(P_{i,j}-P_{i-1,j})}{\Delta\theta^2}\\&+\alpha\frac{H^3_{i,j+1/2}(P_{i,j+1}-P_{i,j}) - H^3_{i,j-1/2}(P_{i,j}-P_{i,j-1})}{\Delta Y^2} = -\varepsilon\sin[(i-1)\Delta\theta]\end{aligned} \tag{6.10}$$

where $H_{i,j} = 1 + \varepsilon\cos[(i-1)\Delta\theta]$.

When we set

$$A = H^3_{i+0.5,j} = [0.5 * (H_{i+1,j} + H_{i,j})]^3$$

$$B = H^3_{i-0.5,j} = [0.5 * (H_{i-1,j} + H_{i,j})]^3$$

$$C = \alpha\left(\frac{\Delta\theta}{\Delta Y}\right)^2 [0.5 * (H_{i,j+1} + H_{i,j})]^3$$

$$D = \alpha\left(\frac{\Delta\theta}{\Delta Y}\right)^2 [0.5 * (H_{i,j-1} + H_{i,j})]^3$$

$$E = A + B + C + D$$

then Equation 6.10 becomes

$$(P_1)_{i,j} = \frac{1}{E}\left[A(P_1)_{i+1,j} + B(P_1)_{i-1,j} + C(P_1)_{i,j+1} + D(P_1)_{i,j-1} + \Delta\theta^2 \varepsilon \sin(i-1)\Delta\theta\right] \tag{6.11}$$

By using the same method to Equation 6.9, this can be rewritten as

$$(P_2)_{i,j} = \frac{1}{E}\left[A(P_2)_{i+1,j} + B(P_2)_{i-1,j} + C(P_2)_{i,j+1} + D(P_2)_{i,j-1} - \Delta\theta^2 \cos(i-1)\Delta\theta\right] \tag{6.12}$$

When $\omega^* = 0$, we set

$$P = pc^2 \Big/ \left(12\eta R_b^2 \, {}^{d\varepsilon}\!/_{dt}\right) \tag{6.13}$$

Therefore, Equation 6.2 can be written in a dimensionless form as

$$\frac{\partial}{\partial\theta}\left(H^3 \frac{\partial P}{\partial\theta}\right) + \alpha \frac{\partial}{\partial Y}\left(H^3 \frac{\partial P}{\partial Y}\right) = \cos\theta \tag{6.14}$$

When $\omega^* = 0$, there is only radial movement. Then, Equations 6.9 and 6.14 are the same. Accordingly, their solutions are also the same, that is, $P_{i,j} = (P_2)_{i,j}$. Therefore, the computational task of Equation 6.14 can be reduced greatly.

Note: during the solving of $(P_1)_{i,j}$ and $(P_2)_{i,j}$, if the node pressures of $(P_1)_{i,j}$ or $(P_2)_{i,j}$ are negative, we do not set them to zero first. After superimposing $(P_1)_{i,j}$ and $(P_2)_{i,j}$ into Equation 6.7, we then set the negative pressures to zero. In this way, we can obtain the different pressures P(I,J) with each different q. By using the corresponding boundary conditions to integrate P(I,J), we can obtain the dimensionless load carrying capacity S of the oil film. Then, the actual dimension load-carrying capacity P of the oil film can finally be obtained.

$$P = \frac{6BD\eta|\omega^*|}{(c/R_b)^2}S \tag{6.15}$$

$$P = \frac{12BD\eta}{(c/R_b)^2}|\frac{d\varepsilon}{dt}|S \quad \omega^* = 0 \tag{6.16}$$

6.2 Numerical Method for Trace of Journal Center

6.2.1 Introduction

Generally, there are three methods to determine the trace of a dynamic journal. They are the mobility method, the Hahn method and the Holland method. Here, the Hahn method is adopted, which superposes the hydrodynamic pressure and the squeeze pressure. After integrating P(I,J), the relationship between load and $\frac{d\varepsilon}{dt}, \frac{d\delta}{dt}$ can be determined. Therefore, the instantaneous velocity can be calculated according to the known dynamic load. Then, the trace of the journal center can be obtained by using the marching method.

From the above method, $\phi = \phi(\varepsilon,q)$ and $S = S(\varepsilon,q)$ can be known. Before we can determine the position of the journal, the lubricant viscosity η, the diameter of the journal d, the diameter of the bearing D, the bearing width B, the journal velocity ω_j, the bearing velocity ω_b and the magnitude and direction angle of load F must be given first. Assuming that the initial position of the journal is (ε,δ), we can calculate the angle of the load carrying capacity, and $\phi = 180 - (\delta - \gamma)$ according to the load direction angle. That F is equal to P can be deduced from the equilibrium relationship between the load and the oil film force. Therefore, ϕ and P are the two known parameters. Then, q can be obtained from $\phi = \phi(\varepsilon,q)$ because ε and ϕ are known. Also, S can be obtained from $S = S(\varepsilon,q)$. We substitute S into Equations 6.15 or 6.16 to determine ω^*. Finally, we substitute ω^* into Equation 6.4 to determine $\frac{d\delta}{dt}$, and substitute ω^* and q into $q = \frac{2}{\omega^*}\frac{d\varepsilon}{dt}$ to determine $\frac{d\varepsilon}{dt}$. In this way, the two essential parameters $\frac{d\varepsilon}{dt}$ and $\frac{d\delta}{dt}$ can be known and then can be used to determine the instantaneous axis velocity. The new position of the journal center in the time interval ΔT can be expressed as

$$\begin{aligned} \varepsilon_i &= \varepsilon_{i-1} + \Delta\varepsilon = \varepsilon_{i-1} + \dot{\varepsilon}\Delta t \\ \delta_i &= \delta_{i-1} + \Delta\delta = \delta_{i-1} + \dot{\delta}\Delta t \end{aligned} \tag{6.17}$$

According to the above method, a series of journal positions can be obtained. Then the trace of dynamic journal bearing can be obtained.

6.2.2 Calculation Steps

1. Take a load F_0 and its direction angle γ_0 as the start point and the initial point (ε_0,δ_0).
2. Calculate ϕ according to γ and δ because $\phi = 180 - (\delta - \gamma)$.
3. Use the bivariate interpolation method to calculate q according to the corresponding ε and ϕ; and use the interpolation method to determine the corresponding load carrying capacity P with the obtained q. Then obtain $\dot{\varepsilon}, \dot{\delta}$.

   There are four situations according to the different oil film load carrying capacity angle ϕ:

   a. As $0° < \phi < 180°$, $\omega^* > 0$. Then, interpolate q from ε and ϕ according to AM1 (K,L) and S1(K,L). Also, obtain the load carrying capacity S. If $|q| > 25$, ω^* can be thought to be zero. Because F is equal to P, we have $\omega^* = \frac{F}{S}\frac{(c/R_b)^2}{6\eta bD}$.

   Therefore, $\dot{\varepsilon}$ and $\dot{\delta}$ can be obtained by Equation 6.18.

$$\begin{aligned} \dot{\delta} &= 0.5 * (\omega_j + \omega_b - \omega^*) \\ \dot{\varepsilon} &= q\omega^*/2 \end{aligned} \tag{6.18}$$

   b. As $180° < \phi < 360°$, $\omega^* < 0$. Interpolate q from ε and ϕ according to AM2(K,L) and S2(K,L). After the load carrying capacity S is obtained, we have $\omega^* = -\frac{F}{S}\frac{(c/R_b)^2}{6\eta bD}$. Then, $\dot{\varepsilon}$ and $\dot{\delta}$ can be obtained.

   c. As $\phi = 180°$, $\omega^* = 0$; that is, $\dot{\varepsilon} > 0$. Therefore, the load carrying capacity S can be determined according to S3(K). Then, $\dot{\varepsilon}$ and $\dot{\delta}$ can be obtained from $\dot{\varepsilon} = \frac{F}{S}\frac{(c/R_b)^2}{12\eta BD}$ and $\dot{\delta} = \frac{\omega_j + \omega_b}{2}$.

   d. When $\phi = 0°$, $\omega^* = 0$; that is, $\dot{\varepsilon} < 0$. Therefore, the load carrying capacity S can be determined according to S4(K). Then, $\dot{\varepsilon}$ and $\dot{\delta}$ can be obtained from $\dot{\varepsilon} = -\frac{F}{S}\frac{(c/R_b)^2}{12\eta BD}$ and $\dot{\delta} = \frac{\omega_j + \omega_b}{2}$.
4. By using $\varepsilon_{i+1} = \varepsilon_i + \Delta\varepsilon = \varepsilon_i + \dot{\varepsilon}\Delta t$, $\delta_{i+1} = \delta_i + \Delta\delta = \delta_i + \dot{\delta}\Delta t$, Δt, $\dot{\varepsilon}$ and $\dot{\delta}$, calculate ε_{i+1} and δ_{i+1} of the next point.
5. Calculate the errors $|\varepsilon_N - \varepsilon_0|$ and $|\delta_N - \delta_0|$ between the start point and the end point in the present loop. If the errors meet the given requirements, the iteration stops. Otherwise, take ε_N and δ_N as the initial start point for a new loop to iterate again until the requirements are met.

6.3 Calculation Program for Dynamic Journal Bearing

6.3.1 Introduction

The calculation program for a dynamic journal bearing includes two main programs and six subroutines. The main programs are used to calculate the load carrying capacity and to solve the trace of the journal. The subroutines are the film thickness subroutine SUBH, the pressure subroutines SUBP1 and SUBP2, the resultant pressure subroutine SUBM, the maximum pressure point calculation subroutine SUBMAX, the interpolation calculation for q subroutine SUBQ and the interpolation calculation for dimensionless load carrying capacity S subroutine SUBSUM1.

6.3.1.1 Assignment Parameters

The assignment parameters are the lubricant viscosity EDA = 0.02 Pa.s, the diameter of the journal DD = 0.02992 m, the diameter of the bearing D = 0.03, the width of the bearing B = 0.024 m, the journal velocity Wj = 3000 rpm and the bearing velocity Wb = 2000 rpm. The assignment parameters can be modified according to the user's need, but the program needs to be recompiled and relinked before running.

6.3.1.2 Input Array Files

The load carrying capacity angle array φ of each case AM1(K,L) and AM2(K,L) in ANGLE.DAT

The load carrying capacity SUM1 of each case S1(K,L), S2(K,L), S3(K) and S4(K) in LOAD.DAT

6.3.1.3 Output Parameters

Pressure distribution P(I,J) in PRESSURE.DAT

Film thickness H(I,J) in FILM.DAT

Calculation of each time EPSON, E, AIMAX, PMAX, SUM1, AI in RESULT.DAT

Axis position eccentricity ε, positive angle δ between the fixed axis X and the biggest film thickness in TRACE.DAT

Axial velocity of each time eccentricity ε, $d\delta/dt$, $d\varepsilon/dt$ in VELOCITY.DAT

6.3.2 Calculation Diagram

The calculation diagram for a dynamic journal bearing is shown in Figure 6.2.

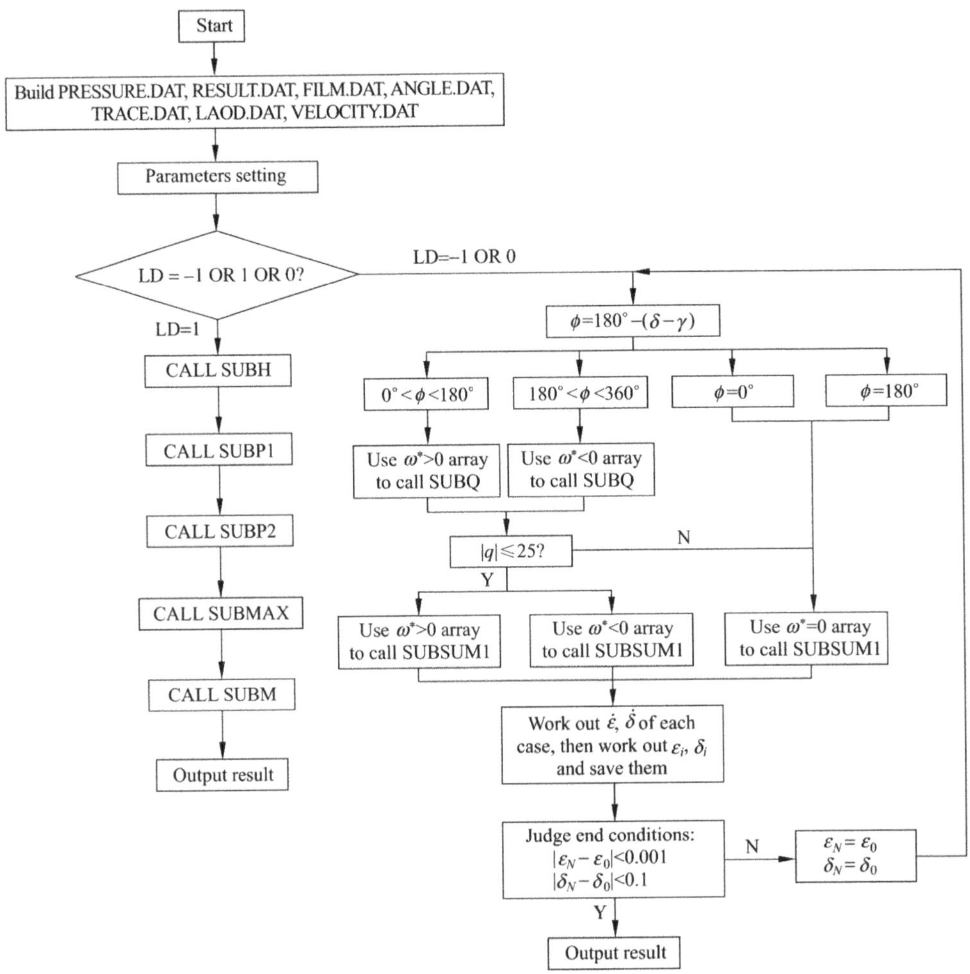

Figure 6.2 Calculation diagram for a dynamic journal bearing.

6.3.3 Calculation Program

```
PROGRAM DYNAMICBEARING
     REAL*4,ALLOCATABLE::DEE(:),DFF(:),VDDET(:),VDECC(:)
     DIMENSION X(121),Y(121),H(121,121),P1(121,121),PD
(121,121),P2(121,121),PE(121,121)
     DIMENSION P(121,121),P3(121,121),QQ(41),DE(121),DF(121)
     DIMENSION HXY(121,121),HXF(121,121),HXB(121,121),HYF
(121,121),HYB(121,121)
     DIMENSION AM(11,41),AM1(11,41),AM2(11,41),S(11,41),S1
(11,41),S2(11,41),S3(11),S4(11),EPS(11)
```

```
      CHARACTER*1 O,O1,O2
      DATA N,B,D,DD,EDA,PI/
121,0.024,0.03,0.02992,0.02,3.14159265/,O1,O2/1HY,1Hy/
      DATA DDET,DECC,DET,EPSON,RI,SUM,WB,WJ,WL/0.104717E
+03,0.927026E-03,30.0,0.656199E
+00,40.0,1500.0,2000.0,3000.0,1000.0/
      DATA EPS/0.05,0.1,0.2,0.3,0.4,0.5,0.6,0.7,0.8,0.9,0.995/
      OPEN(8,FILE='PRESSURE.DAT',STATUS='UNKNOWN')
      OPEN(9,FILE='FILM.DAT',STATUS='UNKNOWN')
      OPEN(13,FILE='RESULT.DAT',STATUS='UNKNOWN')
      OPEN(14,FILE='ANGLE.DAT',STATUS='UNKNOWN')
      OPEN(15,FILE='TRACE.DAT',STATUS='UNKNOWN')
      OPEN(16,FILE='LOAD.DAT',STATUS='UNKNOWN')
      OPEN(17,FILE='VELOCITY.DAT',STATUS='UNKNOWN')
      WRITE(*,*)'Pressure calculation:LD=1; Rotating load:LD=-1;
Suddenly applied load:LD=0. Please input:LD='
      READ(*,*)LD
      WRITE(*,*)' demonstrate the example according to the given
data (Input Y or N)?'
      READ(*,'(A)')O
      IF(O.EQ.O1.OR.O.EQ.O2)GOTO 10
      WRITE(*,*)' Please input the journal velocity:'
      READ(*,*)WB
      WRITE(*,*)' Please input the bearing velocity:'
      READ(*,*)WJ
      IF(LD.EQ.-1)THEN
      WRITE(*,*)' Please input Load and its angular velocity '
      READ(*,*)SUM,WL
      WRITE(*,*)' Please input the initial position of the axis ε,δ'
      READ(*,*)EPSON,DET
      ENDIF
      IF(LD.EQ.1)THEN
      WRITE(*,*)' Please input eccentricity, Axis motion
parameters DDET,DECC'
      READ(*,*)EPSON,DDET,DECC
      ENDIF
      IF(LD.EQ.0)THEN
      WRITE(*,*)' Please input the initial position of the axis ε,δ'
      READ(*,*)EPSON,DET
      WRITE(*,*)' Please input suddenly applied load and its
direction angle '
      READ(*,*)SUM,RI
      ENDIF
10    N1=N-1
      N2=N1/2+1
```

```
PI2=2.0*PI
R=D/2
C=(D-DD)/2.0
PESAI=(D-DD)/D
DX=PI2/N1
DY=1.0/N1
RATIO=PI2*R/B
ALFA=RATIO**2
WB=WB*PI2/60.0
WJ=WJ*PI2/60.0
WL=WL*PI2/60.0
DET=DET*PI/180.0
RI=RI*PI/180.0
W=WB+WJ-2.0*DDET
CC=EPSON
BB=DET
AA=RI
DXDY=PI2/(N1*N1)
DO I=1,N
X(I)=(I-1)*DX
Y(I)=-0.5+(I-1)*DY
ENDDO
DO I=1,N
DO J=1,N
P(I,J)=0.0
P1(I,J)=0.0
P2(I,J)=0.0
P3(I,J)=0.0
ENDDO
ENDDO
IF(LD.EQ.1)THEN
SUM=0.0
GOTO 70
ENDIF
DO K=1,11
WRITE(*,*)K
EPSON=EPS(K)
WRITE(*,*)EPSON
CALL SUBH
CALL SUBP1
CALL SUBP2
DO L=1,41
QQ(L)=-20.0+L-1
DO I=1,N
DO J=1,N
```

```
P3(I,J)=P1(I,J)+QQ(L)*P2(I,J)
IF(P3(I,J).LE.0.0)P3(I,J)=0.0
P(I,J)=P3(I,J)
ENDDO
ENDDO
CALL SUBM
IF(PX.GT.0.0)THEN
AI=180.0-180.0*ATAN(PY/PX)/PI
ELSE
AI=-180.0*ATAN(PY/PX)/PI
ENDIF
AM1(K,L)=AI
S1(K,L)=SUM1
DO I=1,N
DO J=1,N
P3(I,J)=-(P1(I,J)+QQ(L)*P2(I,J))
IF(P3(I,J).LE.0.0)P3(I,J)=0.0
P(I,J)=P3(I,J)
ENDDO
ENDDO
CALL SUBM
IF(PX.GT.0.0)THEN
AI=180.0-180.0*ATAN(PY/PX)/PI
ELSE
AI=360.0-180.0*ATAN(PY/PX)/PI
ENDIF
AM2(K,L)=AI
S2(K,L)=SUM1
DO I=1,N
DO J=1,N
P3(I,J)=P2(I,J)
IF(P3(I,J).LE.0.0)P3(I,J)=0.0
P(I,J)=P3(I,J)
ENDDO
ENDDO
CALL SUBM
S3(K)=SUM1
DO I=1,N
DO J=1,N
P3(I,J)=-P2(I,J)
IF(P3(I,J).LE.0.0)P3(I,J)=0.0
P(I,J)=P3(I,J)
ENDDO
ENDDO
CALL SUBM
```

```
      S4(K)=SUM1
      ENDDO
      ENDDO
100   FORMAT(122(E12.6,1X))
      WRITE(14,100)AM1,AM2
      WRITE(16,100)S1,S2,S3,S4
      DO L=1,41
      QQ(L)=-20.0+L-1
      ENDDO
      EPSON=CC
      DET=BB
      DDT=PI/(WJ*60.0)
      DE(1)=CC
      DF(1)=BB
      KK=1
30    DO II=1,120
      KK=KK+1
      IF(WL.EQ.0.0)THEN
      RI=AA
      ELSE
      RI=(KK-2)*WL*DDT
      ENDIF
      RI=MOD(RI,PI2)
      ZZ=DET
      DET=MOD(ZZ,PI2)
      RI=DET-RI
      AII=ABS(180.0-180.0*RI/PI)
      IF(EPSON.LE.0.05)THEN
      K=1
      AKL=0.0
      ENDIF
      IF(EPSON.GE.0.995)THEN
      K=11
      AKL=0.0
      ENDIF
      IF((EPSON.GT.0.05).AND.(EPSON.LT.0.995))THEN
      AKL=10*EPSON+1
      K=FLOOR(AKL)
      AKL=ABS(K-AKL)
      ENDIF
      IF(ABS(AII-180.0).LE.0.05)THEN
      CALL SUBSUM1
      DECC=(SUM*PESAI**2)/(12.0*EDA*SUM1*B*D)
      DDET=0.5*(WB+WJ)
      ENDIF
```

```
IF((ABS(AII).LE.0.05).OR.(ABS(AII-360.0).LE.0.05))THEN
CALL SUBSUM1
DECC=(-SUM*PESAI**2)/(12.0*EDA*SUM1*B*D)
DDET=0.5*(WB+WJ)
ENDIF
IF((AII.GT.0.05).AND.(AII.LT.179.95))THEN
DO I=1,11
DO J=1,41
AM(I,J)=AM1(I,J)
ENDDO
ENDDO
CALL SUBQ
IF(ABS(Q).LE.25.0)THEN
DO I=1,11
DO J=1,41
S(I,J)=S1(I,J)
ENDDO
ENDDO
ENDIF
CALL SUBSUM1
IF(Q.GT.25.0)THEN
DECC=(SUM*PESAI**2)/(12.0*EDA*SUM1*B*D)
DDET=0.5*(WB+WJ)
ENDIF
IF(Q.LT.-25.0)THEN
DECC=(-SUM*PESAI**2)/(12.0*EDA*SUM1*B*D)
DDET=0.5*(WB+WJ)
ENDIF
IF(ABS(Q).LE.25.0)THEN
W=(SUM*PESAI**2)/(6.0*EDA*SUM1*B*D)
DDET=0.5*(WB+WJ-W)
DECC=0.5*Q*W
ENDIF
ENDIF
IF((AII.GT.180.05).AND.(AII.LT.359.95))THEN
DO I=1,11
DO J=1,41
AM(I,J)=AM2(I,J)
ENDDO
ENDDO
CALL SUBQ
IF(ABS(Q).LE.25.0)THEN
DO I=1,11
DO J=1,41
S(I,J)=S2(I,J)
```

```
      ENDDO
      ENDDO
      ENDIF
      CALL SUBSUM1
      IF(Q.GT.25.0)THEN
      DECC=(-SUM*PESAI**2)/(12.0*EDA*SUM1*B*D)
      DDET=0.5*(WB+WJ)
      ENDIF
      IF(Q.LT.-25.0)THEN
      DECC=(SUM*PESAI**2)/(12.0*EDA*SUM1*B*D)
      DDET=0.5*(WB+WJ)
      ENDIF
      IF(ABS(Q).LE.25.0)THEN
      W=(SUM*PESAI**2)/(6.0*EDA*SUM1*B*D)
      DDET=0.5*(WB+WJ+W)
      DECC=-0.5*Q*W
      ENDIF
      ENDIF
      EPSON=EPSON+DECC*DDT
    DET=ZZ+DDET*DDT
      IF(EPSON.LT.0.0)THEN
      EPSON=-EPSON
      DET=DET+PI
      ENDIF
      DE(II+1)=EPSON
      DF(II+1)=DET
      ALLOCATE(DEE(KK))
      ALLOCATE(DFF(KK))
      ALLOCATE(VDDET(KK-1))
      ALLOCATE(VDECC(KK-1))
      IF(KK.EQ.2)THEN
      DEE(1)=DE(1)
      DFF(1)=DF(1)
      DEE(KK)=DE(II+1)
      DFF(KK)=DF(II+1)
      WRITE(15,"(' Eccentricity ε Positive angle δ between Fixed
  axis X and the biggest film thickness')")
      WRITE(15,100)DEE(1),DFF(1)*180.0/PI
      WRITE(15,100)DEE(KK),DFF(KK)*180.0/PI
      ELSE
      DEE(KK)=DE(II+1)
      DFF(KK)=DF(II+1)
      WRITE(15,100)DEE(KK),DFF(KK)*180.0/PI
      ENDIF
      VDDET(KK-1)=DDET
```

```
      VDECC(KK-1)=DECC
      IF(KK.EQ.2)THEN
      WRITE(17,"(' Eccentricity ε          dδ/dt        dε/dt')")
      WRITE(17,100)DEE(1),VDDET(1),VDECC(1)
      ELSE
      WRITE(17,100)DEE(KK),VDDET(KK-1),VDECC(KK-1)
      ENDIF
      IF(II.EQ.120)THEN
      ANBA=DEE(KK)
      CBA=DFF(KK)
      ENDIF
      XX=MOD(CBA,PI2)
      Y1=ABS(ANBA-CC)
      Y2=ABS(XX-BB)
      AT=CBA
      DEALLOCATE(DEE)
      DEALLOCATE(DFF)
      DEALLOCATE(VDDET)
      DEALLOCATE(VDECC)
      IF(WL.EQ.0.0)THEN
      AB=ABS(DE(II+1)-DE(II))
      AC=ABS(DF(II+1)-DF(II))
      IF((AB.EQ.0.0).AND.(AC.EQ.0.0))THEN
      WRITE(*,*)'End of TRACE CALCULATION'
      STOP
      ENDIF
      ENDIF
      IF((Y1.LE.0.001).AND.(Y2.LE.0.1))EXIT
      ENDDO
      IF((KK.GE.1081).AND.(AT.GE.2*PI2))THEN
      WRITE(*,*)'End of TRACE CALCULATION'
      STOP
      ENDIF
      EPSON=ANBA
      DET=CBA
      IF((Y1.GT.0.001).OR.(Y2.GT.0.1).OR.(AT.LE.2*PI2))GOTO 30
70    CALL SUBH
      CALL SUBP1
      CALL SUBP2
      IF(W.NE.0)THEN
      Q=2*DECC/W
      IF(W.GT.0.0)THEN
      DO I=1,N
      DO J=1,N
      P3(I,J)=P1(I,J)+Q*P2(I,J)
```

```
IF(P3(I,J).LE.0.0)P3(I,J)=0.0
P(I,J)=P3(I,J)
ENDDO
ENDDO
ELSE
DO I=1,N
DO J=1,N
P3(I,J)=-(P1(I,J)+Q*P2(I,J))
IF(P3(I,J).LE.0.0)P3(I,J)=0.0
P(I,J)=P3(I,J)
ENDDO
ENDDO
ENDIF
CALL SUBMAX
CALL SUBM
E=EPSON*C*1.0E3
HMIN=C*(1.0-EPSON)
ALENDA=(6.0*EDA*W)/(PESAI**2)
SUM1=ABS(ALENDA)*B*D*SUM1
IF(W.GT.0.0)THEN
IF(PX.GT.0.0)THEN
AI=180.0-180.0*ATAN(PY/PX)/PI
ELSE
AI=-180.0*ATAN(PY/PX)/PI
ENDIF
ENDIF
IF(W.LT.0.0)THEN
IF(PX.GT.0.0)THEN
AI=180.0-180.0*ATAN(PY/PX)/PI
ELSE
AI=360.0-180.0*ATAN(PY/PX)/PI
ENDIF
ENDIF
ENDIF
IF(W.EQ.0)THEN
IF(DECC.GT.0.0)THEN
DO I=1,N
DO J=1,N
P3(I,J)=P2(I,J)
IF(P3(I,J).LE.0.0)P3(I,J)=0.0
P(I,J)=P3(I,J)
ENDDO
ENDDO
ELSE
DO I=1,N
```

```
      DO J=1,N
      P3(I,J)=-P2(I,J)
      IF(P3(I,J).LE.0.0)P3(I,J)=0.0
      P(I,J)=P3(I,J)
      ENDDO
      ENDDO
      ENDIF
      CALL SUBMAX
      CALL SUBM
      E=EPSON*C*1.0E3
      HMIN=C*(1.0-EPSON)
      AI=180.0*ATAN(PY/PX)/PI
      ALENDA=(12.0*EDA*DECC)/(PESAI**2)
      SUM1=ABS(ALENDA)*B*D*SUM1
      ENDIF
      WRITE(8,40)Y(1),(Y(I),I=1,N)
      DO I=1,N
      WRITE(8,40)X(I)*180/PI,(P(I,J)*ALENDA,J=1,N)
      ENDDO
      WRITE(9,40)Y(1),(Y(I),I=1,N)
      DO I=1,N
      WRITE(9,40)X(I)*180/PI,(H(I,J)*C,J=1,N)
      ENDDO
40    FORMAT(122(E12.6,1X))
      IF(LL.EQ.0) WRITE(13,*)'ε e ψmax, Pmax W,ψ'
      LL=LL+1
      WRITE(13,*)EPSON,E,AIMAX,PMAX,SUM1,AI
      CONTAINS
      SUBROUTINE SUBH
      DO I=1,N
      DO J=1,N
      H(I,J)=1.0+EPSON*COS((I-1)*DX)
      ENDDO
      ENDDO
      DO I=1,N
      I1=I-1
      I2=I+1
      IF(I.EQ.1)I1=N1
      IF(I.EQ.N)I2=2
      DO J=2,N1
      HXF(I,J)=(0.5*(H(I2,J)+H(I,J)))**3
      HXB(I,J)=(0.5*(H(I1,J)+H(I,J)))**3
      HYF(I,J)=ALFA*(0.5*(H(I,J+1)+H(I,J)))**3
      HYB(I,J)=ALFA*(0.5*(H(I,J-1)+H(I,J)))**3
      HXY(I,J)=1.0/(HXF(I,J)+HXB(I,J)+HYF(I,J)+HYB(I,J))
```

```
      ENDDO
      ENDDO
      RETURN
      END SUBROUTINE SUBH
      SUBROUTINE SUBP1
      DO I=1,N
      DO J=1,N
      PD(I,J)=P1(I,J)
      ENDDO
      ENDDO
      IK=0
      TEMP0=DX**2*EPSON
 10   C1=0.0
      DO I=1,N1
      I1=I-1
      I2=I+1
      IF(I1.EQ.0)I1=N1
      IF(I2.EQ.N)I2=1
      DO J=2,N1
      P1(I,J)=(HXF(I,J)*P1(I2,J)+HXB(I,J)*P1(I1,J)+HYF(I,J)*P1
  (I,J+1)+HYB(I,J)*P1(I,J-1)+TEMP0*SIN((I-1)*DX))*HXY(I,J)
      C1=C1+ABS(P1(I,J)-PD(I,J))
      PD(I,J)=P1(I,J)
      ENDDO
      ENDDO
      DO J=2,N1
      P1(N,J)=P1(1,J)
      PD(N,J)=PD(1,J)
      ENDDO
      IK=IK+1
      IF(C1.GT.1.E-20.AND.IK.LE.20000)GOTO 10
      RETURN
      END SUBROUTINE SUBP1
      SUBROUTINE SUBP2
      DO I=1,N
      DO J=1,N
      PE(I,J)=P2(I,J)
      ENDDO
      ENDDO
      IK=0
      TEMP1=DX**2
 20   C1=0.0
      DO I=1,N1
      I1=I-1
      I2=I+1
```

```
      IF(I1.EQ.0)I1=N1
      IF(I2.EQ.N)I2=1
      DO J=2,N1
      P2(I,J)=(HXF(I,J)*P2(I2,J)+HXB(I,J)*P2(I1,J)+HYF(I,J)*P2
(I,J+1)+HYB(I,J)*P2(I,J-1)-TEMP1*COS((I-1)*DX))*HXY(I,J)
      C1=C1+ABS(P2(I,J)-PE(I,J))
      PE(I,J)=P2(I,J)
      ENDDO
      ENDDO
      DO J=2,N1
      P2(N,J)=P2(1,J)
      PE(N,J)=PE(1,J)
      ENDDO
      IK=IK+1
      IF(C1.GT.1.E-20.AND.IK.LE.20000)GOTO 20
      RETURN
      END SUBROUTINE SUBP2
      SUBROUTINE SUBMAX
      TEMP0=PI2/N1
      PMAX=P(1,1)
      DO I=1,N
      DO J=1,N
      IF(P(I,J).GE.PMAX)THEN
      PMAX=P(I,J)
      I_MAX=I
      J_MAX=J
      ENDIF
      ENDDO
      ENDDO
      AIMAX=((I_MAX-1)*TEMP0*180)/PI
      RETURN
      END SUBROUTINE SUBMAX
      SUBROUTINE SUBM
      PX=0.0
      PY=0.0
      TEMP=PI/60.0
      DO I=1,N
      AI=(I-1)*TEMP
      DO J=1,N
      PX=PX-P(I,J)*COS(AI)*DXDY
      PY=PY+P(I,J)*SIN(AI)*DXDY
      ENDDO
      ENDDO
      SUM1=SQRT(PX*PX+PY*PY)
      RETURN
```

```
      END SUBROUTINE SUBM
      SUBROUTINE SUBQ
      Q=0.0
      QI1=0.0
      QI2=0.0
      IF(AII.LT.AM(K,1))THEN
      QI1=QQ(1)*(AII-AM(K,2))/(AM(K,1)-AM(K,2))+QQ(2)*(AII-AM
(K,1))/(AM(K,2)-AM(K,1))
      ENDIF
      IF(AII.GT.AM(K,41))THEN
      QI1=QQ(41)*(AII-AM(K,40))/(AM(K,41)-AM(K,40))+QQ(40)*
(AII-AM(K,41))/(AM(K,40)-AM(K,41))
      ENDIF
      IF((AII.GT.AM(K,1)).AND.(AII.LT.AM(K,41)))THEN
      DO L=1,40
      IF((AII.GE.AM(K,L)).AND.(AII.LE.AM(K,L+1)))THEN
      J=L
      ENDIF
      ENDDO
      QI1=QQ(J)*(AII-AM(K,J+1))/(AM(K,J)-AM(K,J+1))+QQ(J+1)*
(AII-AM(K,J))/(AM(K,J+1)-AM(K,J))
      ENDIF
      IF(AKL.NE.0.0)THEN
      IF(AII.LT.AM(K+1,1))THEN
      QI2=QQ(1)*(AII-AM(K+1,2))/(AM(K+1,1)-AM(K+1,2))+QQ(2)*
(AII-AM(K+1,1))/(AM(K+1,2)-AM(K+1,1))
      ENDIF
      IF(AII.GT.AM(K+1,41))THEN
      QI2=QQ(41)*(AII-AM(K+1,40))/(AM(K+1,41)-AM(K+1,40))+QQ
(40)*(AII-AM(K+1,41))/(AM(K+1,40)-AM(K+1,41))
      ENDIF
      IF((AII.GT.AM(K+1,1)).AND.(AII.LT.AM(K+1,41)))THEN
      DO L=1,40
      IF((AII.GE.AM(K+1,L)).AND.(AII.LE.AM(K+1,L+1)))THEN
      I=L
      ENDIF
      ENDDO
      QI2=QQ(I)*(AII-AM(K+1,I+1))/(AM(K+1,I)-AM(K+1,I+1))+QQ(I
+1)*(AII-AM(K+1,I))/(AM(K+1,I+1)-AM(K+1,I))
      ENDIF
      Q=-QI1*(EPSON-0.1*(K+1))*10.0+QI2*(EPSON-0.1*K)*10.0
      ENDIF
      IF(AKL.EQ.0.0)THEN
      Q=QI1
      ENDIF
      RETURN
```

```
      END SUBROUTINE SUBQ
      SUBROUTINE SUBSUM1
      SUM1=0.0
      SUMY1=0.0
      SUMY2=0.0
      IF(ABS(AII).LE.0.05)THEN
      SUM1=S4(K)*(EPSON-EPS(K+1))/(EPS(K)-EPS(K+1))+S4(K+1)*
(EPSON-EPS(K))/(EPS(K+1)-EPS(K))
      ENDIF
      IF(ABS(AII-180.0).LE.0.05)THEN
      SUM1=S3(K)*(EPSON-EPS(K+1))/(EPS(K)-EPS(K+1))+S3(K+1)*
(EPSON-EPS(K))/(EPS(K+1)-EPS(K))
      ENDIF
      IF((AII.GT.0.05).AND.(AII.LT.179.95))THEN
      IF(Q.LT.-25.0)THEN
      SUM1=S4(K)*(EPSON-EPS(K+1))/(EPS(K)-EPS(K+1))+S4(K+1)*
(EPSON-EPS(K))/(EPS(K+1)-EPS(K))
      ENDIF
      IF(Q.GT.25.0)THEN
      SUM1=S3(K)*(EPSON-EPS(K+1))/(EPS(K)-EPS(K+1))+S3(K+1)*
(EPSON-EPS(K))/(EPS(K+1)-EPS(K))
      ENDIF
      ENDIF
      IF((AII.GT.180.05).AND.(AII.LT.359.95))THEN
      IF(Q.LT.-25.0)THEN
      SUM1=S3(K)*(EPSON-EPS(K+1))/(EPS(K)-EPS(K+1))+S3(K+1)*
(EPSON-EPS(K))/(EPS(K+1)-EPS(K))
      ENDIF
      IF(Q.GT.25.0)THEN
      SUM1=S4(K)*(EPSON-EPS(K+1))/(EPS(K)-EPS(K+1))+S4(K+1)*
(EPSON-EPS(K))/(EPS(K+1)-EPS(K))
      ENDIF
      ENDIF
      IF(ABS(Q).LE.25.0)THEN
      IF(Q.LT.-20.0)THEN
      SUMY1=S(K,1)*(Q-QQ(2))/(QQ(1)-QQ(2))+S(K,2)*(Q-QQ(1))/(QQ
(2)-QQ(1))
      SUMY2=S(K+1,1)*(Q-QQ(2))/(QQ(1)-QQ(2))+S(K+1,2)*(Q-QQ
(1))/(QQ(2)-QQ(1))
      SUM1=SUMY1*(EPSON-EPS(K+1))/(EPS(K)-EPS(K+1))+SUMY2*
(EPSON-EPS(K))/(EPS(K+1)-EPS(K))
      ENDIF
      IF(Q.GT.20.0)THEN
      SUMY1=S(K,41)*(Q-QQ(40))/(QQ(41)-QQ(40))+S(K,40)*(Q-QQ
(41))/(QQ(40)-QQ(41))
      SUMY2=S(K+1,41)*(Q-QQ(40))/(QQ(41)-QQ(40))+S(K+1,40)*(Q-
QQ(41))/(QQ(40)-QQ(41))
```

```
    SUM1=SUMY1*(EPSON-EPS(K+1))/(EPS(K)-EPS(K+1))+SUMY2*
(EPSON-EPS(K))/(EPS(K+1)-EPS(K))
    ENDIF
    IF((Q.GE.-20.0).AND.(Q.LE.20.0))THEN
    QZ=Q+21
    J=FLOOR(QZ)
    SUMY1=S(K,J+1)*(Q-QQ(J))/(QQ(J+1)-QQ(J))+S(K,J)*(Q-QQ(J
+1))/(QQ(J)-QQ(J+1))
    SUMY2=S(K+1,J+1)*(Q-QQ(J))/(QQ(J+1)-QQ(J))+S(K+1,J)*(Q-QQ
(J+1))/(QQ(J)-QQ(J+1))
    SUM1=SUMY1*(EPSON-EPS(K+1))/(EPS(K)-EPS(K+1))+SUMY2*
(EPSON-EPS(K))/(EPS(K+1)-EPS(K))
    ENDIF
    ENDIF
    RETURN
    END SUBROUTINE SUBSUM1
    END
```

6.3.4 Calculation Results

According to the parameters of the working conditions and the form of the time dependent load given in the present program, the traces of the journal are shown in Figures 6.3–6.6. In Figure 6.7, the pressure distribution is given for the instantaneous velocity and eccentricity.

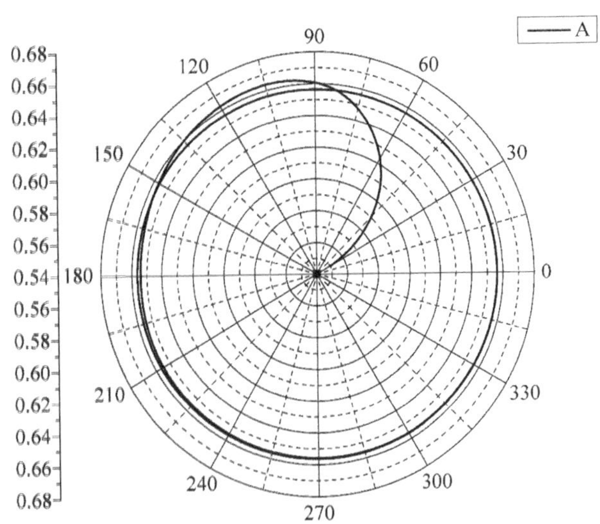

Figure 6.3 Trace of journal with a rotating load (initial parameters $\varepsilon_0 = 0.55, \delta_0 = 30°$, W = 1500).

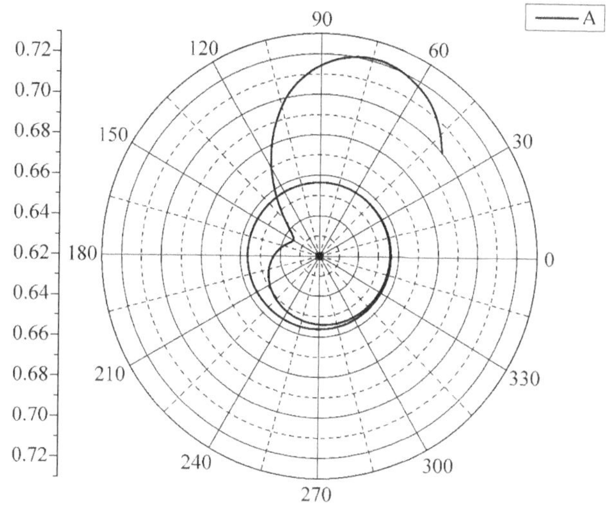

Figure 6.4 Trace of journal with a rotating load (initial parameters $\varepsilon_0 = 0.7, \delta_0 = 40°$, W = 1500).

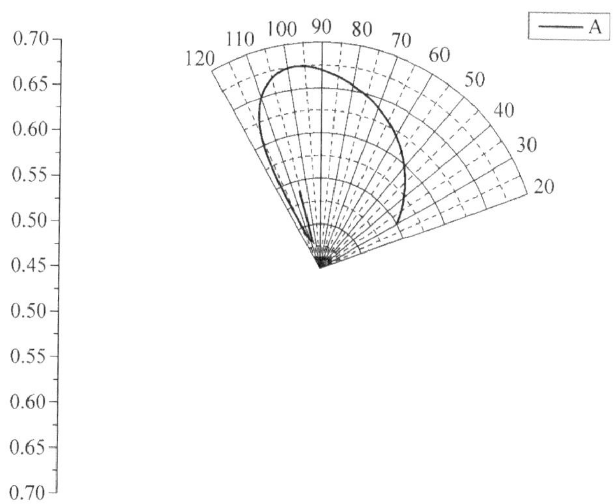

Figure 6.5 Trace of journal with a sudden load (initial parameters $\varepsilon_0 = 0.55$, $\delta_0 = 30°$, W = 1500, $\gamma = 40°$).

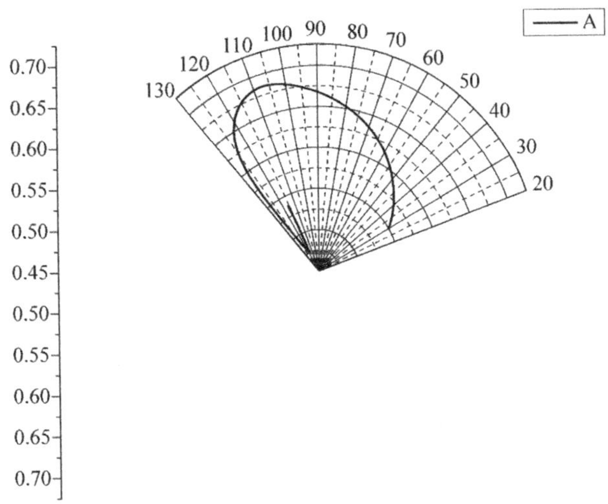

Figure 6.6 Trace of journal with a sudden load (initial parameters $\varepsilon_0 = 0.55, \delta_0 = 30°$, $W = 1500, \gamma = 50°$).

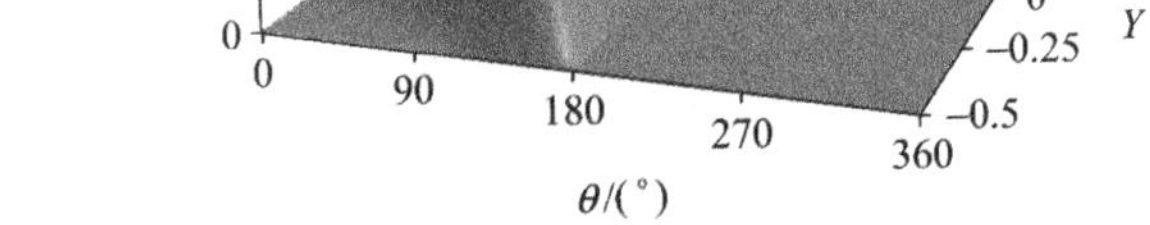

Figure 6.7 Instantaneous pressure distribution of a dynamic journal bearing.

7

Numerical Method and Program for Gas Lubrication

7.1 Basic Equations

7.1.1 General Reynolds Equation of Gas Lubrication

The main feature of gas lubrication is that gas is compressible so that its density is often treated as a variable. Therefore, the Reynolds equation of gas lubrication becomes

$$\frac{\partial}{\partial x}\left(\frac{\rho h^3}{\eta}\frac{\partial p}{\partial x}\right)+\frac{\partial}{\partial y}\left(\frac{\rho h^3}{\eta}\frac{\partial p}{\partial y}\right)=6U\frac{\partial}{\partial x}(\rho h)+6V\frac{\partial}{\partial y}(\rho h)+12\frac{\partial(\rho h)}{\partial t} \tag{7.1}$$

Usually, the viscosity of gas is low. For example, at 20 °C, the viscosity of air is only one-4000th of the viscosity of spindle oil. Therefore, the frictional power loss of gas lubrication can usually be negligible. Furthermore, unlike a liquid, the viscosity of gas slightly increases with an increase of temperature. Therefore, the thermal effect is not very significant unless at a very high speed. Therefore, gas lubrication can usually be considered isothermal and the viscosity of gas is usually considered as a constant.

The relationship of gas density with temperature and pressure is often thought to be

$$\frac{p}{\rho}=gRT \tag{7.2}$$

where T is absolute temperature, R is the gas constant and g is acceleration due to gravity.

For an isothermal process, substituting Equation 7.2 into the Reynolds Equation 7.1, we have

$$\frac{\partial}{\partial x}\left(ph^3\frac{\partial p}{\partial x}\right)+\frac{\partial}{\partial y}\left(ph^3\frac{\partial p}{\partial y}\right)=6\eta U\frac{\partial(ph)}{\partial x}+6\eta V\frac{\partial(ph)}{\partial y}+12\eta\frac{\partial(ph)}{\partial t} \tag{7.3}$$

Numerical Calculation of Lubrication: Methods and Programs, First Edition. Ping Huang.

In numerical calculations, the dimensionless Reynolds equation of gas lubrication is often written as

$$\frac{\partial}{\partial X}\left(PH^3\frac{\partial P}{\partial X}\right)+A^2\frac{\partial}{\partial Y}\left(PH^3\frac{\partial P}{\partial Y}\right)=\Lambda_x\frac{\partial(PH)}{\partial X}+\Lambda_y\frac{\partial(PH)}{\partial Y}+\sigma\frac{\partial(PH)}{\partial T} \tag{7.4}$$

where P is dimensionless pressure ($P=p/p_a$, where p_a is ambient pressure), H is dimensionless film thickness ($H=h/h_0$, where h_0 is minimum film thickness), X and Y are dimensionless coordinates ($X=x/l$ and $Y=y/b$, where l is length and b is width), A is the ratio of length to width ($A=l/b$), Λ is the bearing number ($\Lambda_x=6\eta Ul/p_a/h_0^2$ and $\Lambda_y=6\eta VAl/p_a/h_0^2$), η is gas viscosity, U is sliding velocity, T is dimensionless time ($T=t/\omega_0$), ω_0 is the characteristic resonance frequency and σ is the squeeze parameter ($\sigma=12\eta\omega_0 l^2/p_a/h_0^2$).

7.2 Numerical Method of Gas Lubrication

7.2.1 Basic Equations of Steady and Isothermal Gas Lubrication

The Reynolds equation for steady and isothermal gas lubrication can be written as

$$\frac{\partial}{\partial x}\left(ph^3\frac{\partial p}{\partial x}\right)+\frac{\partial}{\partial y}\left(ph^3\frac{\partial p}{\partial y}\right)=6\eta U\frac{\partial(ph)}{\partial x} \tag{7.5}$$

Therefore, the dimensionless Reynolds equation is

$$\frac{\partial}{\partial X}\left(H^3\frac{\partial P^2}{\partial X}\right)+\alpha\frac{\partial}{\partial Y}\left(H^3\frac{\partial P^2}{\partial Y}\right)=\Lambda\frac{\partial(PH)}{\partial X} \tag{7.6}$$

where $\Lambda=\dfrac{6\eta Ul}{p_a h_0^2}$ is the bearing number or compressibility coefficient; and $\alpha=(l/b)^2$.

If we omit the left term with Y in Equations 7.5 or 7.6, we can obtain a one-dimensional Reynolds equation of steady and isothermal gas lubrication.

For a journal bearing, the Reynolds equation can be expressed by the cylindrical coordinates. The dimensionless parameters are $P=p/p_a$, $H=h/c$, $\theta=x/R$ and $Y=y/b$. Therefore, the Reynolds equation becomes

$$\frac{\partial}{\partial \theta}\left(H^3\frac{\partial P^2}{\partial \theta}\right)+\alpha\frac{\partial}{\partial Y}\left(H^3\frac{\partial P^2}{\partial Y}\right)=\Lambda\frac{\partial(PH)}{\partial \theta} \tag{7.7}$$

where $\Lambda=\dfrac{6\eta UR}{p_a c^2}$ is the bearing number or compressibility coefficient.

It should be pointed out that, for gas lubrication, the pressure boundary condition cannot be set to zero, as in liquid lubrication. The absolute boundary condition must be used. The dimensionless pressure condition will be $P=1$. For common situations, it corresponds to a pressure of one atmosphere, that is, 1.013×10^5 Pa.

7.2.2 *Numerical Method*

7.2.2.1 Discrete Equations of One-Dimensional Gas Lubrication

By using the equidistant difference formula and omitting the term in the y direction of Equation 7.6, we have the discrete Reynolds equation of one-dimensional gas lubrication.

$$\frac{H_{i+1/2}^3 P_{i+1}^2 - (H_{i+1/2}^3 + H_{i-1/2}^3)P_i^2 + H_{i-1/2}^3 P_{i-1}^2}{\Delta X^2} = -\frac{P_{i+1}H_{i+1} - P_{i-1}H_{i-1}}{\Delta X} \quad (7.8)$$

The corresponding iteration equation is

$$P_i^2 = \frac{\Delta X(P_{i+1}H_{i+1} - P_{i-1}H_{i-1}) + H_{i+1/2}^3 P_{i+1}^2 + H_{i-1/2}^3 P_{i-1}^2}{H_{i+1/2}^3 + H_{i-1/2}^3} \quad (7.9)$$

7.2.2.2 Discrete Equations of Two-Dimensional Gas Lubrication

By incorporating the equidistant difference formula into Equation 7.6, we have the numerical Reynolds equation of gas lubrication for surface contact.

$$\begin{aligned}
&\frac{H_{i+1/2,j}^3 P_{i+1,j}^2 - (H_{i+1/2,j}^3 + H_{i-1/2,j}^3)P_{i,j}^2 + H_{i-1/2,j}^3 P_{i-1,j}^2}{\Delta X^2} \\
&+\alpha \frac{H_{i,j+1/2}^3 P_{i,j+1}^2 - (H_{i,j+1/2}^3 + H_{i,j-1/2}^3)P_{i,j}^2 + H_{i,j-1/2}^3 P_{i,j-1}^2}{\Delta Y^2} \\
&= -\frac{P_{i+1,j}H_{i+1,j} - P_{i-1,j}H_{i-1,j}}{\Delta X}
\end{aligned} \quad (7.10)$$

The corresponding iteration equation is

$$P_{i,j}^2 = \frac{\Delta X(P_{i+1,j}H_{i+1,j} - P_{i-1,j}H_{i-1,j}) + H_{i+1/2,j}^3 P_{i+1,j}^2 + H_{i-1/2,j}^3 P_{i-1,j}^2 + \alpha(\Delta X/\Delta Y)^2(H_{i,j+1/2}^3 P_{i,j+1}^2 + H_{i,j-1/2}^3 P_{i,j-1}^2)}{H_{i+1/2,j}^3 + H_{i-1/2,j}^3 + \alpha(\Delta X/\Delta Y)^2(H_{i,j+1/2}^3 + H_{i,j-1/2}^3)} \quad (7.11)$$

7.2.2.3 Discrete Equations of Journal Bearing Gas Lubrication

By using the equidistant difference formula, the discrete Reynolds equation of gas lubrication of a journal bearing is as follows

$$\frac{H^3_{i+1/2,j}P^2_{i+1,j} - (H^3_{i+1/2,j} + H^3_{i-1/2,j})P^2_{i,j} + H^3_{i-1/2,j}P^2_{i-1,j}}{\Delta\theta^2}$$
$$+\alpha\frac{H^3_{i,j+1/2}P^2_{i,j+1} - (H^3_{i,j+1/2} + H^3_{i,j-1/2})P^2_{i,j} + H^3_{i,j-1/2}P^2_{i,j-1}}{\Delta Y^2} \tag{7.12}$$
$$= -\frac{P_{i+1,j}H_{i+1,j} - P_{i-1,j}H_{i-1,j}}{\Delta\theta}$$

The corresponding iteration equation is

$$P^2_{i,j} =$$
$$\frac{\Delta\theta(P_{i+1,j}H_{i+1,j} - P_{i-1,j}H_{i-1,j}) + H^3_{i+1/2,j}P^2_{i+1,j} + H^3_{i-1/2,j}P^2_{i-1,j} + \alpha(\Delta\theta/\Delta Y)^2(H^3_{i,j+1/2}P^2_{i,j+1} + H^3_{i,j-1/2}P^2_{i,j-1})}{H^3_{i+1/2,j} + H^3_{i-1/2,j} + \alpha(\Delta\theta/\Delta Y)^2(H^3_{i,j+1/2} + H^3_{i,j-1/2})} \tag{7.13}$$

When the iterative method is used to solve the above nonlinear equations, the initial pressure of each node must be given first, usually to set them equal to the atmospheric pressure. Then, the dimensionless form will be $P_{ij} = 1$.

7.3 Calculation Program for Gas Lubrication

7.3.1 Calculation Program and Solutions of One-Dimensional Gas Lubrication

Similar to hydrodynamic lubrication, only three parts of the iterative procedure are changed: (1) the new pressure iteration formula is used; (2) liquid viscosity is changed to air viscosity, that is, $\eta = 1.79 \times 10^{-6}$ Pa·s; (3) the nonzero boundary condition $P = 1$ is used. Similarly, in the program, KG = 1 corresponds to the linear slider of gas lubrication, and KG = 2 corresponds to the curve slider of gas lubrication. The calculation diagram is also similar to Figure 2.2. The calculation program is

```
      PROGRAM GASSLIDER
      DIMENSION X(121),H(121),P(121)
      COMMON /COM1/X1,X2,H1,H2,U,EDA,AL,ALOAD,DX
      DATA N,U,X1,X2,H1,H2,EDA,AL/121,1.0,0.0,1.0,1.0,0.5,
 1.79E-5,0.01/
      OPEN(7,FILE='SLIDER.DAT',STATUS='UNKNOWN')
      WRITE(*,*)'If KG=1: Straight slider; KG=2: Curve slider;
 Input KG='
      READ(*,*)KG
      IF(KG.EQ.2)THEN
      X1=-1.0
      X2=1.0
      ELSE
```

```
      KG=1
      ENDIF
      CALL SUBH(KG,N,X,H)
      CALL SUBP(N,X,H,P)
      CALL OUTPUT(KG,N,X,H,P)
      STOP
      END
      SUBROUTINE OUTPUT(KG,N,X,H,P)
      DIMENSION X(N),H(N),P(N)
      COMMON /COM1/X1,X2,H1,H2,U,EDA,AL,ALOAD,DX
      X0=0.0
      DO I=1,N
      X0=X0+P(I)*X(I)
      ENDDO
      X0=X0*AL
      ALOAD=ALOAD*DX*AL*6.0*U*EDA*AL/H2**2
      WRITE(*,*)N,ALOAD,X0
      DO I=1,N
      IF(KG.EQ.1)THEN
      P0=-(-1.0/(H(I)*H2)+H1*H2/(H1+H2)/(H2*H(I))**2+1.0/
(H1+H2))/(H1/H2-1.0)*H2
      WRITE(7,40) X(I),H(I),P(I),P0
      ELSE
      WRITE(7,40) X(I),H(I),P(I)
      ENDIF
      END DO
40    FORMAT(1X,4(E12.6,1X))
      RETURN
      END
      SUBROUTINE SUBH(KG,N,X,H)
      DIMENSION X(N),H(N)
      COMMON /COM1/X1,X2,H1,H2,U,EDA,AL,ALOAD,DX
      DX=1./(N-1.0)
      DO I=1,N
      IF(KG.EQ.1) THEN
      X(I)=X1-(I-1)*DX*(X1-X2)
      H(I)=H1/H2-(H1/H2-1.0)*X(I)
      ELSE
      X(I)=X1-(I-1)*DX*(X1-X2)
      H(I)=1.0+(H1/H2-1.0)*X(I)*X(I)
      ENDIF
      ENDDO
      RETURN
      END
      SUBROUTINE SUBP(N,X,H,P)
```

```
      DIMENSION X(N),H(N),P(N)
      COMMON /COM1/X1,X2,H1,H2,U,EDA,AL,ALOAD,DX
      DO I=2,N-1
      P(I)=0.5
      ENDDO
      P(1)=1.0
      P(N)=1.0
      IK=0
10    C1=0.0
      ALOAD=0.0
      DO I=2,N-1
      A1=(0.5*(H(I+1)+H(I)))**3
      A2=(0.5*(H(I)+H(I-1)))**3
      PD=P(I)
      P(I)=(-DX*(P(I+1)*H(I+1)-P(I-1)*H(I-1))+A1*P(I+1)**2
+A2*P(I-1)**2)/(A1+A2)
      IF(P(I).LT.0.0)P(I)=0.0
      P(I)=SQRT(P(I))
      P(I)=0.5*PD+0.5*P(I)
      C1=C1+ABS(P(I)-PD)
      ALOAD=ALOAD+P(I)
      ENDDO
      ERO=C1/ALOAD
      IK=IK+1
      IF(ERO.GT.1.E-7)GOTO 10
      RETURN
      END
```

The numerical solutions of gas lubrication for a one-dimensional slider are shown in Figure 7.1.

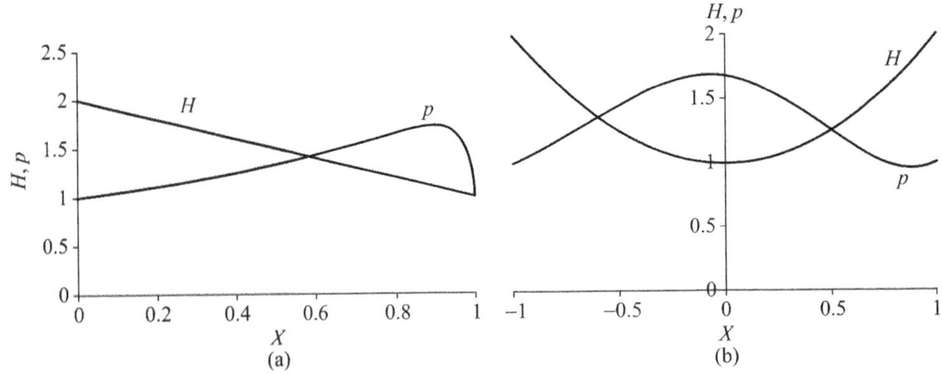

Figure 7.1 Numerical solutions of gas lubrication for a one-dimensional slider.

7.3.2 *Numerical Program and Solutions of Two-Dimensional Gas Lubrication*

Similar to hydrodynamic lubrication, the numerical program and solutions of two-dimensional gas lubrication are given as follows. In the program, KG = 1 is for a plane slider while KG = 2 is for a curved surface slider. The calculation diagram of the program is similar to Figure 3.1, but the iterative formula and the boundary conditions are different.

```
      PROGRAM GASSURFACE
      DIMENSION X(121),Y(121),H(121,121),P(121,121)
      COMMON /COM1/DX,ALFA,X1,X2,H1,H2,ALOAD,U,EDA,ALX,ALY,ALENDA
      DATA N,EDA,ALX,ALY,U,H1,H2,X1,X2/121,1.79E-
5,0.03,0.024,1.0,0.1,0.05,0.0,1.0/
      OPEN(8,FILE='PRESSURE.DAT',STATUS='UNKNOWN')
      OPEN(9,FILE='FILM.DAT',STATUS='UNKNOWN')
      WRITE(*,*)'If KG=1: Plane surface; KG=2: Curve surface;
Input KG='
      READ(*,*)KG
      CALL SUBH(KG,N,X,Y,H)
      CALL SUBP(N,X,Y,H,P)
      CALL OUTPUT(N,X,Y,H,P)
      STOP
      END
      SUBROUTINE SUBH(KG,N,X,Y,H)
      DIMENSION X(N),Y(N),H(N,N)
      COMMON /COM1/DX,ALFA,X1,X2,H1,H2,ALOAD,U,EDA,ALX,ALY,
ALENDA
      IF(KG.EQ.2)THEN
      X1=-1.0
      X2=1.0
      ELSE
      KG=1
      ENDIF
      DX=1.0/(N-1.0)
      ALFA=(ALX/ALY)**2
      DO I=1,N
      X(I)=X1-(I-1)*DX*(X1-X2)
      IF(KG.EQ.1)Y(I)=-0.5-(I-1)*DX*(X1-X2)
      IF(KG.EQ.2)Y(I)=-1.0-(I-1)*DX*(X1-X2)
      ENDDO
      DO I=1,N
      DO J=1,N
      IF(KG.EQ.1)H(I,J)=H1/H2-X(I)*(H1/H2-1.0)
      IF(KG.EQ.2)H(I,J)=1.0+(X(I)*X(I)+Y(J)*Y(J))*(H1/H2-1.0)
      ENDDO
```

```
      ENDDO
      RETURN
      END
      SUBROUTINE SUBP(N,X,Y,H,P)
      DIMENSION X(N),Y(N),H(N,N),P(N,N)
      COMMON /COM1/DX,ALFA,X1,X2,H1,H2,ALOAD,U,EDA,ALX,ALY,
ALENDA
      DATA PA/1.013E5/
      ALENDA=6.0*EDA*U*ALX/PA/(H2*1.0E-5)**2
      DO I=1,N
      P(I,1)=1.0
      P(I,N)=1.0
      P(1,I)=1.0
      P(N,I)=1.0
      ENDDO
      DO I=2,N-1
      DO J=2,N-1
      P(I,J)=1.5
      ENDDO
      ENDDO
      IK=0
10    C1=0.0
      ALOAD=0.0
      DO I=2,N-1
      I1=I-1
      I2=I+1
      DO J=2,N-1
      J1=J-1
      J2=J+1
      PD=P(I,J)
      A1=(0.5*(H(I2,J)+H(I,J)))**3
      A2=(0.5*(H(I,J)+H(I1,J)))**3
      A3=ALFA*(0.5*(H(I,J2)+H(I,J)))**3
      A4=ALFA*(0.5*(H(I,J)+H(I,J1)))**3
      P(I,J)=(-DX*ALENDA*(P(I+1,J)*H(I+1,J)-P(I-1,J)*H(I-1,J))+A1*P
 (I2,J)**2+A2*P(I1,J)**2+A3*P(I,J2)**2+A4*P(I,J1)**2)/(A1+A2+A3+A4)
      IF(P(I,J).LT.0.0)P(I,J)=0.0
      P(I,J)=SQRT(P(I,J))
      P(I,J)=0.7*PD+0.3*P(I,J)
      IF(P(I,J).LT.0.0)P(I,J)=0.0
      C1=C1+ABS(P(I,J)-PD)
      ALOAD=ALOAD+P(I,J)
      ENDDO
      ENDDO
      IK=IK+1
```

```
      C1=C1/ALOAD
      WRITE(*,*)IK,C1,ALOAD
      IF(C1.GT.1.E-7)GOTO 10
      RETURN
      END
      SUBROUTINE OUTPUT(N,X,Y,H,P)
      DIMENSION X(N),Y(N),H(N,N),P(N,N)
      COMMON /COM1/DX,ALFA,H1,H2,ALOAD,U,EDA,ALX,ALY,ALENDA
      ALENDA=6.0*U*EDA*ALX/H2**2
      ALOAD=ALOAD*ALENDA*DX*DX*ALX*ALY/(N-1.0)/(N-1.0)
      WRITE(8,40)Y(1),(Y(I),I=1,N)
      DO I=1,N
      WRITE(8,40)X(I),(P(I,J),J=1,N)
      ENDDO
      WRITE(9,40)Y(1),(Y(I),I=1,N)
      DO I=1,N
      WRITE(9,40)X(I),(H(I,J),J=1,N)
      ENDDO
40    FORMAT(122(E12.6,1X))
      STOP
      END
```

The numerical solutions of gas lubrication for a two-dimensional slider are shown in Figure 7.2.

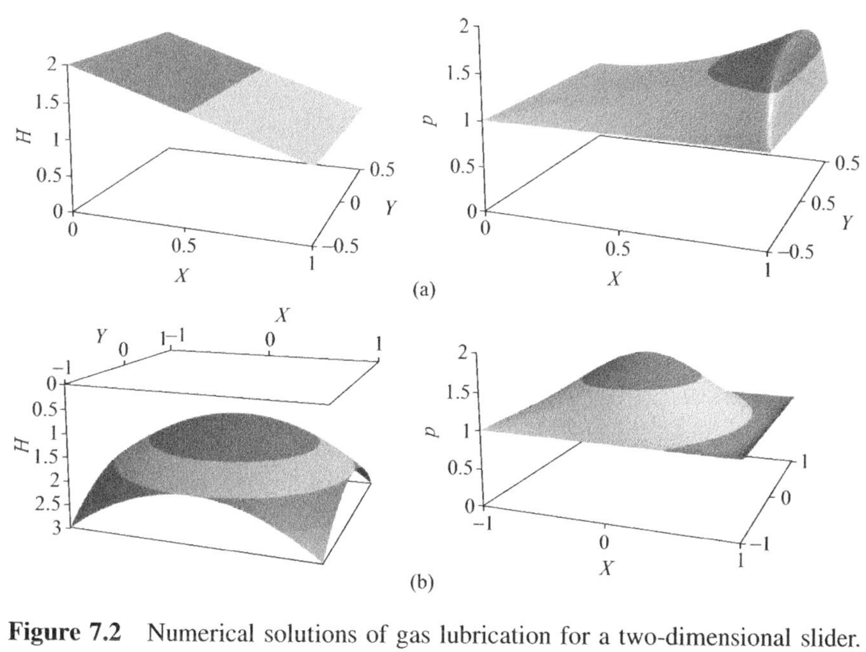

Figure 7.2 Numerical solutions of gas lubrication for a two-dimensional slider.

7.3.3 *Numerical Program and Solutions of Journal Bearing Gas Lubrication*

7.3.3.1 Calculation Working Conditions

The calculation program of steady-state gas lubrication for a journal bearing includes one main program and two subroutines, which are the subroutine SUBH to calculate the film thickness and the subroutine SUBP to calculate the pressure.

Pre-assignment parameters and their values	
Bearing width	B = 60.0 E–3 m
Journal radius	R = 25.0 E–3 m
Radius clearance	C0 = 5.0 E–5 m
Work rotational speed	AN = 6.0 E4 m/s
Environmental pressure	PA = 1.013 E5 Pa
Gas dynamic viscosity of gas	EDA = 1.79 E–5 Pa·s
Eccentricity	EPSON = 0.7

The pre-assignment parameters can be modified depending on the user's requirement, but the program needs to be re-compiled and re-linked before running.

The output parameters are the pressure distribution P(I, J) in the file PRESSURE.DAT and the film thickness H (I, J) in the file FILM.DAT. The calculation diagram is similar to Figure 4.3, but the iterative formula and boundary conditions are different.

7.3.3.2 Calculation Program

```
      PROGRAM GASJOURNAL
      DIMENSION H(61,21),P(61,21)
      DATA B,R,C0,AN,PA,EDA,EPSON/60.0E-3,25.0E-3,5.0E-
5,6.0E4,1.013E5,1.79E-5,0.7/
      OPEN(9,FILE='PRESSURE.DAT',STATUS='UNKNOWN')
      OPEN(8,FILE='FILM.DAT',STATUS='UNKNOWN')
      PI=3.1415926
      N=61
      M=21
      DX=2.0*PI/FLOAT(N-1)
      DY=1./FLOAT(M-1)
      OMEGA=AN*2.0*PI/60.0
      U=OMEGA*R
      ALENDA=6.0*EDA*U*R/PA/C0**2
      ALFA=(R/B*DX/DY)**2
      CALL SUBH(N,M,DX,EPSON,H)
      CALL SUBP(N,M,DX,EPSON,ALFA,ALENDA,H,P)
```

```
      CALL OUTPUT(N,M,DX,DY,H,P)
      STOP
      END
      SUBROUTINE SUBH(N,M,DX,EPSON,H)
      DIMENSION H(N,M)
      DO I=1,N
      SETA=(I-1.0)*DX
      DO J=1,M
      H(I,J)=1.0+EPSON*COS(SETA)
      ENDDO
      ENDDO
      RETURN
      END
      SUBROUTINE SUBP(N,M,DX,EPSON,ALFA,ALENDA,H,P)
      DIMENSION H(N,M),P(N,M)
      DO I=2,N-1
      DO J=2,M-1
      P(I,J)=1.1
      ENDDO
      ENDDO
      DO I=1,N
      P(I,1)=1.0
      P(I,M)=1.0
      ENDDO
      DO J=1,M
      P(1,J)=1.0
      P(N,J)=1.0
      ENDDO
      IK=0
10    C1=0.0
      ALOAD=0.0
      DO I=2,N-1
      I1=I-1
      I2=I+1
      DO J=2,M-1
      J1=J-1
      J2=J+1
      PD=P(I,J)
      A1=(0.5*(H(I2,J)+H(I,J)))**3
      A2=(0.5*(H(I,J)+H(I1,J)))**3
      A3=ALFA*(0.5*(H(I,J2)+H(I,J)))**3
      A4=ALFA*(0.5*(H(I,J)+H(I,J1)))**3
      P(I,J)=(-DX*ALENDA*(P(I+1,J)*H(I+1,J)-P(I-1,J)*H(I-1,
J))+A1*P(I2,J)**2+A2*P(I1,J)**2+A3*P(I,J2)**2+A4*P(I,J1)**2)/
(A1+A2+A3+A4)
```

```
      IF(P(I,J).LT.0.0)P(I,J)=0.0
      P(I,J)=SQRT(P(I,J))
      P(I,J)=0.7*PD+0.3*P(I,J)
      IF(P(I,J).LT.0.0)P(I,J)=0.0
      C1=C1+ABS(P(I,J)-PD)
      ALOAD=ALOAD+P(I,J)
      ENDDO
      ENDDO
      IK=IK+1
      C1=C1/ALOAD
      WRITE(*,*)IK,C1,ALOAD
      IF(C1.GT.1.E-7)GOTO 10
      RETURN
      END
      SUBROUTINE OUTPUT(N,M,DX,DY,H,P)
      DIMENSION Y(21),H(N,M),P(N,M)
      DO J=1,M
      Y(J)=(J-1.)*DY-0.5
      ENDDO
      WRITE(8,40)Y(1),(Y(J),J=1,M)
      WRITE(9,40)Y(1),(Y(J),J=1,M)
      DO I=1,N
      AX=(I-1.0)*360.0/(N-1.0)
      WRITE(8,40)AX,(H(I,J),J=1,M)
      WRITE(9,40)AX,(P(I,J),J=1,M)
      ENDDO
40    FORMAT(22(E12.6,1X))
      STOP
      END
```

7.3.3.3 Calculation Results

With the given program, the given working parameters and the given film thickness (Figure 7.3a), the pressure distribution can be obtained as shown in Figure 7.3b. The results show that there exists a small region of negative pressure ($P < 1$).

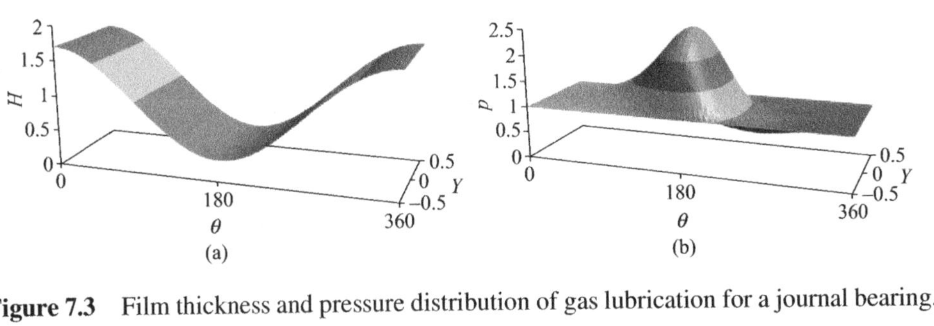

Figure 7.3 Film thickness and pressure distribution of gas lubrication for a journal bearing.

8

Numerical Method and Program for Rarefied Gas Lubrication

8.1 Basic Equations

The general equations for gas lubrication are based on compressible gas dynamics equations for a continuous medium. The compressible Reynolds equation is often used to solve these problems. However, under near-vacuum conditions or in a very small clearance, the common model based on gas dynamics theory no longer applies.

The conventional hydrodynamic theory cannot explain a flow under near-vacuum conditions or in a very small clearance because of the rarefaction effect. Such an effect will appear in microelectro-mechanical systems (MEMS), the magnetic head/disk of a hard disk and so on.

Rarefied gas flows were first studied in a high-altitude environment. When a high-velocity object flies at high altitude, it is in rarified conditions. Aerospace is related to rarefied gas dynamics and heat transfer. With the development of nanotechnology, rarified gas problems are found in many other fields, such as plasma material processing, micro-etching, fine chemicals, vacuum systems and so on. The following discussion focuses on the rarefied gas lubrication effect in the magnetic head and hard disk of the computer.

In order to describe the extent of deviation from a continuous medium, Knudsen defined a dimensionless parameter K_n, that is, the Knudsen number. Its expression is

$$K_n = \frac{\lambda}{h_0} \tag{8.1}$$

where λ is the molecular mean free path of the gas and h_0 is the characteristic size of the gas flow, which is usually a channel height or tube diameter.

Numerical Calculation of Lubrication: Methods and Programs, First Edition. Ping Huang.

For an ideal gas in equilibrium, the molecular mean free path is related to pressure, which can be expressed as

$$\lambda = \frac{1}{\sqrt{2}\pi d^2 n} = \frac{kT}{\sqrt{2}\pi d^2 p} \tag{8.2}$$

where d is the molecular diameter, n is the number density of the gas, p is pressure, k is the Boltzmann constant and T is absolute temperature.

When the molecular mean free path of gas changes only a little, the smaller the characteristic size h_0, the larger is the Knudsen number K_n. Therefore, the problem may deviate largely from the classical gas lubrication problem if the device is small, that is, on a micron or nanometer scale.

K_n expresses the extent of ratification. A large K_n means a significant rarified effect. According to the value of K_n, the gas flow region can be divided into the following four regimes.

1. When $K_n \leq 0.01$, the flow regime is considered to be a continuous medium region. In this region, the flow has no-slip boundary conditions and obeys the Navier–Stokes equation.
2. When $0.01 < |K_n| \leq 0.1$, the flow regime is a slip flow. It is usually assumed that a no-slip boundary condition no longer applies. A mean free path from the bottom of the Knudsen layer begins to play a control role between wall and fluid. In a slip flow regime, the flow is under the control of Navier–Stokes equations and the velocity slip and temperature jump conditions are used to establish the rarified effect model through the wall slip.
3. When $0.1 < |K_n| \leq 10$, it is a transition flow regime. In this regime, the Navier–Stakes equations are no longer valid. The effect of molecules colliding with each other must be considered.
4. When $K_n > 10$, the fluid is considered to be in a free molecular flow. In this regime, each collision between molecules can be ignored compared with the role of collisions between molecules and the boundary.

The compressible gas lubrication Reynolds equation based on the continuum medium assumption and simplified approach in the previous chapter is no longer valid when the gas film thickness is of the same order as the gas molecular mean free path. The influence of the gas rarefaction effect must be considered.

Fukui and Kanedo (1988) analyzed ultra-thin gas film lubrication based on a linearized Boltzmann equation to obtain a generalized lubrication Reynolds equation suitable for rarefied gas lubrication. Its dimensionless form is

$$\frac{\partial}{\partial X}\left(PH^3Q\frac{\partial P}{\partial X}\right) + A^2\frac{\partial}{\partial Y}\left(PH^3Q\frac{\partial P}{\partial Y}\right) = \Lambda_x\frac{\partial(PH)}{\partial X} + \Lambda_y\frac{\partial(PH)}{\partial Y} + \sigma\frac{\partial(PH)}{\partial T} \tag{8.3}$$

where P is dimensionless pressure ($P=p/p_0$), p_0 is ambient pressure, H is dimensionless film thickness ($H=h/h_0$), h_0 is minimum film thickness, X and Y are dimensionless coordinates ($X=x/l$; $Y=y/b$), A is the ratio of length l and head width b ($A=l/b$), Λ is bearing number ($\Lambda_x=6\eta Ul/p_0/h_0^2$; $\Lambda_y=6\eta Vb/p_0/h_0^2$), η is gas viscosity, U is sliding velocity, Q is a dimensionless coefficient of flow, T is dimensionless time ($T=t/\omega_0$), ω_0 is the characteristic resonance frequency and σ is the squeeze constant ($\sigma=12\eta\omega_0 l^2/p_0/h_0^2$).

In Equation 8.3, the dimensionless poiseuille flow factor including the gas rarefaction effect Q can be expressed as

$$Q=\frac{Q_P}{Q_{Con}}=f(D,\alpha) \tag{8.4}$$

where $D=D_0PH$, D_0 is the inverse Knudsen number determined by minimum film thickness h_0 and environmental pressures p_a [$D_0=p_0h_0/\left(\eta\sqrt{2RT_0}\right)$], R is the gas constant (287 J/kg·K), T_0 is room temperature (293 K), Q_{con} is the continuum poiseuille flow factor (taking the value of $D/6$) and α is the accommodation coefficient.

The poiseuille flow factor Q_p is a function of the inverse Knudsen number D. Fukui and Kaneko (1990) also give polynomial formulas by fitting numerical calculation data for the flow coefficient with the inverse Knudsen number D. The formulas are

$$Q_p=\begin{cases} D/6+1.0162+1.0653/D-2.1354/D^2 & D\leq 5 \\ 0.13852D+1.25087+0.15653/D-0.00969/D^2 & 0.15\leq D<5 \\ -2.22919D+2.10673+0.01653/D-0.0000694/D^2 & 0.01\leq D<0.15 \end{cases} \tag{8.5}$$

Further, an analytical formula fit analytically with the above numerical results by the authors is proposed, which has a wider range

$$Q_p=\frac{D}{6}+1.0162+0.40134\ln\left(1+\frac{1.2477}{D}\right) \tag{8.6}$$

8.2 Numerical Method of Rarefied Gas Lubrication

8.2.1 Rarefied Gas Lubrication Model

Figure 8.1 shows a typical gas-lubricated slider.

The gas film thickness $h(x, y)$ is related to the slider attitude, that is, the pitch angle θ. It can be expressed as

$$h(x,y)=h_0+(l-x)\sin\theta \tag{8.7}$$

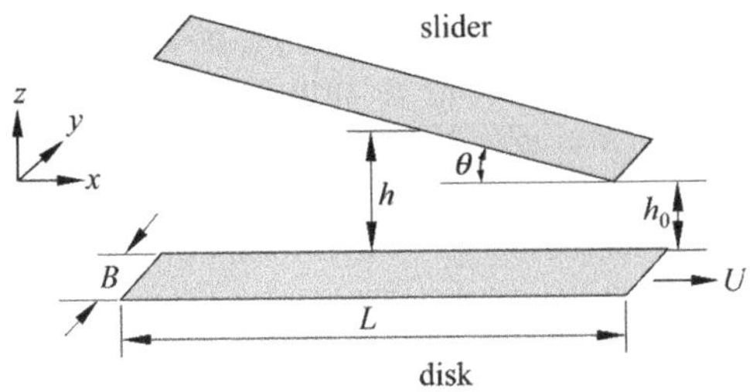

Figure 8.1 Gas lubrication slider model.

where h_0 is the minimum gas film thickness, l is the length of the slider and b is the width.

For a steady problem, the time term is not considered. Therefore, the dimensionless Reynolds equation can be simplified as

$$\frac{\partial}{\partial X}\left(PH^3Q\frac{\partial P}{\partial X}\right)+A^2\frac{\partial}{\partial Y}\left(PH^3Q\frac{\partial P}{\partial Y}\right)=\Lambda_x\frac{\partial(PH)}{\partial X} \tag{8.8}$$

The dimensionless pressure boundary condition of the above equation is $P|_s=1.0$

8.2.2 *Treatment of the Ultra-Thin Gas Film Lubrication Equation*

In ultra-thin gas lubrication, the bearing number Λ is very large, usually more than 1 000 000. This makes the solution severely unstable. In order to solve Equation 8.8, the following two problems must be solved first.

1. The bearing consists of two shear flow items, Λ_x and Λ_y. When h_0 is small, they are far greater than the other items. Therefore, if the traditional lubrication calculation is still used, the large number will make the iterative process unstable.
2. In the incompressible Reynolds equation, the shear flow does not contain the pressure, so the pressure must be solved from the left-hand items of Equation 8.8. However, due to gas compressibility, the right-hand items now contain pressure p. This provides a new possibility for solving the equation.

Then, the dimensionless Reynolds equation for steady ultra-thin gas lubrication can be rewritten as

$$\Lambda\frac{\partial}{\partial X}(PH)=\frac{\partial}{\partial X}\left(QH^3\frac{\partial P^2}{\partial X}\right)+A^2\frac{\partial}{\partial Y}\left(QH^3\frac{\partial P^2}{\partial Y}\right) \tag{8.9}$$

8.3 Discretization and Iteration of Modified Reynolds Equation

8.3.1 Discrete Equation

By using the upwind scheme, Equation 8.9 can be discretized as follows

$$2\Lambda(P_{i,j}H_{i,j}-P_{i-1,j}H_{i-1,j})/\Delta X_i=\left[q_{i+1/2,j}(P_{i+1,j}^2-P_{i,j}^2)-q_{i-1/2,j}(P_{i,j}^2-P_{i-1,j}^2)\right]/\Delta X_i^2$$
$$+A^2\left[q_{i,j+1/2}(P_{i,j+1}^2-P_{i,j}^2)-q_{i,j-1/2}(P_{i,j}^2-P_{i,j-1}^2)\right]/\Delta Y_j^2 \tag{8.10}$$

where $q=QH^3$ is the volume flow coefficient, $q_{i+1/2,j}=\dfrac{2q_{i+1,j}q_{i,j}}{q_{i+1,j}+q_{i,j}}$.

After finishing, we have

$$P_{i,j}=\frac{2P_{i-1,j}\Lambda H_{i-1,j}/\Delta X_i+\left(q_{i+1/2,j}P_{i+1,j}^2+q_{i-1/2,j}P_{i-1,j}^2\right)/\Delta X_i^2+A^2\left(q_{i,j+1/2}P_{i,j+1}^2+q_{i,j-1/2}P_{i,j-1}^2\right)/\Delta Y_j^2}{2\Lambda H_{i,j}/\Delta X_i+P_{i,j}\left(q_{i+1/2,j}+q_{i-1/2,j}\right)/\Delta X_i^2+A^2P_{i,j}\left(q_{i,j+1/2}+q_{i,j-1/2}\right)/\Delta Y_j^2} \tag{8.11}$$

8.3.2 Iteration Method

The super relaxation iteration method is used to solve Equation 8.11 as follows

$$\hat{P}_{i,j}=\alpha\bar{P}_{i,j}+(1-\alpha)\tilde{P}_{i,j}, \tag{8.12}$$

where $\bar{P}$, $\tilde{P}$ and $\hat{P}$ are the variables before, during and after the current iteration, respectively, and α is a relaxation parameter, $0<\alpha\le 1$.

In order to obtain the converged solution, the relaxation parameter α starts with a small value, for example, 0.1 and gradually increases to 1.0 during the iteration. In order to monitor the process, the following strict and direct convergence criterion is used.

$$E_{i,j}\le\varepsilon \tag{8.13}$$

where ε is the error convergence precision, taking values from 0.01 to 0.00001; and E_{ij} are node residuals which can be obtained by the following expression.

$$E_{i,j}=2\Lambda\left(P_{i,j}H_{i,j}-P_{i-1,j}H_{i-1,j}\right)/\Delta X_i$$
$$-\left[q_{i+1/2,j}\left(P_{i+1,j}^2-P_{i,j}^2\right)-q_{i-1/2,j}\left(P_{i,j}^2-P_{i-1,j}^2\right)\right]/\Delta X_i^2$$
$$-A^2\left[q_{i,j+1/2}\left(P_{i,j+1}^2-P_{i,j}^2\right)-q_{i,j-1/2}\left(P_{i,j}^2-P_{i,j-1}^2\right)\right]/\Delta Y_j^2 \tag{8.14}$$

If the convergence condition is satisfied, the iteration process ends. Otherwise, we change the iteration factor α to iterate the pressure again until it meets the convergence conditions.

8.4 Calculation Program for Rarefied Gas Lubrication of Slider

8.4.1 Procedures Introduction

The calculation program consists of a main program and three subroutines, including SUBH calculating the gas film thickness, SUBP calculating the pressure and OUTPUT giving the output results.

8.4.1.1 Parameter Pre-Assignment

Length of slider	AL = 4.0E-6 m
Width of slider	B = 3.3E-6 m
Minimum film thickness	H0 = 5.0E-9 m
Velocity of the disk	U = 25.0 m/s
Gas kinetic viscosity	EDA = 1.8060E-5 Pa·s
Ambient pressure	PA = 1.0135E5 Pa
Tilt angle	ALFA = 0.01 rad
Gas constant	R = 287.03
Absolute temperature	T0 = 293.0 K

Parameter pre-assignment can be modified depending on the user's situation, but the program needs to be recompiled and relinked before running.

8.4.1.2 Output Parameters

```
Pressure distribution   P(I,J)        PRESSURE.DAT
Gas film thickness      H(I,J)        FILM.DAT
```

8.4.2 Calculation Diagram

The calculation diagram of rarified gas lubrication is shown in Figure 8.2.

8.4.3 Calculation Program

```
      PROGRAM RARIFIEDGAS
      IMPLICIT REAL*8 (A&hyphen;H,O-Z)
      DIMENSION P(161,161),H(161,161),X(161),Y(161),F(161,161),
     QW(161,161)
```

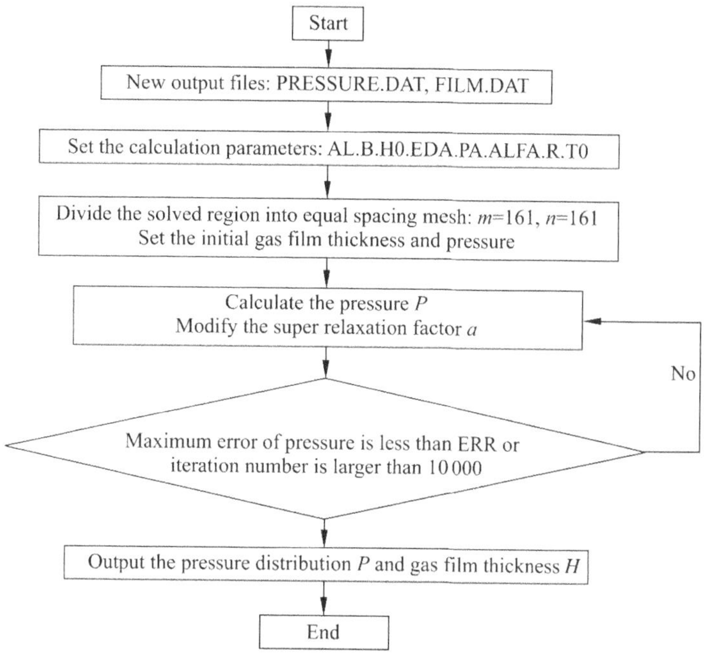

Figure 8.2 Calculation diagram of rarified gas lubrication.

```
      DATA AL,B,H0,U,EDA,PA,ALFA,R,T0/4.0E-6,3.3E-6,5.0E-
9,25.0,1.8060E-5,1.0135E5,0.01,287.03,293.0/
DATA N,M/161,161/
      OPEN(8,FILE='FILM.DAT',STATUS='UNKNOWN')
      OPEN(9,FILE='PRESSURE.DAT',STATUS='UNKNOWN')
      PI=3.1415926
      BETA1=0.01
      ALENDA=6.0*EDA*U*AL/(H0**2*PA)
      DELTA=AL*DSIN(ALFA)/H0
      A=AL/B
      D0=PA*H0/EDA/DSQRT(2.0*R*T0)
      DX=1.0/FLOAT(N-1)
      DY=1.0/FLOAT(M-1)
      CALL SUBH(N,M,DX,DY,DELTA,X,Y,H)
      CALL SUBP(N,M,DX,DY,D0,A,ALENDA,BETA1,U,X,Y,H,P,F,QW)
      CALL OUTPUT(N,M,A,ALFA,D0,H0,PA,AL,B,U,X,Y,H,P)
      STOP
      END
      SUBROUTINE SUBH(N,M,DX,DY,DELTA,X,Y,H)
      IMPLICIT REAL*8 (A-H,O-Z)
      DIMENSION X(N),Y(M),H(N,M)
      DO I=1,N
```

```
X(I)=(I-1)*DX
ENDDO
DO J=1,M
Y(J)=(J-1)*DY-0.5
ENDDO
DO I=1,N
DO J=1,M
H(I,J)=1.0+DELTA*(1.0-X(I))
ENDDO
ENDDO
RETURN
END
SUBROUTINE SUBP(N,M,DX,DY,D0,A,ALENDA,BETA1,U,X,Y,H,P,F,QW)
IMPLICIT REAL*8 (A-H,O-Z)
DIMENSION X(N),Y(M),H(N,M),P(N,M),F(N,M),QW(N,M)
DO I=1,N
DO J=1,M
P(I,J)=1.0
F(I,J)=0.0
QW(I,J)=1.0
ENDDO
ENDDO
DO I=2,N-1
DO J=2,M-1
P(I,J)=P(I-1,J)*H(I-1,J)/H(I,J)
ENDDO
ENDDO
K=0
DO WHILE(K<10000)
DO I=1,N
DO J=1,M
DR=D0*P(I,J)*H(I,J)
QC=DR/6.0
QP=QC+1.0162+0.40134*DLOG(1.0+1.2477/DR)
QW(I,J)=QP/QC
ENDDO
ENDDO
ERR=0.0
DO J=2,M-1
DO I=2,N-1
C0=2.0*ALENDA*H(I,J)/DX
C5=2.0*ALENDA*P(I-1,J)*H(I-1,J)/DX
TMP1=QW(I+1,J)*H(I+1,J)**3
TMP2=QW(I,J)*H(I,J)**3
QHP1=0.5*(TMP1+TMP2)
```

```
TMP1=QW(I-1,J)*H(I-1,J)**3
QHM1=0.5*(TMP1+TMP2)
C1=P(I,J)*(QHP1+ QHM1)/DX**2
C3=(P(I+1,J)**2*QHP1+P(I-1,J)**2*QHM1)/DX**2
TMP1=QW(I,J+1)*H(I,J+1)**3
QHP1=0.5*(TMP1+TMP2)
TMP1=QW(I,J-1)*H(I,J-1)**3
QHM1=0.5*(TMP1+TMP2)
C2=A**2*P(I,J)*(QHP1+ QHM1)/DY**2
C4=A**2*(P(I,J+1)*P(I,J+1)*QHP1+P(I,J-1)*P(I,J-1)*QHM1)/
DY**2
TMP=(C5+C3+C4)/(C0+C1+C2)
P(I,J)=P(I,J)+BETA1*(TMP-P(I,J))
F(I,J)=P(I,J)*(C0+C1+C2)-C3-C4-C5
IF(ABS(F(I,J)).GT.ERR)ERR=ABS(F(I,J))
ENDDO
ENDDO
K=K+1
WRITE(*,"('K=',I6,6X,'ERR=',E12.6)")K,ERR
IF(ERR.LT.1.E-5) EXIT
IF(K.GT.500)BETA1=0.1
IF(K.GT.2000)BETA1=0.5
IF(K.GT.3000)BETA1=0.75
IF(K.GT.5000)BETA1=0.95
ENDDO
RETURN
END
SUBROUTINE OUTPUT(N,M,A,ALFA,D0,H0,PA,AL,B,U,X,Y,H,P)
IMPLICIT REAL*8 (A-H,O-Z)
DIMENSION X(N),Y(M),H(N,M),P(N,M)
N2=N/2
SM=0.0
X0=0.0
AM0=0.0
PMAX=0.0
PMIN=0.0
DO I=1,N
DO J=1,M
P(I,J)=P(I,J)-1.0
SM=SM+P(I,J)
AM0=AM0+P(I,J)*(X(N2)-X(I))
IF(P(I,J).GT.PMAX)THEN
PMAX=P(I,J)
IMAX=I
JMAX=J
```

```
      ENDIF
      IF(P(I,J).LT.PMIN)PMIN=P(I,J)
      H(I,J)=H0*H(I,J)
      ENDDO
      ENDDO
      X0=AM0/SM+X(N2)
      AM0=AM0*PA*AL**2*B/(N-1)/(M-1)
      SM=SM*PA*AL*B/(N-1)/(M-1)
      WRITE(8,40)Y(1),(Y(J),J=1,M)
      WRITE(8,40)(X(I),(H(I,J),J=1,M),I=1,N)
      WRITE(9,40)Y(1),(Y(J),J=1,M)
      WRITE(9,40)(X(I),(P(I,J),J=1,M),I=1,N)
40    FORMAT(162(E12.6,1X))
      RETURN
      END
```

8.4.4 Calculation Results

The working conditions and parameters in the program are: length of slider $l = 4.0 \times 10^{-6}$ m, width of slider $b = 3.3 \times 10^{-6}$ m, minimum height of slider $h_0 = 5.0 \times 10^{-9}$ m, pitch angle $\theta = 0.01$ rad, velocity of disk $U = 25$ m·s^{-1} and gas viscosity $\eta = 1.806 \times 10^{-5}$ Pa·s.

According to the above parameters and considering the rarefication gas effect, the calculation results are shown in Figure 8.3. This indicates that the pressure distribution of the gas film is significantly less than that without considering the rarefication effect.

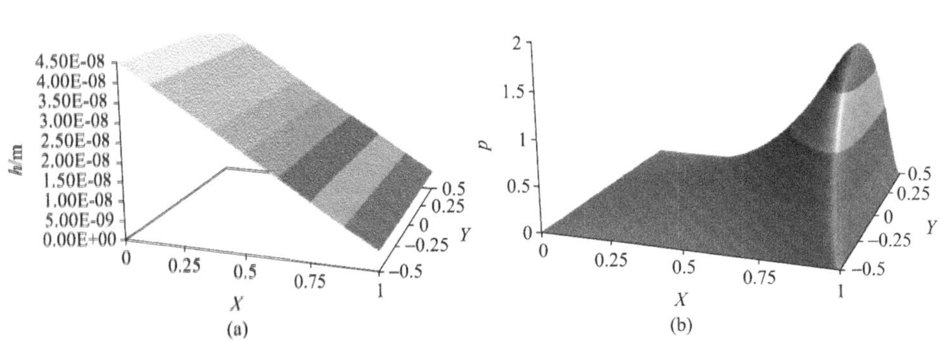

Figure 8.3 Film thickness (a) and pressure distribution (b) of rarefied gas lubrication.

9

Numerical Method and Program for One-dimensional Grease Lubrication

9.1 Basic Equations

9.1.1 Introduction

Grease is a kind of lubricant made by adding thickening agents to oil, to form a semi-solid or jellylike substance. The thickening agents are commonly metallic soaps. The fibers of the soaps form a mesh framework for storing the oil. As the grease has a three-dimensional frame structure composed of fibers, it is not laminar, so that the grease lubrication process shows complex mechanical properties, such as time dependence and visco-plasticity. Figure 9.1 gives the rheological behavior of grease. The main characteristics of grease can be summarized as follows.

1. The viscosity of grease usually increases with a decrease of the shear strain rate. Thus, the relationship of the shear stress and shear rate is nonlinear.
2. As shown in Figure 9.1, grease has a yield shear stress τ_s. Only when the shear stress τ is larger than τ_s will grease present the properties of a fluid. When $\tau \leq \tau_s$, grease is like a solid. It may have a certain amount of elastic deformation. Because grease has a yield shear stress, the grease lubricant film appears to be a nonflowing layer when $\tau \leq \tau_s$. In the flowing layer, the velocities perpendicular to the layer are the same.
3. Grease has a thixotropic property, which means that when grease flows under a certain shear rate, both the shear stress and the viscosity decrease gradually with time increasing. Furthermore, its viscosity will partially recover after the shear process ceases. Thus, grease lubrication is time dependent. When we call grease lubrication "steady," it is assumed to be not time dependent.

Numerical Calculation of Lubrication: Methods and Programs, First Edition. Ping Huang.

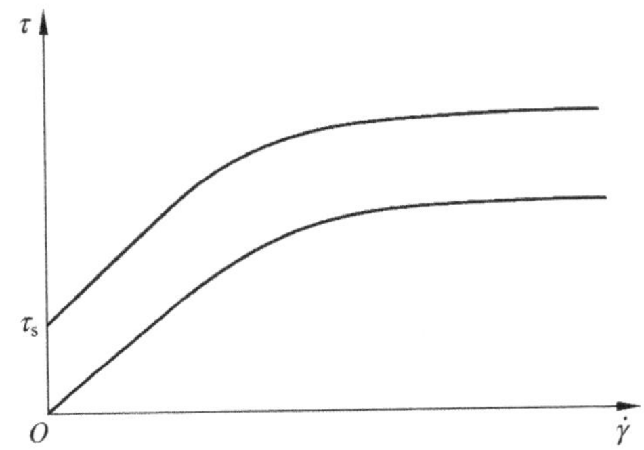

Figure 9.1 Rheological property of grease.

9.1.2 *Constitutive Equations of Grease*

Three kinds of constitutive equations are commonly used to describe the rheological property of grease. They are:

The Ostwald model

$$\tau = \phi \dot{\gamma}^n \tag{9.1}$$

The Bingham model

$$\tau = \tau_s + \phi \dot{\gamma} \tag{9.2}$$

The Herschel–Bulkley model

$$\tau = \tau_s + \phi \dot{\gamma}^n \tag{9.3}$$

where n is the rheological index, usually $n < 1$; and ϕ is a plastic viscosity similar to the viscosity of oil.

The Herschel–Bulkley model is much closer to experimental results and is especially more accurate at low velocity. In addition, when $n = 1$, the Herschel–Bulkley model turns into the Bingham model; or when $\tau_s = 0$, it turns into the Ostwald model. Therefore, the Herschel–Bulkley model is universal. Strictly speaking, the three rheological parameters τ_s, ϕ and n should be functions of temperature and pressure. For isothermal lubrication, we do not need to consider the influence of temperature. Furthermore, the rheological index n is usually assumed not to be related to pressure p; but the yield shear stress τ_s and plastic viscosity ϕ vary with pressure p according to the following relationships

$$\begin{aligned} \tau_s &= \tau_{s0} e^{ap} \\ \phi &= \phi_0 e^{ap} \end{aligned} \tag{9.4}$$

where τ_{s0} is the initial yield shear stress, ϕ_0 is the initial plastic viscosity of grease at atmosphere pressure and α is the viscosity–pressure coefficient of the basic oil in the grease.

9.1.3 Reynolds Equation

The equations of grease lubrication are similar to those of oil lubrication. The Reynolds equation of one-dimensional grease lubrication can also be derived from the constitutive equation, the equilibrium equation and the continuity equation. However, because the constitutive equation of grease contains the yield shear stress τ_s, the lubricant film is divided into two parts, known as the nonshearing flow part and the shearing flow part. The two parts must be dealt with separately, and so the deriving process is complicated. For an example, the one-dimensional Reynolds equation based on the Herschel–Bulkley model is

$$\frac{dp}{dx} = 2\phi\left[2\left(2+\frac{1}{n}\right)\right]^n \frac{U^n(h-\bar{h})^n}{h^{2n+1}}\left[1-\frac{2\tau_s}{\frac{dp}{dx}h}\right]^{-(n+1)}\left[1+\frac{n}{n+1}\frac{2\tau_s}{\frac{dp}{dx}h}\right]^{-n} \tag{9.5}$$

where $\bar{h}$ is the film thickness at $dp/dx = 0$.

If we set $\tau_s = 0$, Equation 9.5 will become the Reynolds equation of the Ostwald model, as follows

$$\frac{dp}{dx} = \text{sign}\left(h-\bar{h}\right)\frac{2\phi}{h^{2n+1}}U^n\left[2\left(2+\frac{1}{n}\right)\right]^n\left|h-\bar{h}\right|^n \tag{9.6}$$

where sign $\left(h-\bar{h}\right)$ is the symbol taking the same sign as $h-\bar{h}$.

If we set $\tau_s = 0$ and $n = 1$, Equation 9.5 turns into the Reynolds equation of a Newtonian fluid

$$\frac{dp}{dx} = 12\phi U\frac{h-\bar{h}}{h^3} \tag{9.7}$$

where ϕ is equal to the dynamic viscosity η.

It should be noted that the Reynolds equation can be derived according to different rheological models, but its applications are not exactly the same. In order to obtain practical solutions to meet the requirements, one must carefully choose a rheological model according to the working conditions.

9.2 Numerical Method of One-Dimensional Grease Lubrication

Because the Reynolds equation for grease lubrication using the Ostwald model is a first order differential equation, we can use the front or back discrete differential formulas to express it. Therefore, we have

$$\frac{p_i - p_{i-1}}{x_i - x_{i-1}} = \text{sign}(h - \bar{h}) \frac{2\phi}{h_i^{2n+1}} U^n \left[2\left(2 + \frac{1}{n}\right)\right]^n |h - \bar{h}|^n \tag{9.8}$$

Then, its iterative form is

$$P_i = P_{i-1} + \text{sign}(H_i - \bar{H}) \frac{1}{H_i^{2n+1}} |H_i - \bar{H}|^n \Delta X \tag{9.9}$$

where P is the dimensionless pressure, $p = 2\phi l \left\{ \frac{U}{h_2^2} \left[2\left(2 + \frac{1}{n}\right)\right] \right\}^n P.$

Although Equation 9.6 seems only a first order difference equation, it is actually integrated from a quadratic differential equation. Therefore, the equation contains one integral constant and thus it is a second order differential equation. Also, the constant $\bar{H}$ needs to be determined with the pressure boundary condition on one end of the border. When we start to calculate, the pressure of the first node is determined by the boundary condition. Then, we can obtain the pressure of the next node by Equation 9.9. After we finish the iteration, the pressure of the last node usually does not meet the boundary of this end. Therefore, we should adjust $\bar{H}$ until this pressure boundary condition meets the requirement.

9.3 Calculation Program of One-Dimensional Grease Lubrication

9.3.1 Calculation Diagram

Because the Reynolds equation for grease lubrication given by Equation 9.9 has been integrated once, the iterative process is different from the boundary value problem of the quadratic differential equation mentioned before. First, we set the pressure of the first node to be equal to one boundary value, for example, P(1) = 0. Then, we carry out an iteration to calculate the pressure of the next node and then for the whole mesh. If the pressure of the other end does not satisfy the boundary condition, for example, P(N) ≠ 0, we adjust H0 (that is $\bar{H}$) to iterate all the nodes again until P(N) ≈ 0.

For adjusting H0, in order to make the iteration fast, in the present program, we use the dichotomy method. This method is fast at the beginning and has high accuracy when close to convergence. The adjusting approach is that if the increment of H0 is DH and if the sign of DH changes during adjustment, we halve DH and then iterate again. The calculation diagram of one-dimensional grease lubrication is as shown in Figure 9.2.

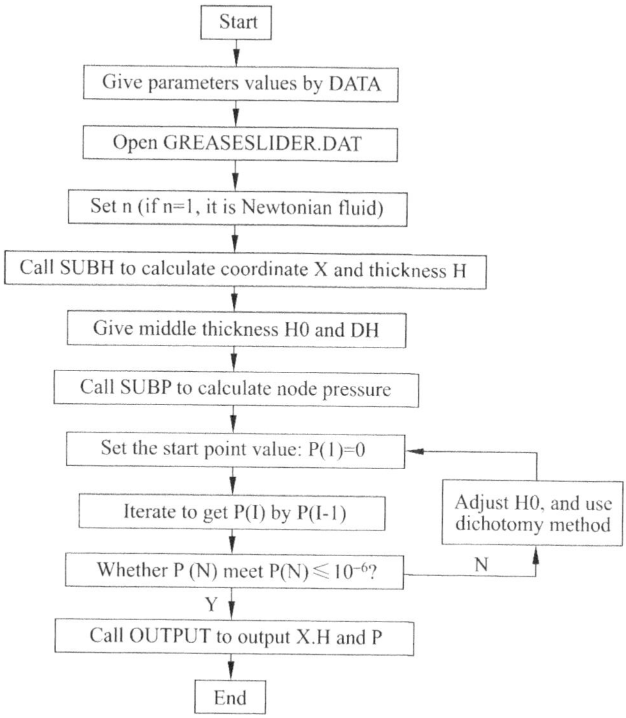

Figure 9.2 Calculation diagram of one-dimensional grease lubrication.

9.3.2 *Calculation Program*

```
PROGRAM GREASESLIDER
DIMENSION X(121),H(121),P(121)
COMMON /COM1/X1,X2,H1,H2,U,EDA,AL,ALOAD,DX,P0
DATA N,U,X1,X2,H1,H2,EDA,AL/121,1.0,0.0,1.0,1.0,0.5,0.08,0.01/
OPEN(7,FILE='GREASESLIDER.DAT',STATUS='UNKNOWN')
WRITE(*,*)'n=?'
READ(*,*)AN
P0=2.0*EDA*AL*(U*(4.0+2.0/AN)/H2**2)**AN
CALL SUBH(N,X,H)
H0=0.5*(H(1)+H(N))
DH=1.E-4*H0
CALL SUBP(N,AN,H0,DH,X,H,P)
CALL OUTPUT(KG,N,AN,X,H,P)
STOP
END
SUBROUTINE SUBH(N,X,H)
DIMENSION X(N),H(N)
```

```
      COMMON /COM1/X1,X2,H1,H2,U,EDA,AL,ALOAD,DX,P0
      DX=1./(N-1.0)
      DO I=1,N
      X(I)=X1-(I-1)*DX*(X1-X2)
      H(I)=H1/H2-(H1/H2-1.0)*X(I)
      ENDDO
      RETURN
      END
      SUBROUTINE SUBP(N,AN,H0,DH,X,H,P)
      DIMENSION X(N),H(N),P(N)
      COMMON /COM1/X1,X2,H1,H2,U,EDA,AL,ALOAD,DX,P0
      P(1)=0.0
   10 DO I=2,N
      IF(H(I)-H0.GT.0.0)THEN
      A=H(I)-H0
      P(I)=P(I-1)+A**AN*DX/H(I)**(2.0*AN+1.0)
      ENDIF
      IF(H(I)-H0.LT.0.0)THEN
      A=H0-H(I)
      P(I)=P(I-1)-A**AN*DX/H(I)**(2.0*AN+1.0)
      ENDIF
      IF(H(I)-H0.EQ.0.0)P(I)=P(I-1)
      ENDDO
      IF(ABS(P(N)).GT.1.E-6)THEN
      IF(P(N).GT.0.0)THEN
      H0=H0+DH
      IF(KG.EQ.0)DH=DH/2.0
      KG=1
      ENDIF
      IF(P(N).LT.0.0)THEN
      H0=H0-DH
      IF(KG.EQ.1)DH=DH/2.0
      KG=0
      ENDIF
      WRITE(*,*)P(N)
      GOTO 10
      ENDIF
      RETURN
      END
      SUBROUTINE OUTPUT(KG,N,AN,X,H,P)
      DIMENSION X(N),H(N),P(N)
      COMMON /COM1/X1,X2,H1,H2,U,EDA,AL,ALOAD,DX,P0
      X0=0.0
      ALOAD=0.0
      DO I=1,N
```

```
      ALOAD=ALOAD+P(I)
      X0=X0+P(I)*X(I)
      ENDDO
      X0=X0*AL
      ALOAD=P0*AL*ALOAD*DX
      WRITE(*,*)AN,ALOAD,X0
      DO I=1,N
      WRITE(7,40) X(I),H(I),P(I)
      END DO
40    FORMAT(1X,3(E12.6,1X))
      RETURN
      END
```

9.3.3 Calculation Results

Figure 9.3 shows the pressure distribution of the different indexes n. When $n=1$, it represents a Newtonian fluid. Usually, n of grease is less than 1. In the figure, we set $n=0.8$. There is a significant difference in its dimensionless pressure from that of $n=1$. This indicates that n has a direct influence on the dimensionless pressure and load carrying capacity. For example, when $n=0.8$, the load $w=0.833\times10^{-5}$ N; and when $n=1.0$, $w=1.026\times10^{-5}$ N. Therefore, we cannot directly use the dimensionless pressure to determine the magnitude of the load carrying capacity. In fact, because grease is nonlinear, its nondimensional pressure seems less than that of linear oil. However, because the grease viscosity is much higher than that of oil, the dimensional load of grease is larger than that of oil.

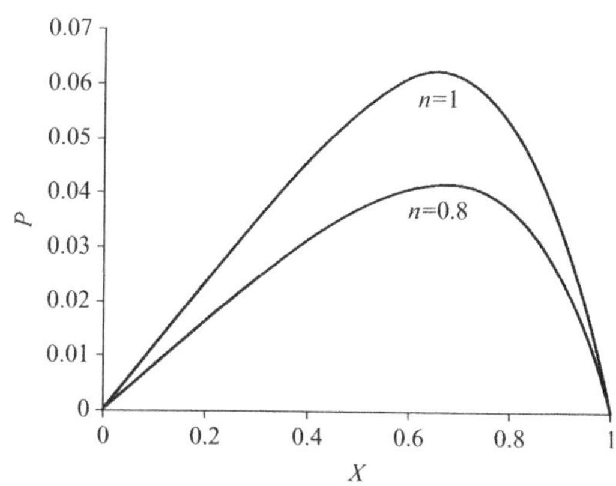

Figure 9.3 Numerical solutions of pressure in one-dimensional grease lubrication.

Part Two

Numerical Method for Energy Equation

10

Energy Equation and its Discrete Form

The temperature of a lubricating film is an important factor influencing its lubricating properties. The temperature significantly changes the viscosity of the lubricant, such that it influences the pressure and load carrying capacity. In addition, the change of clearance shape caused by thermal deformation on the lubrication surfaces can also influence the lubrication properties. Furthermore, a high temperature may cause lubrication failure to damage the material surface. Usually, 120 ~ 140 °C are used as the limits of the local temperature.

10.1 Basic Equations

In a lubrication calculation, the general energy equation can be written as

$$\rho c_p\left(u\frac{\partial T}{\partial x}+v\frac{\partial T}{\partial y}+w\frac{\partial T}{\partial z}\right)=k\left(\frac{\partial^2 T}{\partial x^2}+\frac{\partial^2 T}{\partial y^2}+\frac{\partial^2 T}{\partial z^2}\right)-\frac{T}{\rho}\frac{\partial \rho}{\partial T}\left(u\frac{\partial p}{\partial x}+v\frac{\partial p}{\partial y}+w\frac{\partial p}{\partial z}\right)+\Phi \tag{10.1}$$

where ρ is the lubricant density, c_p is the specific heat of the lubricant, T is the temperature, u, v and w are the speed components of the fluid in the x, y and z directions, k is the lubricant thermal conductivity, p is the pressure and Φ is the dissipation work item, which can be simplified as

$$\Phi=\eta\left[\left(\frac{\partial u}{\partial z}\right)^2+\left(\frac{\partial v}{\partial z}\right)^2\right] \tag{10.2}$$

where η is the viscosity.

Equation 10.1 is very complicated. In order to obtain the temperature distribution in a lubricating film we often need to simplify the energy equation.

Numerical Calculation of Lubrication: Methods and Programs, First Edition. Ping Huang.

10.1.1 Simplified Energy Equation

Considering that T, p and η do not change across the film thickness and ρ is independent of temperature, after being integrated across the film thickness the energy Equation 10.1 can be transformed into the commonly used energy equation of hydrodynamic lubrication as follows

$$q_x\frac{\partial T}{\partial x}+q_y\frac{\partial T}{\partial y}=\frac{\eta U^2}{J\rho c_p h}+\frac{h^3}{12\eta J\rho c_p}\left[\left(\frac{\partial p}{\partial x}\right)^2+\left(\frac{\partial p}{\partial y}\right)^2\right] \tag{10.3}$$

where $q_x=\dfrac{Uh}{2}-\dfrac{h^3}{12\eta}\dfrac{\partial p}{\partial x}$, $q_y=-\dfrac{h^3}{12\eta}\dfrac{\partial p}{\partial y}$, and J is the mechanical equivalent of heat.

In the temperature calculation, the energy equation should be dimensionless. We set

$$X=x/l;Y=y/b;H=h/h_0;\alpha=l/b;P=\frac{h_0^2}{6U\eta_0 l}p;Q=\frac{q}{Uh_0};\eta^*=\frac{\eta}{\eta_0};T^*=\frac{2J\rho c_p h_0^2}{Ul\eta_0}T$$

where η_0 is the entrance viscosity.

Substituting all the parameters above into energy Equation 10.1, we obtain its dimensionless form

$$\frac{\partial T^*}{\partial X}=\frac{1}{Q_x}\left\{-\alpha Q_y\frac{\partial T^*}{\partial Y}+\frac{2\eta^*}{H}+\frac{6H}{\eta^*}\left[\left(\frac{\partial P}{\partial X}\right)^2+\alpha^2\left(\frac{\partial P}{\partial Y}\right)^2\right]\right\} \tag{10.4}$$

where $Q_x=\dfrac{H}{2}-\dfrac{H^3}{2}\dfrac{\partial P}{\partial X}$ and $Q_y=-\dfrac{H^3}{2}\dfrac{\partial P}{\partial Y}$.

From the equation above we know that, in order to get the temperature field, we must obtain the pressure first so that we have the values of $\dfrac{\partial P}{\partial X}$ and $\dfrac{\partial P}{\partial Y}$. Further, pressure is influenced by temperature. Therefore, to solve a thermal hydrodynamic lubrication problem, we must solve the Reynolds equation and the energy equation simultaneously.

10.1.2 Boundary Conditions

It should be pointed out that the temperature analysis is a two-dimensional problem, but Equation 10.4 is a first order differential equation. Therefore, only one boundary condition in the x or y direction needs to be given. Generally, the temperature $T_{1,j}$ at the inlet is known, as shown in Figure 10.1, but the boundary condition in the y direction depends on the circumstance. Usually, for a symmetric problem, we have $\dfrac{\partial T}{\partial Y}=0$ at the centerline. In this way, the other boundary condition in the y direction need not be given anymore. Two-dimensional thermal lubrication and journal bearings thermal lubrication are both handled in this way.

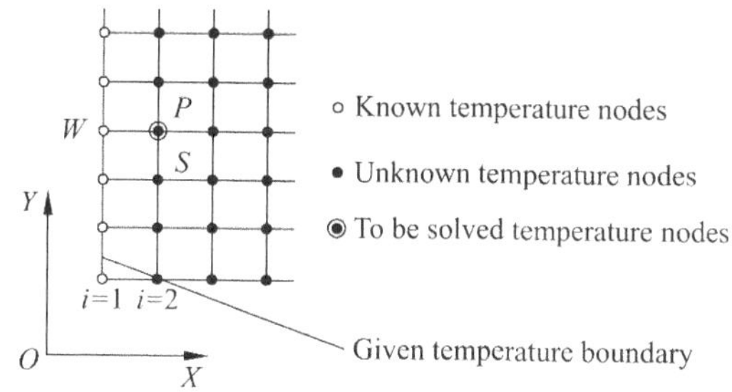

Figure 10.1 Mesh of temperature computation.

10.1.3 Numerical Method

The characteristics of the temperature calculation of a lubricant film are that the lubricant enters the inlet under the supplied temperature. Then, the temperature gradually changes with the flow. Such a problem is called the initial value problem. Usually, we use the marching method to solve the initial value problem. Its basic points are as follows.

As shown in Figure 10.1, we first divide the solution domain. Then, we select one row at which the temperature is known in the y direction as the initial value, for example, we take $i=1$ where oil is supplied by a tank. In this way, for each node of Row 1, the temperature T_{1j}^* is known.

After T_{1j}^* is known, with the discrete differential formula or the three-point parabola formula we can calculate the value of $\left(\frac{\partial T^*}{\partial Y}\right)_{1j}$. Because the pressure has been solved from the Reynolds equation, we can also determine $\left(\frac{\partial P}{\partial X}\right)_{1j}$ and $\left(\frac{\partial P}{\partial Y}\right)_{1j}$ as well as Q_{1j} and Q_{1j}. Substituting these values into Equation 10.4, we can obtain $\left(\frac{\partial T^*}{\partial X}\right)_{1j}$. Thus, we can obtain the temperature T_{2j}^* of each node of Row 2.

After T_{2j}^* have been determined, we repeat the steps above to obtain T_{3j}^* and so on. We can finally obtain the temperature of all the nodes in this way.

It should be noted that the following points should be paid attention if we use the marching method to solve the temperature. First, the forward direction must be consistent with the direction of lubricant flow. If along the x direction, we should meet the condition $Q_x > 0$. However, when the supplied oil is a high pressure or a great pressure gradient in the inlet area, $Q_x < 0$ may occur in the convergence clearance. This means that there is a counter flow. For such a situation, we cannot simply use the marching method to solve the temperature field. Furthermore, If $Q_x = 0$, $\frac{\partial T^*}{\partial X}$ derived from Equation 10.4 will be infinite. At such a point, the above method will fail.

10.2 Influence of Temperature on Lubricant Performance

10.2.1 Viscosity–Temperature Equation

When considering the influence of temperature on viscosity, the Barus equation is usually used.

$$\eta = \eta_0 \exp[-\beta(T - T_0)] \tag{10.5}$$

where η_0 is the viscosity at T_0; and β is the viscosity–temperature coefficient, which can be usually regarded as 0.03 $°C^{-1}$.

Although the following Rolelands equation is slightly more complicated than the Barus equation, it is more realistic.

$$\eta = \eta_0 \exp\left\{ (\ln \eta_0 + 9.67)\left[\left(\frac{T - 138}{T_0 - 138} \right)^{-1.1} - 1 \right] \right\} \tag{10.6}$$

10.2.2 Density–Temperature Equation

The influence of temperature on density is mainly because the thermal expansion changes the volume of the lubricant, leading to a decrease in the density. In general lubrication conditions, the influence of temperature on density is so small that it is often ignored. If there is a need to consider the influence of temperature on the density of a lubricant, we can take the density–temperature equation. If the thermal expansion coefficient is α_T, the density–temperature equation can be expressed as

$$\rho = \rho_0[1 - \alpha_T(T - T_0)] \tag{10.7}$$

where ρ is the density at temperature T, ρ_0 is the density at temperature T_0 and the unit of α_T is $°C^{-1}$.

Usually the value of α_T of oil can be expressed by the following two relationships. When the viscosity is lower than 3000 mPa s, we have $\lg\eta \leq 3.5$. Then, α_T is equal to

$$\alpha_T = \left(10 - \frac{9}{5} \lg \eta \right) \times 10^{-4} \tag{10.8}$$

If the viscosity is higher than 3000 mPa s, we have $\lg\eta > 3.5$. Then, α_T is equal to

$$\alpha_T = \left(5 - \frac{3}{8} \lg \eta \right) \times 10^{-4} \tag{10.9}$$

where the unit of the viscosity η is mPa s.

10.3 Numerical Method for Thermal Hydrodynamic Lubrication

10.3.1 Methods and Program for One-dimensional Thermal Hydrodynamic Lubrication

10.3.1.1 Basic Equations

The basic one-dimensional energy equation can be obtained from either Equation 10.3 or Equation 10.4 by omitting the y direction terms. Then, we have

$$\frac{\partial T^*}{\partial X} = \frac{1}{Q_x}\left\{\frac{2\eta^*}{H} + \frac{6H}{\eta^*}\left(\frac{\partial P}{\partial X}\right)^2\right\} \tag{10.10}$$

where $Q_x = \frac{H}{2} - \frac{H^3}{2}\frac{\partial P}{\partial X}$.

Discretizing Equation 10.10, we have

$$\frac{T_i^* - T_{i-1}^*}{X_i - X_{i-1}} = \frac{1}{Q_x}\left\{\frac{2\eta_i^*}{H_i} + \frac{6H_i}{\eta_i^*}\left(\frac{P_i - P_{i-1}}{X_i - X_{i-1}}\right)^2\right\} \tag{10.11}$$

where $Q_x = \frac{H_i}{2} - \frac{H_i^3}{2}\frac{P_i - P_{i-1}}{X_i - X_{i-1}}$.

The viscosity–temperature equation (Barus equation)

$$\eta = \eta_0 \exp[-\beta(T - T_0)] \tag{10.12}$$

Or, the dimensionless form is

$$\eta_i^* = \exp[-\beta(T_i - T_0)] \tag{10.13}$$

10.3.1.2 Calculation Diagram

In order to obtain the temperature, we need the viscosity at each node. As the viscosity is directly influenced by the temperature, we need to carry out an iteration. The calculation diagram for one-dimensional temperature without pressure coupling is shown in Figure 10.2.

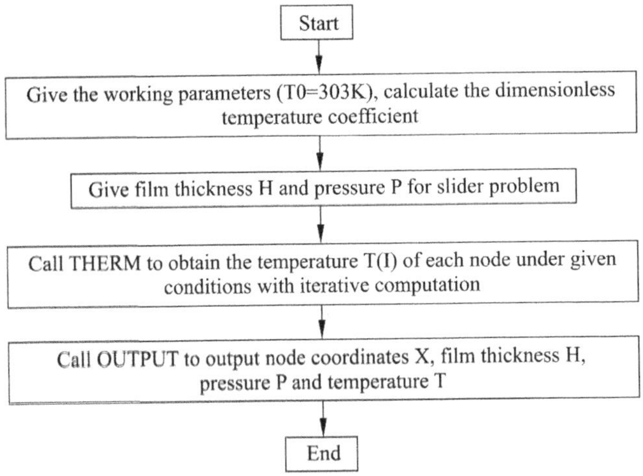

Figure 10.2 Calculation diagram for one-dimensional temperature without pressure coupling.

10.3.1.3 Calculation Program

```
PROGRAM LINETHERM
   DIMENSION X(200),P(200),H(200),T(200)
   DATA U,AL,EDA0,RO,C,AJ,H1,H2/1.0,0.01,0.05,890.0,1870.0,
4.184,5.5E-6,5.E-6/
   OPEN(8,FILE='OUT.DAT',STATUS='UNKNOWN')
   N=129
   A=U*AL*EDA0/2.0/AJ/RO/C/H2**2
   T0=303.0/A
   DX=1./(N-1.0)
   HH=H1/H2
   DH=HH-1.0
   DO I=1,N
   X(I)=(I-1)*DX
   H(I)=HH-DH*X(I)
   P(I)=-(-1.0/(H(I))+HH/(HH+1.0)/H(I)**2+1.0/(HH+1.0))/DH
   T(I)=T0
   ENDDO
   P(1)=0.0
   P(N)=0.0
   CALL THERM(N,A,DX,T0,X,P,H,T)
   CALL OUTPUT(N,A,T0,X,H,P,T)
   STOP
   END
```

```
      SUBROUTINE THERM(N,A,DX,T0,X,P,H,T)
      DIMENSION X(N),P(N),H(N),T(N)
   10 ERT=0.0
      DO I=2,N
      TOLD=T(I)
      EDA=EXP(-0.03*A*(T(I)-T0))
      QX=0.5*H(I)-0.5*H(I)**3*((P(I)-P(I-1))/DX)
      T(I)=T(I-1)+DX*(2.0*EDA/H(I)+6.0*H(I)/EDA*((P(I)-P(I-1))/
DX)**2)/QX
      T(I)=0.5*(TOLD+T(I))
      ERT=ERT+ABS(T(I)-TOLD)
      ENDDO
      ERT=A*ERT/(303.0*(N-1))
      WRITE(*,*)ERT
      IF(ERT.GT.1.E-6)GOTO 10
      RETURN
      END
      SUBROUTINE OUTPUT(N,A,T0,X,H,P,T)
      DIMENSION X(N),H(N),P(N),T(N)
      DO I=1,N
      T(I)=A*(T(I)-T0)
      END DO
      DO I=1,N
      WRITE(8,30)X(I),H(I),P(I),T(I)
      ENDDO
   30 FORMAT(4(1X,E12.6))
      RETURN
      END
```

10.3.1.4 Calculation Results

As given in the program, the working parameters are: velocity $U = 1\ \mathrm{m\,s^{-1}}$, slider length $Al = 0.01$ m, initial viscosity of lubricant $EDA0 = 0.05$ Pa s, lubricant density $RO = 890\ \mathrm{kg\,m^{-3}}$, specific heat of lubricant $C = 1870\ \mathrm{J\,kg^{-1}\,K^{-1}}$, mechanical equivalent of heat $AJ = 4.184\ \mathrm{J\,cal^{-1}}$, maximum thickness $H1 = 5.5 \times 10^{-6}$ m and minimum film thickness $H2 = 5 \times 10^{-6}$ m. The temperature calculated is linear to the coordinates, and the maximum temperature rise is about 5 °C. Theoretically, because viscosity influences pressure, the viscosity obtained from the temperature–viscosity equation should be substituted into the Reynolds equation to get the pressure, and then we solve the temperature again until convergency is obtained. If the temperature is not too high, we can omit this step. The calculation results of temperature for slider lubrication at a given pressure are shown in Figure 10.3.

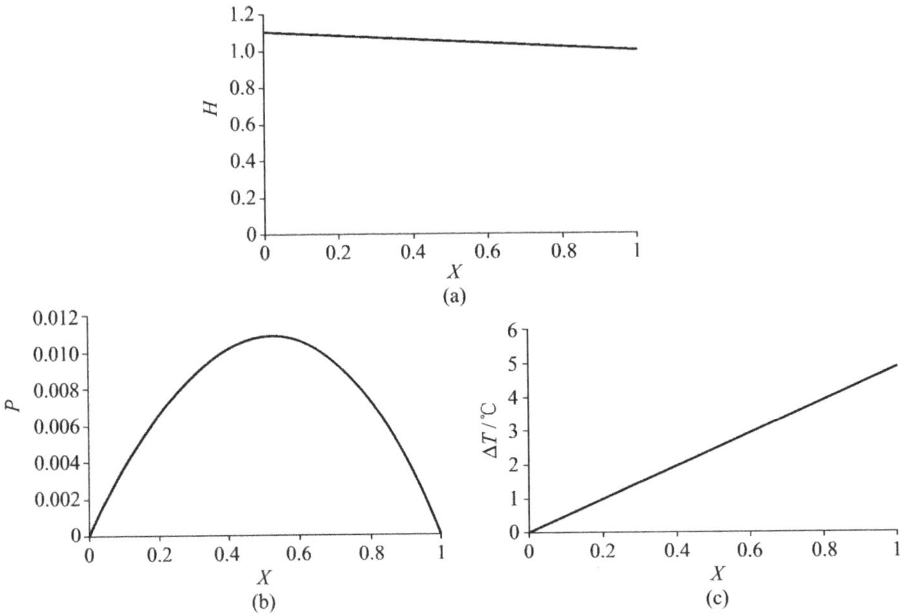

Figure 10.3 Temperature calculation results for slider lubrication at a given pressure. (a) Film thickness, H, (b) surface pressure, P, (c) temperature rise, ΔT.

10.3.2 Numerical Method and Program for Two-dimensional Thermal Hydrodynamic Lubrication

10.3.2.1 Basic Equations

The energy equation is

$$\frac{\partial T^*}{\partial X} = \frac{1}{Q_x}\left\{-\alpha Q_y \frac{\partial T^*}{\partial Y} + \frac{2\eta^*}{H} + \frac{6H}{\eta^*}\left[\left(\frac{\partial P}{\partial X}\right)^2 + \alpha^2\left(\frac{\partial P}{\partial Y}\right)^2\right]\right\} \tag{10.14}$$

where $Q_x = \frac{H}{2} - \frac{H^3}{2}\frac{\partial P}{\partial X}$ and $Q_y = -\frac{H^3}{2}\frac{\partial P}{\partial Y}$.

Discretizing Equation 10.13, we have

$$\frac{T^*_{i,j} - T^*_{i-1,j}}{X_i - X_{i-1}} = \frac{1}{Q_x}\left\{-\alpha Q_y \frac{T^*_{i,j} - T^*_{i,j-1}}{Y_j - Y_{j-1}} + \frac{2\eta^*_{i,j}}{H_{i,j}} + \frac{6H_{i,j}}{\eta^*_{i,j}}\left[\left(\frac{P_i - P_{i-1}}{X_i - X_{i-1}}\right)^2 + \alpha^2\left(\frac{P_{i,j} - P_{i,j-1}}{Y_j - Y_{j-1}}\right)^2\right]\right\} \tag{10.15}$$

where $Q_x = \frac{H_{i,j}}{2} - \frac{H_{i,j}^3}{2}\frac{P_{i,j} - P_{i-1,j}}{X_i - X_{i-1}}$, $Q_y = -\frac{H_{i,j}^3}{2}\frac{P_{i,j} - P_{i,j-1}}{Y_j - Y_{j-1}}$, $\alpha = \frac{l}{b}$, l is the length of the slider and b is the width of the slider.

The viscosity–temperature equation (Barus equation)

$$\eta = \eta_0 \exp[-\beta(T - T_0)] \tag{10.16}$$

Or

$$\eta_{i,j}^* = \exp[-\beta(T_{i,j} - T_0)] \tag{10.17}$$

10.3.2.2 Calculation Diagram

In order to obtain the temperature, we need the viscosity at each node. As the viscosity is directly influenced by the temperature, we need to carry out an iteration. The calculation diagram for two-dimensional temperature without pressure coupling is shown in Figure 10.4.

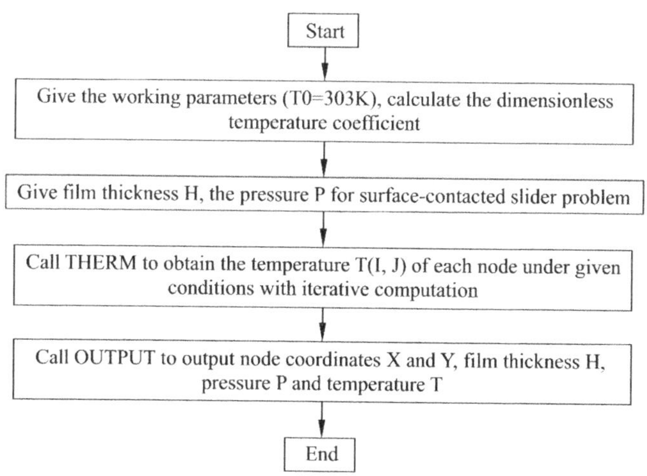

Figure 10.4 Calculation diagram for two-dimensional temperature without pressure coupling.

10.3.2.3 Calculation Program

```
PROGRAM SURFACETHERM
   DIMENSION X(200),Y(200),P(20000),H(20000),T(20000)
   DATA U,ALX,ALY,EDA0,RO,C,AJ,H1,H2/1.0,0.01,0.01,0.05,890.0,
1870.0,4.184,1.1E-6,1.E-6/
   OPEN(7,FILE='FILM.DAT',STATUS='UNKNOWN')
   OPEN(8,FILE='PRESSURE.DAT',STATUS='UNKNOWN')
   OPEN(9,FILE='TEM.DAT',STATUS='UNKNOWN')
   N=129
   M=65
   A=U*ALX*EDA0/2.0/AJ/RO/C/H2**2
   T0=303.0/A
   DX=1./(N-1.0)
   DY=1./(M-1.0)
   HH=H1/H2
   DH=HH-1.0
   ALFA1=ALX/ALY
   CALL INIT(N,M,DX,DY,HH,DH,T0,X,Y,H,P,T)
   CALL THERM(N,M,A,ALFA1,DX,DY,T0,X,Y,P,H,T)
   CALL OUTPUT(N,M,A,T0,X,Y,H,P,T)
   STOP
   END
   SUBROUTINE INIT(N,M,DX,DY,HH,DH,T0,X,Y,H,P,T)
   DIMENSION X(N),Y(M),H(N,M),P(N,M),T(N,M)
   DO I=1,N
   X(I)=(I-1)*DX
   ENDDO
   DO J=1,M
   Y(J)=-0.5+(J-1)*DY
   ENDDO
   DO I=1,N
   DO J=1,M
   H(I,J)=HH-DH*X(I)
   P(I,J)=-(-1.0/(H(I,J))+HH/(HH+1.0)/H(I,J)**2+1.0/(HH+1.0))/DH*
 (1.0-4.0*Y(J)*Y(J))
   T(I,J)=T0
   ENDDO
   ENDDO
   DO I=1,N
   P(I,1)=0.0
   P(I,M)=0.0
   ENDDO
   DO J=1,M
```

```
      P(1,J)=0.0
      P(N,J)=0.0
      ENDDO
      RETURN
      END
      SUBROUTINE THERM(N,M,A,ALFA1,DX,DY,T0,X,Y,P,H,T)
      DIMENSION X(N),Y(M),H(N,M),P(N,M),T(N,M)
10    ERT=0.0
      DO I=2,N
      DO J=M/2+1,1,-1
      TOLD=T(I,J)
      EDA=EXP(-0.03*A*(T(I,J)-T0))
      DPDX=(P(I,J)-P(I-1,J))/DX
      IF(J.EQ.M/2+1)THEN
      DPDY=0.0
      DTDY=0.0
      ELSE
      DPDY=(P(I,J+1)-P(I,J))/DY
      DTDY=(T(I,J+1)-T(I,J))/DY
      ENDIF
      QX=0.5*H(I,J)-0.5*H(I,J)**3*DPDX
      QY=-0.5*H(I,J)**3*DPDY
      AA=-0.5*ALFA1*QY*DTDY
      AB=2.0*EDA/H(I,J)
      AC=6.0*H(I,J)/EDA*(DPDX**2+ALFA1**2*DPDY**2)
      BA=QX/DX-ALFA1*QY/DY
      BB=QX/DX*T(I-1,J)-ALFA1*QY/DY*T(I,J+1)
      T(I,J)=(BB+AB+AC)/BA
      T(I,J)=0.7*TOLD+0.3*T(I,J)
      ERT=ERT+ABS(T(I,J)-TOLD)
      ENDDO
      ENDDO
      ERT=A*ERT/(303.0*(N-1)*(M-1))
      WRITE(*,*)ERT
      IF(ERT.GT.1.E-8)GOTO 10
      DO I=2,N
      DO J=1,M/2
      T(I,M-J+1)=T(I,J)
      ENDDO
      ENDDO
      RETURN
      END
      SUBROUTINE OUTPUT(N,M,A,T0,X,Y,H,P,T)
      DIMENSION X(N),Y(M),H(N,M),P(N,M),T(N,M)
```

```
      DO I=1,N
      DO J=1,M
      T(I,J)=A*(T(I,J)-T0)
      END DO
      ENDDO
      WRITE(7,30)X(1),(Y(J),J=1,M)
      WRITE(8,30)X(1),(Y(J),J=1,M)
      WRITE(9,30)X(1),(Y(J),J=1,M)
      DO I=1,N
      WRITE(7,30)X(I),(H(I,J),J=1,M)
      WRITE(8,30)X(I),(P(I,J),J=1,M)
      WRITE(9,30)X(I),(T(I,J),J=1,M)
      ENDDO
   30 FORMAT(130(1X,E12.6))
      RETURN
      END
```

10.3.2.4 Calculation Results

As given in the program, the working parameters are:

Velocity	$U = 1\,m\,s^{-1}$
Slider length	$ALX = 0.01\,m$
Slider width	$ALY = 0.01\,m$
Initial viscosity of lubricant	$EDA0 = 0.05\,Pa \cdot s$
Lubricant density	$RO = 890\,kg\,m^{-3}$
Specific heat of lubricant	$C = 1870\,J\,kg^{-1}\,K^{-1}$
Mechanical equivalent of heat	$AJ = 4.184\,J \cdot cal^{-1}$
Maximum film thickness	$H1 = 1.1 \times 10^{-6}\,m$
Minimum film thickness	$H2 = 1 \times 10^{-6}\,m$

The results show that the calculated temperature is no longer linear to the coordinate and the maximum temperature rise is about 50 °C. Theoretically, because viscosity influences pressure, the viscosity obtained from the temperature–viscosity equation should be substituted into the Reynolds equation to get the pressure, and then we solve the temperature again until convergency is obtained. Here we only give the temperature calculation under the current pressure without iteration. The temperature calculation results for two-dimensional slider lubrication is shown in Figure 10.5.

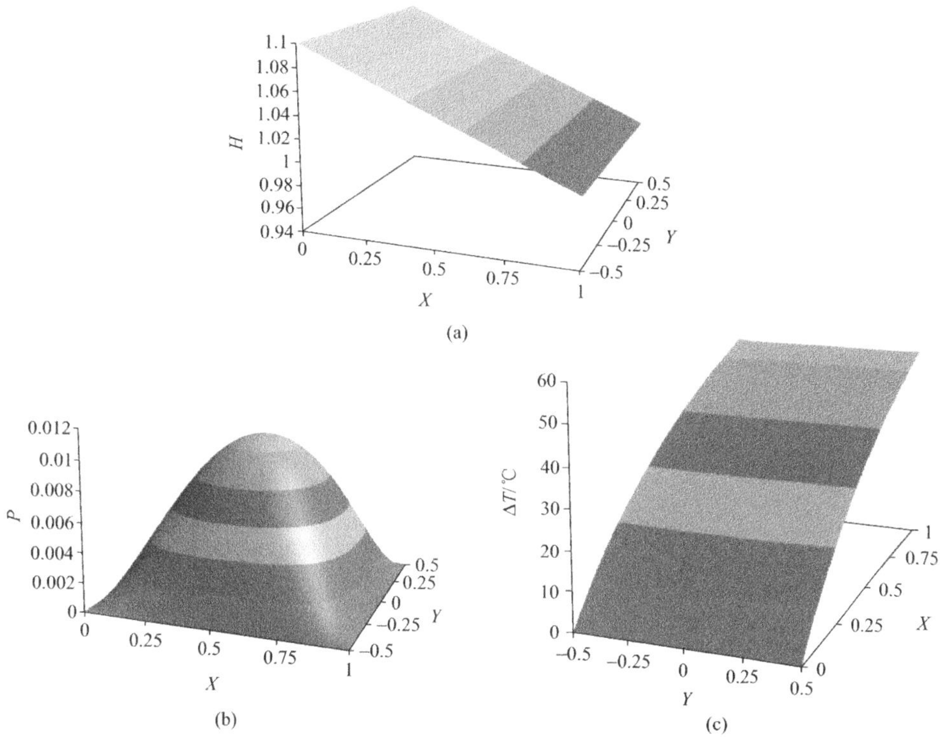

Figure 10.5 Temperature calculation results for two-dimensional slider lubrication. (a) Film thickness H, (b) surface pressure, P, (c) temperature rise, ΔT.

11

Numerical Method and Program for Incompressible and Steady Thermal Hydrodynamic Lubrication of Journal Bearing

11.1 Basic Equations

11.1.1 Reynolds Equation

By using cylindrical coordinates, that is, let $x = R\theta$, $dx = Rd\theta$, the Reynolds equation becomes

$$\frac{\partial}{R^2\partial\theta}\left(\frac{h^3}{\eta}\frac{\partial p}{\partial\theta}\right) + \frac{\partial}{\partial y}\left(\frac{h^3}{\eta}\frac{\partial p}{\partial y}\right) = 6U\frac{dh}{Rd\theta} \tag{11.1}$$

Or in the dimensionless form

$$\frac{\partial}{\partial\theta}\left(\frac{H^3}{\eta^*}\frac{\partial P}{\partial\theta}\right) + \alpha^2\frac{\partial}{\partial Y}\left(\frac{H^3}{\eta^*}\frac{\partial P}{\partial Y}\right) = \frac{dH}{d\theta} \tag{11.2}$$

where H is dimensionless film thickness $[H = h/c = (1 + \varepsilon\cos\theta)]$, c is bearing clearance, ε is eccentricity, α is the ratio of R and b ($\alpha = \frac{R}{b}$, R is bearing radius, b is bearing width), Y is a dimensionless coordinate ($Y = y/b$) and P is dimensionless pressure ($P = pc^2/6U\eta_0 R$).

Numerical Calculation of Lubrication: Methods and Programs, First Edition. Ping Huang.

11.1.2 Energy Equation

$$\frac{\partial T^*}{\partial \theta} = \frac{1}{Q_x}\left\{-\alpha Q_y \frac{\partial T^*}{\partial Y} + \frac{2\eta^*}{H} + \frac{6H}{\eta^*}\left[\left(\frac{\partial P}{\partial \theta}\right)^2 + \alpha^2\left(\frac{\partial P}{\partial Y}\right)^2\right]\right\} \tag{11.3}$$

where $Q_x = \frac{H}{2} - \frac{H^3}{2}\frac{\partial P}{\partial \theta}$ and $Q_y = -\frac{H^3}{2}\frac{\partial P}{\partial Y}$.

11.1.3 Viscosity–Temperature Equation

The Barus equation is used as follows

$$\eta = \eta_0 \exp[-\beta(T - T_0)] \tag{11.4}$$

In most cases, the Roelands equation is better

$$\eta = \eta_0 \exp\left\{(\ln \eta_0 + 9.67)\left[\left(\frac{T - 138}{T_0 - 138}\right)^{-1.1} - 1\right]\right\} \tag{11.5}$$

11.2 Numerical Method

11.2.1 Discrete Reynolds Equation

The discrete Reynolds equation was given in Chapter 4 (see Equation 4.9). If we add the viscosity term, it becomes

$$\frac{H^3_{i+1/2}/\eta^*_{i+1/2,j}(P_{i+1,j} - P_{i,j}) - H^3_{i-1/2}/\eta^*_{i-1/2,j}(P_{i,j} - P_{i-1,j})}{\Delta\theta^2} + \alpha^2 H_i^3 \frac{(P_{i+1,j} - P_{i,j})/\eta^*_{i,j+1/2} - (P_{i,j} - P_{i-1,j})/\eta^*_{i,j-1/2}}{\Delta Y^2} = -\varepsilon \sin\theta_i \tag{11.6}$$

where η^* is the dimensionless viscosity $H_{i+1/2} = \left[1 + \varepsilon\cos\left(\frac{\theta_i + \theta_{i+1}}{2}\right)\right] \approx \frac{H_{i+1} + H_i}{2}$ and $H_{i-1/2} = \left[1 - \varepsilon\cos\left(\frac{\theta_i + \theta_{i-1}}{2}\right)\right] \approx \frac{H_i + H_{i-1}}{2}$.

The pressure boundary conditions were also given in Chapter 4.

In the axial direction: $P_{i,1} = 0$ and the symmetrical condition $\left.\frac{\partial P}{\partial Y}\right|_{Y=0} = 0$ is used, that is, to solve half of the area.

In the circumferential direction: $P_{1,j} = 0$, $P_{i,j} = 0$ and $P_{i+1,j} - P_{i,j} = 0$. By iteration, at $P < 0$ nodes we set P to be equal to 0 so that we can ultimately determine the outlet position.

11.2.2 Discrete Energy Equation

The dimensionless energy Equation 11.3 can be written in the discrete form as follows.

$$\frac{T^*_{i,j}-T^*_{i-1,j}}{\theta_i-\theta_{i-1}}=\frac{1}{Q_x}\left\{-\alpha Q_y\frac{T^*_{i,j}-T^*_{i,j-1}}{Y_j-Y_{j-1}}+\frac{2\eta^*_{i,j}}{H_{i,j}}+\frac{6H_{i,j}}{\eta^*_{i,j}}\left[\left(\frac{P_i-P_{i-1}}{\theta_i-\theta_{i-1}}\right)^2+\alpha^2\left(\frac{P_{i,j}-P_{i,j-1}}{Y_j-Y_{j-1}}\right)^2\right]\right\} \tag{11.7}$$

where $Q_x=\frac{H_{i,j}}{2}-\frac{H^3_{i,j}}{2}\frac{P_{i,j}-P_{i-1,j}}{\theta_i-\theta_{i-1}}$ and $Q_y=-\frac{H^3_{i,j}}{2}\frac{P_{i,j}-P_{i,j-1}}{Y_j-Y_{j-1}}$.

As pointed out before, because the energy equation is a one-dimensional partial differential equation, one boundary condition can only be given in the two directions θ and Y. The temperature boundary conditions will be as follows.

In the circumferential direction, the mixed oil temperatures are considered at the start point of θ, namely

$$T^{k+1}_{1,j}=\frac{T^k_{1,j}+T^k_{N,j}}{2} \tag{11.8}$$

In the axial direction, the symmetrical condition of Y is used, that is, solving half of the area.

$$\left.\frac{\partial T}{\partial Y}\right|_{Y=0}=0 \tag{11.9}$$

11.2.3 Temperature–Viscosity Equation

Here, the Roelands Viscosity–Temperature equation is used.

$$\eta_{i,j}=\exp\left\{(\ln\eta_0+9.67)\left[\left(\frac{T_{i,j}-138}{T_0-138}\right)^{-1.1}-1\right]\right\} \tag{11.10}$$

11.3 Calculation Program

11.3.1 Calculation Diagram

Because the temperature calculation should use the node viscosity which is affected by temperature, an iteration is needed. Furthermore, the symmetry condition is used in the Y direction and the temperatures of the inlet and outlet

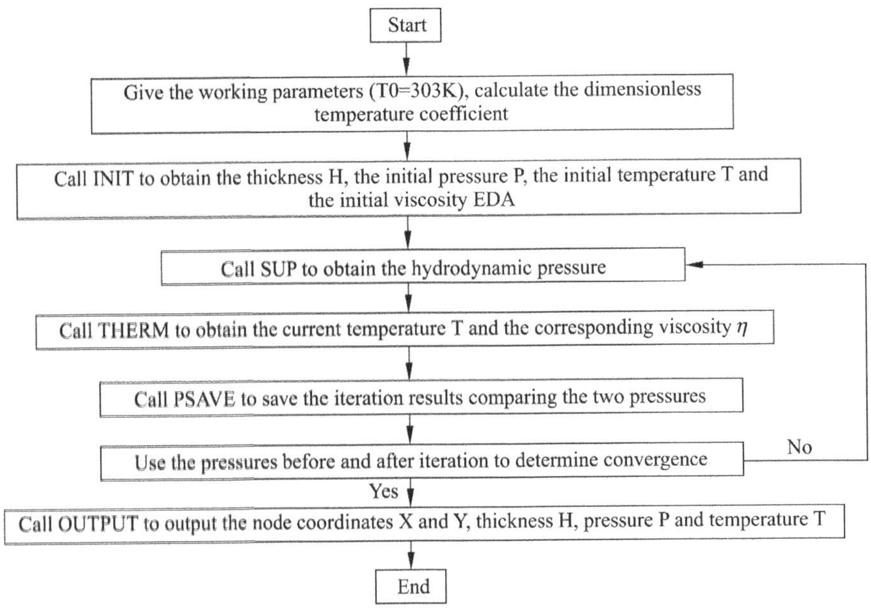

Figure 11.1 Calculation diagram.

of oil are determined by the mixed condition. The calculation diagram is shown in Figure 11.1.

11.3.2 Calculation Program

```
PROGRAM JOURNALTHERM
      DIMENSION X(200),Y(200),P(20000),H(20000),T(20000),EDA
(20000),POLD(20000)
      DATA EDA0,RO,C,AJ,B,R,RATIO,AN,EPSON/
0.05,890.0,1870.0,4.184,1.0E-1,6.0E-2,0.003,3.0E3,0.7/
      OPEN(7,FILE='FILM.DAT',STATUS='UNKNOWN')
      OPEN(8,FILE='PRESSURE.DAT',STATUS='UNKNOWN')
      OPEN(9,FILE='TEM.DAT',STATUS='UNKNOWN')
      OPEN(10,FILE='EDA.DAT',STATUS='UNKNOWN')
      PI=3.1415926
      N=61
      M=41
      DX=2.0*PI/FLOAT(N-1)
      DY=1./FLOAT(M-1)
      ALFA1=R/B
      C0=RATIO*R
      U=R*AN*2.0*PI/60.0
```

```
      ALFA=(R/B*DX/DY)**2
      A=U*R*EDA0/(2.0*AJ*RO*C*C0**2)
      T0=303.0
      KG=0
      CALL INIT(N,M,DX,DY,A,T0,EPSON,X,Y,H,T,EDA)
10    CALL SUBP(N,M,DX,EPSON,ALFA,H,P,EDA)
      CALL THERM(N,M,A,ALFA1,DX,DY,EDA0,T0,X,Y,P,H,T,EDA)
      IF(KG.EQ.0)THEN
      CALL PSAVE(KG,N,M,P,POLD,ERO)
      KG=1
      GOTO 10
      ENDIF
      CALL PSAVE(KG,N,M,P,POLD,ERO)
      WRITE(*,*)'EROP,KG=',ERO,KG
      KG=KG+1
      IF(ERO.GT.1.E-6.AND.KG.LT.10)GOTO 10
      CALL OUTPUT(N,M,A,T0,X,Y,H,P,T,EDA)
      STOP
      END
      SUBROUTINE INIT(N,M,DX,DY,A,T0,EPSON,X,Y,H,T,EDA)
      DIMENSION X(N),Y(M),H(N,M),T(N,M),EDA(N,M)
      DO I=1,N
      X(I)=(I-1.0)*DX
      ENDDO
      DO J=1,M
      Y(J)=-0.5+(J-1)*DY
      DO I=1,N
      H(I,J)=1.0+EPSON*COS(X(I))
      T(I,J)=T0/A
      EDA(I,J)=1.0
      ENDDO
      ENDDO
      RETURN
      END
      SUBROUTINE SUBP(N,M,DX,EPSON,ALFA,H,P,EDA)
      DIMENSION H(N,M),P(N,M),EDA(N,M)
      DO I=2,N-1
      DO J=2,M-1
      P(I,J)=0.5
      ENDDO
      ENDDO
      DO J=1,M
      P(1,J)=0.0
      P(N,J)=0.0
      ENDDO
```

```
      DO I=1,N
      P(I,1)=0.0
      P(I,M)=0.0
      ENDDO
      IK=0
10    C1=0.0
      ALOAD=0.0
      DO I=2,N-1
      I1=I-1
      I2=I+1
      DO J=2,M-1
      PD=P(I,J)
      J1=J-1
      J2=J+1
      A1=(0.5*(H(I2,J)+H(I,J)))**3/(0.5*(EDA(I2,J)+EDA(I,J)))
      A2=0.5*(H(I,J)+H(I1,J)))**3/(0.5*(EDA(I,J)+EDA(I1,J)))
      A3=ALFA*H(I,J)**3/(0.5*(EDA(I,J2)+EDA(I,J)))
      A4=ALFA*H(I,J)**3/(0.5*(EDA(I,J)+EDA(I,J1)))
      A5=A1*P(I2,J)+A2*P(I1,J)+A3*P(I,J2)+A4*P(I,J1)
      A6=A1+A2+A3+A4
      P(I,J)=(-DX*(H(I2,J)-H(I1,J))+A5)/A6
      P(I,J)=0.7*PD+0.3*P(I,J)
      IF(P(I,J).LT.0.0)P(I,J)=0.0
      C1=C1+ABS(P(I,J)-PD)
      ALOAD=ALOAD+P(I,J)
      ENDDO
      ENDDO
      IK=IK+1
      C1=C1/ALOAD
      IF(C1.GT.1.E-6)GOTO 10
      RETURN
      END
      SUBROUTINE PSAVE(KG,N,M,P,POLD,ERO)
      DIMENSION P(N,M),POLD(N,M)
      IF(KG.EQ.0)GOTO 10
      ERO=0.0
      EROMAX=-1.0
      W=0.0
      DO I=1,N
      DO J=1,M
      AE=ABS(P(I,J)-POLD(I,J))
      ERO=ERO+AE
      IF(AE.GT.EROMAX)THEN
      II=I
      JJ=J
```

```
      EROMAX=AE
      ENDIF
      W=W+ABS(P(I,J))
      ENDDO
      ENDDO
      ERO=ERO/W
10    DO I=1,N
      DO J=1,M
      POLD(I,J)=P(I,J)
      ENDDO
      ENDDO
      RETURN
      END
      SUBROUTINE THERM(N,M,A,ALFA1,DX,DY,EDA0,T0,X,Y,P,H,T,EDA)
      DIMENSION X(N),Y(M),H(N,M),P(N,M),T(N,M),EDA(N,M)
      KG=0
10    ERT=0.0
      DO I=2,N
      DO J=1,M/2+1
      TOLD=T(I,J)
      EDA(I,J)=EXP((ALOG(EDA0)+9.67)*(((A*T(I,J)-138.0)/
(T0-138.0))**(-1.1)-1.0))
      IF(I.NE.N)DPDX=0.5*(P(I+1,J)-P(I-1,J))/DX
      IF(I.EQ.N)DPDX=0.5*(P(1,J)-P(I-1,J))/DX
      QX=0.5*H(I,J)-0.5*H(I,J)**3*DPDX
      DPDY=ALFA1*(P(I,J+1)-P(I,J))/DY
      IF(J.EQ.M/2+1)DPDY=0.0
      QY=-0.5*H(I,J)**3*DPDY
      AA=QX/DX*T(I-1,J)-ALFA1*QY*T(I,J+1)/DY
      AB=2.0*EDA(I,J)/H(I,J)
      AC=6.0*H(I,J)/EDA(I,J)*(DPDX**2+DPDY**2)
      BB=QX/DX-ALFA1*QY/DY
      T(I,J)=(AA+AB+AC)/BB
      IF(A*T(I,J).GE.403.)THEN
      WRITE(*,*)'T OVER THE LIMIT 100'
      WRITE(*,*)I,J,T(I,J)
      STOP 00001
      ENDIF
      T(I,J)=0.7*TOLD+0.3*T(I,J)
      ERT=ERT+ABS(T(I,J)-TOLD)/303.
      ENDDO
      ENDDO
      ERT=A*ERT/((N-1)*(M-1))
      KG=KG+1
      DO J=1,M
```

```
      T(1,J)=0.5*(T(1,J)+T(N,J))
      ENDDO
      WRITE(*,*)'ERT,KG=',ERT,KG
      IF(ERT.GT.1.E-3)GOTO 10
      DO I=1,N
      DO J=1,M/2
      T(I,M-J+1)=T(I,J)
      EDA(I,M-J+1)=EDA(I,J)
      ENDDO
      ENDDO
      RETURN
      END
      SUBROUTINE OUTPUT(N,M,A,T0,X,Y,H,P,T,EDA)
      DIMENSION X(N),Y(M),H(N,M),P(N,M),T(N,M),EDA(N,M)
      TMAX=0.0
      DO I=1,N
      DO J=1,M
      T(I,J)=A*T(I,J)-T0
      IF(T(I,J).GT.TMAX)TMAX=T(I,J)
      END DO
      ENDDO
      WRITE(*,*)'TAMX=',TMAX
      WRITE(7,30)X(1),(Y(J),J=1,M)
      WRITE(8,30)X(1),(Y(J),J=1,M)
      WRITE(9,30)X(1),(Y(J),J=1,M)
      WRITE(10,30)X(1),(Y(J),J=1,M)
      DO I=1,N
      WRITE(7,30)X(I),(H(I,J),J=1,M)
      WRITE(8,30)X(I),(P(I,J),J=1,M)
      WRITE(9,30)X(I),(T(I,J),J=1,M)
      WRITE(10,30)X(I),(EDA(I,J),J=1,M)
      ENDDO
30    FORMAT(42(1X,E12.6))
      RETURN
      END
```

11.3.3 Calculation Results

The working parameters in the program are: lubricant initial viscosity EDA0 = 0.05 Pa s, lubricant density RO = 890 kg m^{-3}, lubricant specific heat C = 1870 J kg^{-1} K^{-1}, mechanical equivalent of heat AJ = 4.184 J cal^{-1}, bearing width B = 0.1 m, bearing radius R = 0.06 m, bearing clearance RATIO = 0.003, rotational speed AN = 3000 rpm and eccentricity EPSON = 0.7.

Figure 11.2 shows the film thickness, surface pressure distribution and temperature distribution of the journal with the above working conditions by the

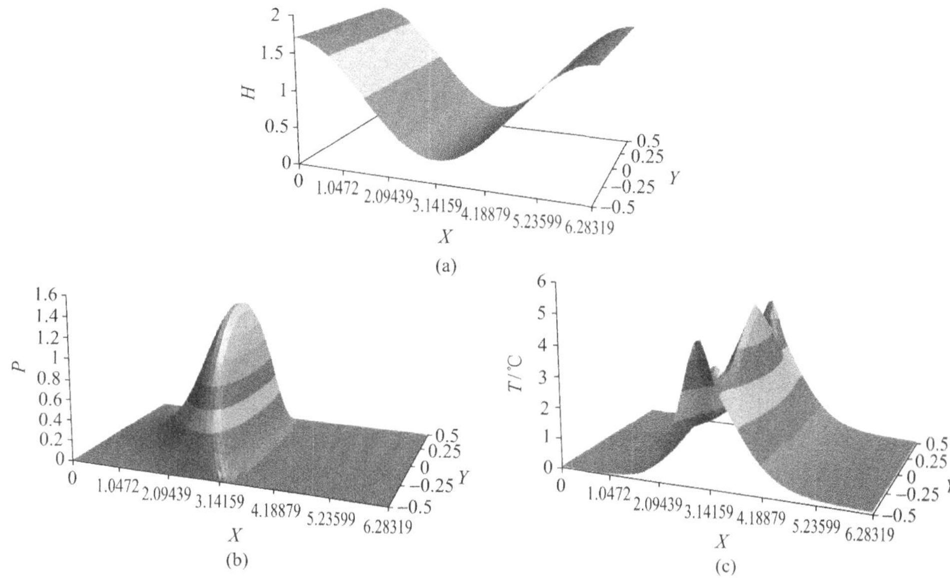

Figure 11.2 Results of thermal hydrodynamic lubrication of journal bearing. (a) Film thickness, H, (b) surface pressure, P, (c) temperature, T.

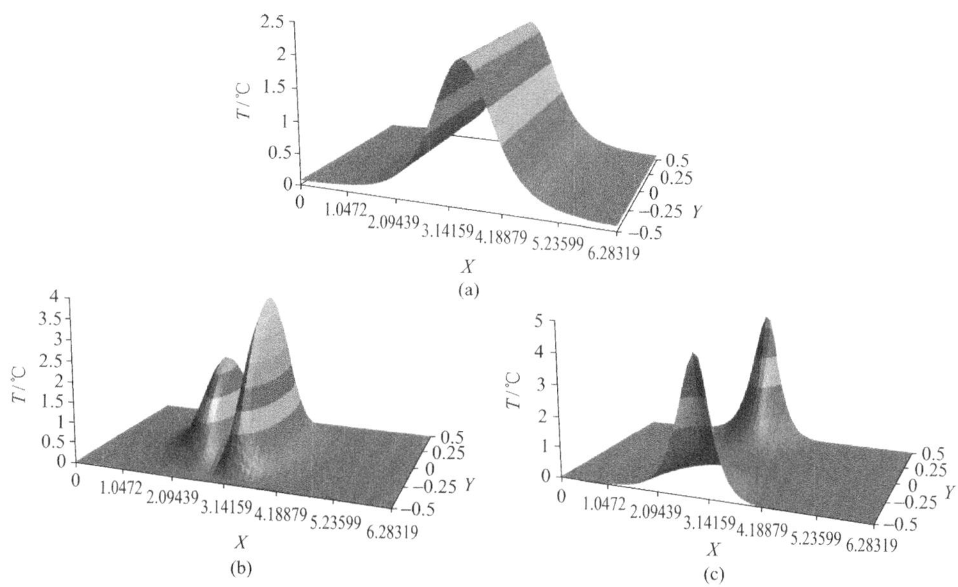

Figure 11.3 Each individual impact on temperature. (a) Temperature is inversely proportional to film thickness H. (b) Temperature is proportional to $\left(\frac{\partial p}{\partial x}\right)^2$. (c) Temperature is proportional to $\pm\left(\frac{\partial p}{\partial y}\right)^2$.

JOURNALTHERM program. The maximum temperature rise obtained is about 6 °C.

From Equation 11.3, it is known that the temperature consists of three main parts. These are:

1. The temperature is inversely proportional to film thickness H. Therefore, the lowest temperature is reached at the maximum film thickness and the highest temperature at the minimum film thickness.
2. The temperature is proportional to $\left(\frac{\partial p}{\partial x}\right)^2$. So, the maximum temperature is reached at the maximum $\pm\left(\frac{\partial p}{\partial x}\right)$. If we only consider this impact, the temperature forms two peaks.
3. The temperature is proportional to $\pm\left(\frac{\partial p}{\partial y}\right)^2$. If we only consider this impact, the temperature is low in the middle and high at both ends.

Considering the three separate effects, with the above working conditions, the calculated temperature distribution will be three figures, which reflect these impacts, as shown in Figure 11.3.

Part Three

Numerical Method for Elastic Deformation and Thermal Elastohydrodynamic Lubrication

12

Numerical Method and Program for Elastic Deformation and Viscosity–Pressure Equation

12.1 Basic Equations of Elastic Deformation

12.1.1 Film Thickness Equation

In elastohydrodynamic lubrication (EHL) calculations, we need to superimpose the elastic deformation equation on the original film thickness. An equivalent problem in which an elastic cylinder contacts with a rigid plane is shown in Figure 12.1. The expression of the film thickness can be written as

$$h(x) = h_c + \frac{x^2}{2R} + v(x) \tag{12.1}$$

where h_c is the center film thickness before deformation (which can be determined based on the load balance condition), R is the equivalent radius of the two surfaces ($\frac{1}{R} = \frac{1}{R_1} \pm \frac{1}{R_2}$, where + is for the outer contact and − is for the inner contact) and $v(x)$ is the elastic deformation generated by pressure.

12.1.2 Elastic Deformation Equation

For the line contact, because the length and radius of the surfaces are much larger than the contact width, the problem can be considered in a plane strain state, that is, its

Numerical Calculation of Lubrication: Methods and Programs, First Edition. Ping Huang.

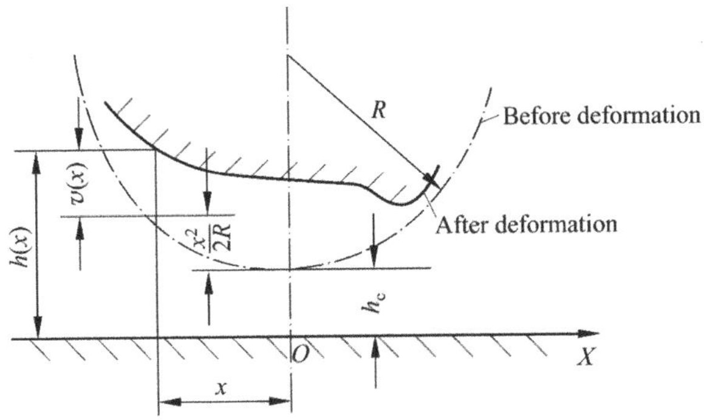

Figure 12.1 Shape of film thickness in line contact.

elastic deformation is equivalent to a pressure distribution acting on an infinitely wide elastic plane, as shown in Figure 12.2.

According to the theory of elasticity, the elastic deformation in the vertical direction can be written as

$$v(x) = -\frac{2}{\pi E}\int_{s_1}^{s_2} p(s)\ln{(s-x)^2}ds + c \tag{12.2}$$

where s is an additional coordinate of the x-axis which indicates the distance between the load $p(s)$ds and the origin of the coordinate, $p(s)$ is the pressure at s, s_1 is the start point of the pressure $p(s)$, s_2 is the end point, E is the equivalent elastic modulus $\frac{1}{E} = \frac{1}{2}\left(\frac{1-\nu_1^2}{E_1} + \frac{1-\nu_2^2}{E_2}\right)$ and c is a constant to be determined. Because the central film thickness h_c in Equation 12.1 can be determined based on the load

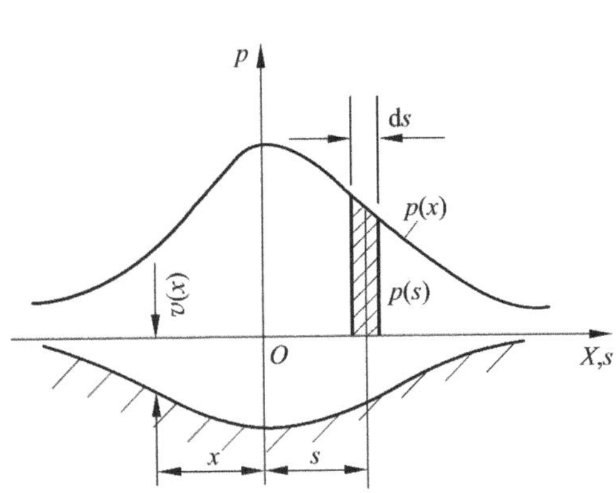

Figure 12.2 Elastic deformation in line contact.

balance condition, c can usually be merged into it. Therefore, it is no longer considered in the following.

12.2 Numerical Methods and Programs of Elastic Deformation

12.2.1 Numerical Method and Program of Elastic Deformation Equation in Line Contact

12.2.1.1 Singular Point Treatment

Because the pressure $p(x)$ can be obtained by solving the Reynolds equation, the elastic deformation $v(x)$ can be solved by numerically integrating the pressure. The integral part of the deformation Equation 12.2 is

$$v = \int_{s_1}^{s_2} p(s)\ln(s-x)^2 ds \tag{12.3}$$

Although there may exist a singular point where x is equal to s in Equation 12.3, that point can be removed in the numerical integration. If the singular point is contained in an integration interval, the mean value theorem can be used in this interval; that is, we take the midpoint between the two nodes so as to avoid the singular point, that is

$$\Delta v = \int_{x}^{x+\Delta x} p(s)\ln(s-x)^2 ds \approx p(x+\Delta x/2)\ln(\Delta x/2)^2 \Delta x \tag{12.4}$$

In this way, the singular point can be removed from the integration interval without significant difference.

For the equant grid, the numerical integration of Equation 12.3 can be simplified as

$$v_i = \Delta x \sum_{j=1}^{N} (a_{i-j} + \ln \Delta x) p_j \tag{12.5}$$

where $\Delta x = s - x$, v_i is the elastic deformation of the ith node, p_j is the pressure of the jth node and $a_{i\text{-}j} = (i\text{-}j+0.5)(\ln|i\text{-}j+0.5|\text{-}1)(\ln(i\text{-}j+0.5)\text{-}1)\text{-}(i\text{-}j-0.5)(\ln|i\text{-}j-0.5|\text{-}1)$, $i\text{-}j = 0, \ldots, N$.

12.2.1.2 Calculation Diagram and Program

The calculation diagram of elastic deformation in line contact is shown in Figure 12.3. In this diagram, N is the number of the nodes, DX is the equant distance of the interval, P is the pressure array and VI is the elastic deformation array.

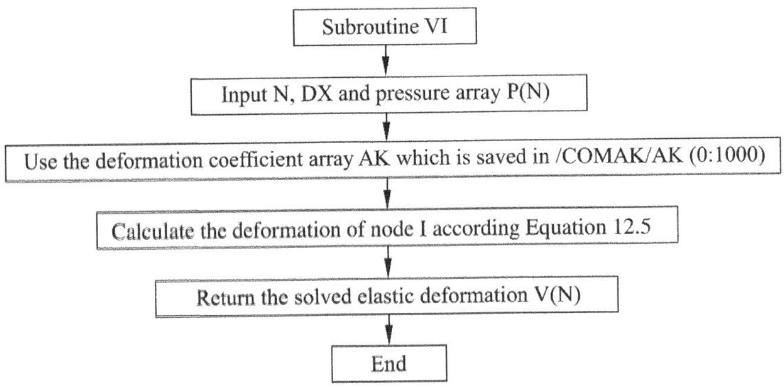

Figure 12.3 Calculation program of elastic deformation in line contact.

The calculation program is

```
SUBROUTINE VI(N,DX,P,V)
      DIMENSION P(N),V(N)
      COMMON /COMAK/AK(0:1100)
      PAI1=0.318309886
      C=ALOG(DX)
      DO 10 I=1,N
      V(I)=0.0
      DO 10 J=1,N
      IJ=IABS(I-J)
10    V(I)=V(I)+(AK(IJ)+C)*DX*P(J)
      DO I=1,N
      V(I)=-PAI1*V(I)
      ENDDO
      RETURN
      END
      SUBROUTINE SUBAK(MM)
      COMMON /COMAK/AK(0:1100)
      DO 10 I=0,MM
10    AK(I)=(I+0.5)*(ALOG(ABS(I+0.5))-1.)-(I-0.5)*(ALOG(ABS(I-0.5))-1.)
      RETURN
      END
```

In the above program, the input variables are the node number N, the distance DX between equidistant nodes, and the pressure P(I) of each node. The output variables are the elastic deformation array V(I). Because AK(I) is independent of pressure, it only needs to be calculated once, although it must be used many times in elastic deformation calculation. In order to save time, AK(I) is calculated once in advance in subroutine SUBAK.

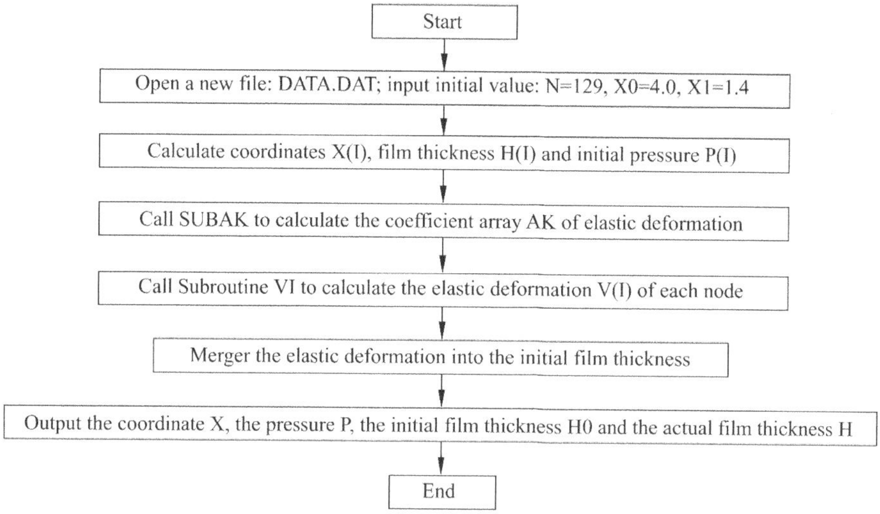

Figure 12.4 Calculation diagram of elastic deformation in line contact.

12.2.1.3 Calculation Diagram and Program of Film Thickness

The calculation diagram (Figure 12.4) for the elastic deformation V(I) is superimposed on the initial film thickness to obtain the deformed film thickness.

In this diagram, N is the node number, X0 is the inlet coordinate, X1 is the outlet coordinate, AK is the coefficient array of elastic deformation, V is the elastic deformation array, X is the coordinate array, P is the pressure array and H is the film thickness array.

The calculation program is

```
DIMENSION P(1000),H0(1000),H(1000),V(1000),X(1000)
      OPEN (8,FILE='DATA.DAT',STATUS='UNKNOWN')
      N=129
      X1=1.4
      X0=-4.0
      DX=(X1-X0)/(N-1.0)
      DO I=1,N
      X(I)=-4.0+(I-1)*DX
      H0(I)=0.5*X(I)*X(I)
      H(I)=H0(I)
      P(I)=0.0
      IF(X(I).GE.-1.0.AND.X(I).LE.1.0)THEN
      P(I)=SQRT(1-X(I)*X(I))
      ENDIF
      ENDDO
```

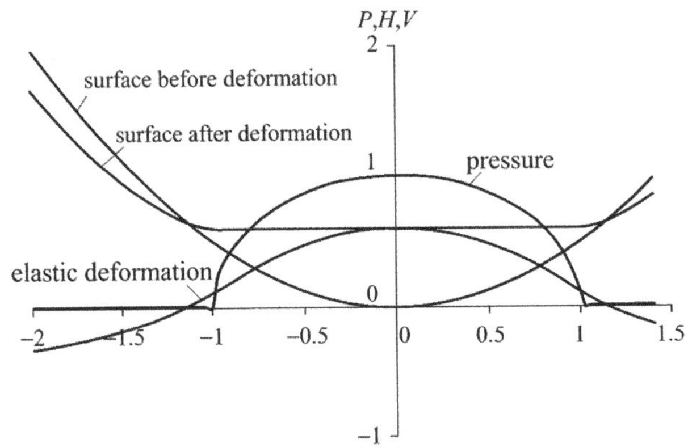

Figure 12.5 Calculation results of elastic deformation in line contact.

```
CALL SUBAK(N)
CALL VI(N,DX,P,V)
DO I=1,N
H(I)=H(I)+V(I)
WRITE(*,*)X(I),P(I),V(I),H0(I),H(I)
WRITE(8,*)X(I),P(I),V(I),H0(I),H(I)
ENDDO
STOP
END
```

Results: The difference in film thickness before and after deformation is shown in Figure 12.5. In the program, the total nodes are 129, the dimensionless coordinate X is from -4.0 to 1.4, the dimensionless film thickness can be expressed as $H(X) = H_c + X^2/2 + V(X)$ and the dimensionless Hertz contact pressure is $P = \sqrt{1 - X^2}$. From the figure, we can see that elastic deformation makes the cylinderical surface a flat one, which is consistent with the theory of elasticity. Because of not using the load balance calculation, there is a gap between the two surfaces in the figure. If in solid contact, the actual gap is zero.

12.2.2 Numerical Method and Program of Elastic Deformation Equation in Point Contact

12.2.2.1 Film Thickness

For a point contact, the solution region is generally a rectangular area, as shown in Figure 12.6, where AB is the inlet border, CD is the outlet border and AD and BC are the leakage sides. In the figure, α, β and γ can be used to decide the solution domain.

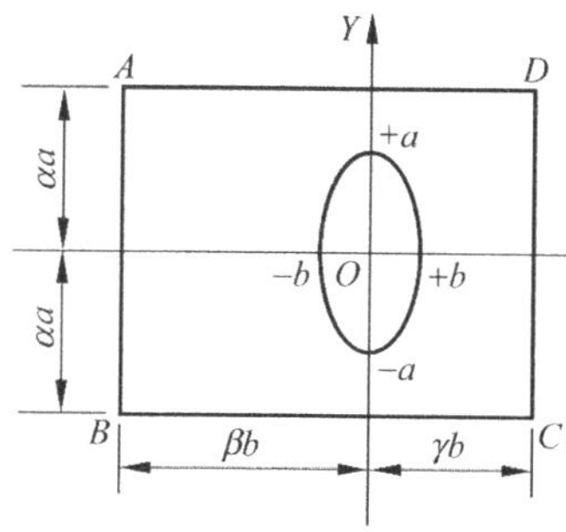

Figure 12.6 Solution region in point contact.

Further, a point contact problem can usually be simplified into an equivalent elastic ellipsoid contacting with a rigid plane. When considering elastic deformation, the film thickness can be expressed as

$$h(x,y) = h_0 + \frac{x^2}{2R_x} + \frac{y^2}{2R_y} + v(x,y) \tag{12.6}$$

where h_0 is the center film thickness to be determined in calculation, R_x and R_y are the equivalent curvature radii in the x and y directions and $v(x,y)$ is the elastic deformation.

12.2.2.2 Elastic Deformation

According to the theory of elasticity, the relationship between the pressure $p(x,y)$ and the deformation $v(x,y)$ can be expressed as

$$v(x,y) = \frac{2}{\pi E} \iint\limits_{\Omega} \frac{p(s,t)}{\sqrt{(x-s)^2 + (y-t)^2}} dsdt \tag{12.7}$$

where s and t are the additional coordinates in the x and y directions, Ω is the solving domain (as shown in Figure 12.7) and E is the equivalent elastic module of the surfaces.

The denominator of the integral part in Equation 12.7 is the distance between the pressure point $(s,\ t)$ and the deformed point $(x,\ y)$. If $x = s$ and $y = t$, the integral is singular. However, it can be removed with the variable substitution method. By transforming the coordinate origin to $(s,\ t)$, that is, $x' = x\text{-}s$ and $y' = y\text{-}t$, Equation 12.7 becomes

$$v(x,y) = \frac{2}{\pi E} \iint\limits_{\Omega} \frac{p'(x',y')}{\sqrt{x'^2 + y'^2}} dx'dy' \tag{12.8}$$

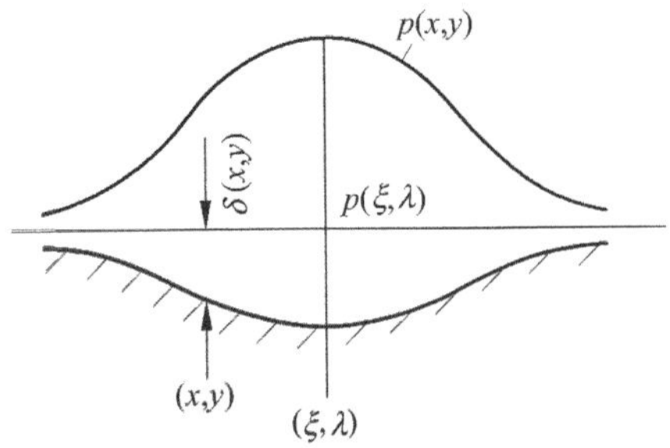

Figure 12.7 Elastic deformation.

Then, using the polar coordinate, that is, $x' = r\cos\theta$ and $y' = r\sin\theta$, we will obtain the following integral equation.

$$v(x,y) = \frac{2}{\pi E}\iint_{\Omega} p''(r,\theta)drd\theta \tag{12.9}$$

The above integral equation no longer has a singular point.

12.2.2.3 Numerical Method

The main difficulty of the numerical calculation of elastic deformation in a point contact is that there is a large amount of computation. If the usual numerical integration method is used, because the pressure distribution $p\,(x, y)$ is different (for it is generated from the Reynolds equation), the deformation of each node will be calculated in each iteration and over the entire solution domain. This makes the computational work very heavy and it needs a large amount of computer storage space. An effective way to overcome this difficulty is to use a deformation matrix, as follows.

Supposing the solution domain is divided into a mesh with m nodes in the x direction and n nodes in the y direction, we can calculate the elastic deformation D_{ij}^{kl} at Node (k,l) when there is a unit pressure acting at Node (i,j) while the remaining nodes have no pressure. Then, the discrete form of the elastic deformation equation becomes

$$v_{kl} = \frac{2}{\pi E}\sum_{i=1}^{n}\sum_{j=1}^{m} D_{ij}^{kl} p_{ij} \tag{12.10}$$

where v_{kl} is the elastic deformation at Node (k,l), p_{ij} is the pressure at Node (i,j) and D_{ij}^{kl} is the elastic deformation coefficient between Node (k,l) and Node (i,j).

Similar to a line contact, we only need to calculate D_{ij}^{kl} once and store it for later calculations. Substituting D_{ij}^{kl} into Equation 12.10 to calculate the deformation repeatly in the iteration process, the amount of computation can be significantly reduced with no significant loss of accuracy.

If we use an isometric grid mesh, we have

$$D_{ij}^{kl} = D_{kj}^{il}$$
$$D_{ij}^{kl} = D_{il}^{kj} \tag{12.11}$$

Further, the storage unit can be reduced from $m^2 \times n^2$ to $m \times n$ and the specific formula of the elastic deformation becomes

$$v_{ij} = \Delta x^2 \sum_{k=1}^{N} \sum_{l=1}^{N} a_{i-k,j-l} p_{kl} \tag{12.12}$$

where $\Delta x = \Delta y$ is the distance between two neighboring nodes and $a_{i\text{-}k,j\text{-}l}$ is the integration coefficient given by the following formula.

$$\begin{aligned} a_{i-k,j-l} = {} & (|j-l|+0.5)\ln\left[f(|i-k|+0.5,|j-l|+0.5)/f(|i-k|-0.5,|j-l|+0.5)\right] \\ & +(|i-k|-0.5)\ln\left[f(|j-l|-0.5,|i-k|-0.5)/f(|j-l|+0.5,|i-k|-0.5)\right] \\ & +(|j-l|-0.5)\ln\left[f(|i-k|-0.5,|j-l|-0.5)/f(|i-k|+0.5,|j-l|-0.5)\right] \\ & +(|i-k|+0.5)\ln\left[f(|j-l|+0.5,|i-k|+0.5)/f(|j-l|-0.5,|i-k|+0.5)\right] \end{aligned} \tag{12.13}$$

where Function $f(x, y) = x + \sqrt{x^2 + y^2}$.

12.2.2.4 Calculation Diagram

The calculation diagram of elastic deformation in the point contact is shown in Figure 12.8. The deformation coefficient AK should be calculated in advance, where N × N is the number of the nodes, DX = DY is the equant distance of the mesh interval, P is the pressure array and VI is the elastic deformation array.

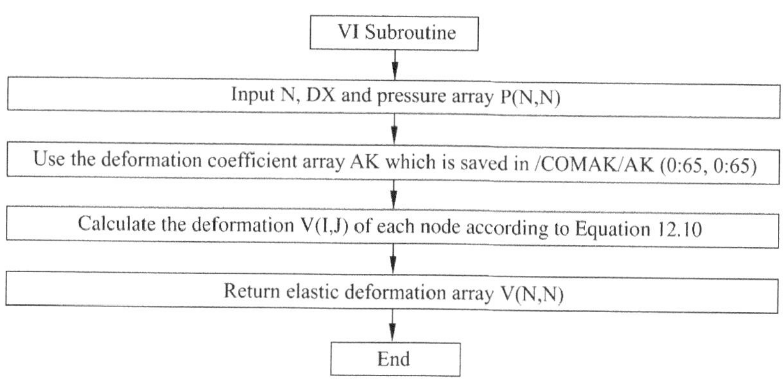

Figure 12.8 Calculation program of elastic deformation in point contact.

12.2.2.5 Program

```
      SUBROUTINE VI (N,DX,P,V)
      DIMENSION P(N,N),V(N,N)
      COMMON /COMAK/AK(0:65,0:65)
      PAI1=0.2026423
      DO 40 I=1,N
      DO 40 J=1,N
      H0=0.0
      DO 30 K=1,N
      IK=IABS(I-K)
      DO 30 L=1,N
      JL=IABS(J-L)
30    H0=H0+AK(IK,JL)*P(K,L)
40    V(I,J)=H0*DX*PAI1
      RETURN
      END
      SUBROUTINE SUBAK(MM)
      COMMON /COMAK/AK(0:65,0:65)
      S(X,Y)=X+SQRT(X**2+Y**2)
      DO 10 I=0,MM
      XP=I+0.5
      XM=I-0.5
      DO 10 J=0,I
      YP=J+0.5
      YM=J-0.5
      A1=S(YP,XP)/S(YM,XP)
      A2=S(XM,YM)/S(XP,YM)
      A3=S(YM,XM)/S(YP,XM)
      A4=S(XP,YP)/S(XM,YP)
      AK(I,J)=XP*ALOG(A1)+YM*ALOG(A2)+XM*ALOG(A3)+YP*ALOG(A4)
10    AK(J,I)=AK(I,J)
      RETURN
      END
```

It should be noted that, in the program, the input node number N is only in one direction. Therefore, the total number of nodes is equal to N × N. DX = DY is the equal node spacing. P (I,J) is the pressure array, which is obtained by solving the Reynolds equation. V(I,J) is the elastic deformation array, which will be output after being calculated. In addition, because $a_{i-k,j-l}$ is not related to the pressure, it can be calculated in advance and be ready to be used to calculate the elastic deformation. Therefore, in order to save computing time, subroutine SUBAK is used to calculate $a_{i-k,j-l}$.

12.2.2.6 Example

The elastic deformation of a ball with radius R contacting with a rigid plane is calculated by using the Hertzian contact stress distribution. The solving domain is between X from -1.2 to1.2 and Y from -1.2 to 1.2. The region is divided into 33×33 isometric nodes. The calculation diagram is shown in Figure 12.9.

In this diagram, VI is the elastic deformation array, P is the pressure array, H0 is the film thickness parameter of the two surfaces and H is the film thickness array.

The calculation program is

```
DIMENSION P(4500),H(4500),V(4500),X(65),Y(65)
OPEN (8,FILE='PRESS.DAT',STATUS='UNKNOWN')
OPEN (10,FILE='FILM.DAT',STATUS='UNKNOWN')
N=33
CALL SUBAK(N)
CALL PCAL(N,X,Y,P,H,V)
STOP
END
SUBROUTINE PCAL(N,X,Y,P,H,V)
DIMENSION P(N,N),H(N,N),X(N),Y(N),V(N,N)
COMMON /COMAK/AK(0:65,0:65)
KL=ALOG(N-1.)/ALOG(2.)-1.99
DX=2.4/(N-1.0)
DO I=1,N
X(I)=-1.2+DX*(I-1)
A=X(I)*X(I)
```

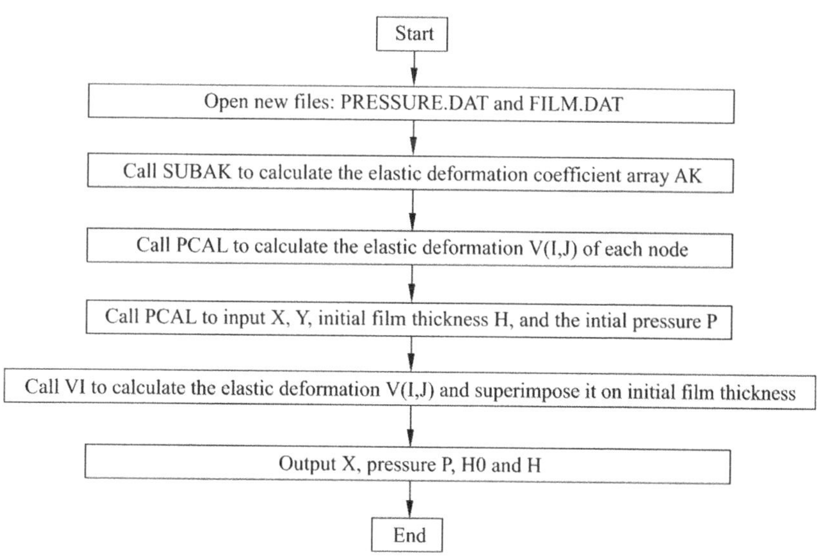

Figure 12.9 Calculation diagram of elastic deformation in point contact.

```
      DO J=1,N
      Y(J)=-1.2+DX*(J-1)
      P(I,J)=0.0
      H(I,J)=0.5*A+0.5*Y(J)*Y(J)
      ENDDO
      ENDDO
      M=0
      DO I=1,N
      DO J=1,N
      A=1.0-X(I)*X(I)-Y(J)*Y(J)
      IF(A.GE.0.0) P(I,J)=SQRT(A)
      ENDDO
      ENDDO
      CALL VI(N,DX,P,V)
      DO 10 I=1,N
      DO 10 J=1,N
      H(I,J)=H(I,J)+V(I,J)
10    CONTINUE
      XP=1.0
      WRITE(8,20)XP,(Y(I),I=1,N)
      WRITE(10,20)XP,(Y(I),I=1,N)
      DO I=1,N
      WRITE(8,20)X(I),(P(I,J),J=1,N)
      WRITE(10,20)X(I),(H(I,J),J=1,N)
      ENDDO
20    FORMAT(1X,34(F6.3,1X))
      STOP
      END
```

Results: The pressure distribution and the calculated film thickness are shown in Figure 12.10. During the calculation of elastohydrodynamic lubrication, h_0 in

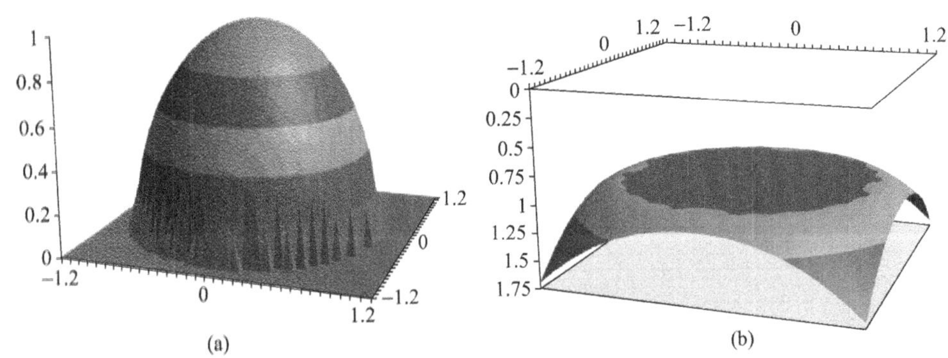

Figure 12.10 Calculation results of elastic deformation in point contact. (a) Pressure distribution. (b) Calculated film thickness.

Equation 12.6 or H0 in the program should be adjusted by the load balance condition.

12.3 Viscosity–Pressure and Density–Pressure Equations

When pressure increases, the distance between the molecules of a liquid or gas decreases and so the intermolecular force increases. Therefore, viscosity increases. Generally, when the pressure of a mineral oil is above 0.02 GPa, viscosity changes significantly with pressure. The viscosity change rate also varies with pressure. The viscosity of lubricant under a pressure of 1 GPa can increase by several orders of magnitude. When the pressure is higher, mineral oil loses its liquid nature and changes into a waxy solid. Therefore, for hydrodynamic lubrication under a heavy load or especially for elastic hydrodynamic lubrication, the viscosity–pressure relationship is an important factor.

12.3.1 Viscosity–Pressure Relationship

There are several equations to describe the relationship of viscosity and pressure. They are:

$$\text{Barus} \quad \eta = \eta_0 e^{\alpha p} \tag{12.14}$$

$$\text{Roelands} \quad \eta = \eta_0 \exp\{(\ln \eta_0 + 9.67)[-1 + (1 + p/p_0)^z]\} \tag{12.15}$$

$$\text{Cameron} \quad \eta = \eta_0 (1 + cp)^{16} \tag{12.16}$$

where η is the viscosity of the lubricant, η_0 is the viscosity at $p = 0$, α is the coefficient of viscosity–pressure, p_0 is the pressure coefficient (usually $p_0 = 5.1 \times 10^9$), z is an index (usually $z = 0.68$) and c is approximately chosen as $\alpha/15$.

When the pressure is >1 GPa, the viscosity is too large with the Barus viscosity–pressure equation, but it is more realistic with the Roelands viscosity–pressure equation.

The coefficient α of viscosity–pressure can be chosen as $2.2 \times 10^{-8}\ \text{m}^2\cdot\text{N}^{-1}$. Tables 12.1 and 12.2 give the coefficients of viscosity–pressure for some lubricants.

Table 12.1 Coefficients α of the viscosity–pressure of some mineral oils ($\times 10^{-8}\ \text{m}^2\cdot\text{N}^{-1}$)

Temperature (°C)	Cycloalkyl			Paraffin base		
	Spindle oil	Light oil	Heavy oil	Light oil	Heavy oil	Cylinder oil
30	2.1	2.6	2.8	2.2	2.4	3.4
60	1.6	2.0	2.3	1.9	2.1	2.8
90	1.3	1.6	1.8	1.4	1.6	2.2

Table 12.2 Coefficients α of the viscosity–pressure of some base oil at 25 °C ($\times 10^{-8}\,m^2 \cdot N^{-1}$)

Lubricant	α	Lubricant	α
Paraffin base	$1.5 \sim 2.4$	Alkyl silicone oil	$1.4 \sim 1.8$
Cycloalkyl	$2.5 \sim 3.6$	Poly ether	$1.1 \sim 1.7$
Aromatic	$4 \sim 8$	Aromatic silicone oil	$3 \sim 5$
Polyolefin	$1.5 \sim 2.0$	Chlorinated paraffins	$0.7 \sim 5$
Diester	$1.5 \sim 2.5$		

12.3.2 Viscosity–Pressure–Temperature Relationship

The viscosity–temperature equation was discussed in Section 10.2. When taking into account together the influences of temperature and pressure on viscosity, the viscosity–temperature and viscosity–pressure equations are usually combined together. The commonly used expressions are as follows.

Barus and Reynolds formula

$$\eta = \eta_0 \exp[\alpha p - \beta(T - T_0)] \tag{12.17}$$

Roelands formula

$$\eta = \eta_0 \exp\left\{ (\ln \eta_0 + 9.67) \left[(1 + 5.1 \times 10^{-9} p)^{0.68} \times \left(\frac{T - 138}{T_0 - 138} \right)^{-1.1} - 1 \right] \right\} \tag{12.18}$$

Equation 12.16 is simpler and easier, but Equation 12.17 is more accurate. Therefore, the latter is often used.

12.3.3 Density–Pressure Relationship

For conveniency, the density–pressure equation can be expressed as

$$\rho = \rho_0 \left(1 + \frac{0.6p}{1 + 1.7p} \right) \tag{12.19}$$

where ρ_0 is the density of lubricant at $p = 0$; and the unit of p is GPa.

Because Equations 12.14–12.19 are analytical, they can be directly used in the numerical calculation. For example, their dimensionless equations can be written as

$$\eta^*_{i,j} = \exp(\alpha p_{i,j}) \tag{12.20}$$

$$\eta_{i,j}^* = \exp\left\{(\ln \eta_0 + 9.67)\left[-1 + \left(1 + p_{i,j}/p_0\right)^z\right]\right\} \tag{12.21}$$

$$\eta_{i,j}^* = (1 + cp_{i,j})^{16} \tag{12.22}$$

$$\eta_{i,j}^* = \exp[\alpha p_{i,j} - \beta(T_{i,j} - T_0)] \tag{12.23}$$

$$\eta_{i,j}^* = \exp\left\{(\ln \eta_0 + 9.67)\left[(1 + 5.1 \times 10^{-9} p_{i,j})^{0.68} \times \left(\frac{T_{i,j} - 138}{T_0 - 138}\right)^{-1.1} - 1\right]\right\} \tag{12.24}$$

and

$$\rho_{i,j}^* = \left(1 + \frac{0.6p_{i,j}}{1 + 1.7p_{i,j}}\right) \tag{12.25}$$

where $\eta_{i,j}^*$ is the dimensionless viscosity of the lubricant ($\eta_{i,j}^* = \eta_{i,j}/\eta_0$) and $\rho_{i,j}^*$ is the dimensionless density of the lubricant ($\rho_{i,j}^* = \rho_{i,j}/\rho_0$).

13

Numerical Method and Program for EHL in Line Contact

13.1 Basic Equations

The basic equations of elastohydrodynamic lubrication (EHL) in line contact include the Reynolds equation, the film thickness equation, the elastic deformation equation, the viscosity–pressure equation and the density–pressure equation.

Reynolds equation

$$\frac{d}{dx}\left(\frac{\rho h^3}{\eta}\frac{dp}{dx}\right) = 12U\frac{d\rho h}{dx} \tag{13.1}$$

where ρ is the lubricant density, η is the lubricant viscosity, U is the average velocity ($U=(u_1+u_2)/2$; u_1 and u_2 are the tangential velocities for the two surfaces), p is pressure, h is film thickness and x is the coordinate along the line of motion.

Film thickness equation

$$h(x) = h_c + \frac{x^2}{2R} + v(x) \tag{13.2}$$

where h_c is the central film thickness (determined by calculation), R is the equivalent radius of the surfaces $\left(\frac{1}{R} = \frac{1}{R_1} + \frac{1}{R_2}\right)$ and $v(x)$ is the elastic deformation.

Elastic deformation equation

$$v(x) = -\frac{2}{\pi E}\int_{x_0}^{x_e} p(s)\ln(s-x)^2 ds + c \tag{13.3}$$

Numerical Calculation of Lubrication: Methods and Programs, First Edition. Ping Huang.

where E is the equivalent elastic modulus $\left[\frac{1}{E}=\frac{1}{2}\left(\frac{1-\nu_1{}^2}{E_1}+\frac{1-\nu_2{}^2}{E_2}\right)\right]$, ν_1 and ν_2 are the Poisson's ratios of the two surfaces, E_1 and E_2 are the elastic modulus of the two surfaces, x_0 is the inlet coordinate, x_e is the outlet coordinate and c is the integral constant.

Viscosity–pressure equation

$$\eta=\eta_0\exp\left\{(\ln\eta_0+9.67)\left[\left(1+5.1\times10^{-9}p\right)^{0.68}-1\right]\right\} \tag{13.4}$$

where η_0 is the lubricant viscosity at $p=0$.

Density–pressure equation

$$\rho=\rho_0\left(1+\frac{0.6p}{1+1.7p}\right) \tag{13.5}$$

where ρ_0 is the lubricant density at $p=0$ and the pressure p is in GPa.

Because elastic deformation in the film-thickness equation and viscosity in the viscosity–pressure equation are dependent on pressure, the general solving way is to give an initial pressure distribution first (such as Hertz contact stress distribution) to calculate the film thickness and viscosity. Then, we substitute them into the Reynolds equation to obtain a new pressure distribution. According to the previous pressure distribution, we iteratively calculate the elastic deformation and film thickness. We repeat the above steps until the pressure difference between two neighboring results is less than the given error. Thus, we can obtain the final pressure distribution and film thickness containing the elastic deformation.

13.2 Numerical Method

13.2.1 Dimensionless Equations

The dimensionless Reynolds equation of EHL in line contact is

$$\frac{d}{dX}\left(\varepsilon\frac{dP}{dX}\right)-\frac{d(\rho^*H)}{dX}=0 \tag{13.6}$$

where $\varepsilon=\frac{\rho^*H^3}{\eta^*\lambda}$, $\lambda=\frac{12\eta_0UR^2}{b^2p_H}$, η^* is the dimensionless lubricant viscosity ($\eta^*=\eta/\eta_0$), ρ^* is the dimensionless lubricant density ($\rho^*=\rho/\rho_0$), P is the dimensionless pressure ($P=p/p_{\mathrm{H}}$), p_{H} is the maximum Hertzian pressure, X is a dimensionless coordinate ($X=x/b$), b is the half-width of the contact area and H is the dimensionless film thickness $\left(H=\frac{hR}{b^2}\right)$.

The boundary conditions of the Reynolds Equation 13.6 are

$$\text{Inlet boundary condition}: P(X_0) = 0$$
$$\text{Outlet boundary conditions}: P(X_e) = 0 \text{ and } \frac{dP(X_e)}{dX} = 0.$$

where X_0 and X_e are the dimensionless coordinates of the inlet and outlet; and X_0 is given but X_e should be determined by the outlet boundary conditions.

Film thickness equation

$$H(X) = H_0 + \frac{X^2}{2} - \frac{1}{\pi}\int_{X_0}^{X_e} \ln|X - X'|p(X')dX' \tag{13.7}$$

where H_0 is the dimensionless film thickness to be determined in calculation. Viscosity–pressure equation

$$\eta^* = \exp\left\{[\ln(\eta_0) + 9.67]\left[-1 + \left(1 + 5.1 \times 10^{-9}P \cdot p_H\right)^{0.68}\right]\right\} \tag{13.8}$$

Density–pressure equation

$$\rho^* = \left(1 + \frac{0.6p}{1 + 1.7p}\right) \tag{13.9}$$

Load equation

$$W = \int_{X_0}^{X_e} Pdx = \frac{\pi}{2} \tag{13.10}$$

13.2.2 Discrete Equations

With the central and the forward discrete differential Equations 1.22 and 1.24, the discrete form of the Reynolds Equation 13.6 can be written as

$$\frac{\varepsilon_{i-1/2}P_{i-1} - (\varepsilon_{i-1/2} + \varepsilon_{i+1/2})P_i + \varepsilon_{i+1/2}P_{i+1}}{\Delta X^2} = \frac{\rho_i^* H_i - \rho_{i-1}^* H_{i-1}}{\Delta X} \tag{13.11}$$

where $\varepsilon_{i\pm1/2} = \frac{1}{2}(\varepsilon_i + \varepsilon_{i\pm1})$ and $\Delta X = X_i - X_{i-1}$.

The boundary condition in the inlet $P(X_0) = 0$; and the boundary of the outlet can be obtained by setting the negative pressure equal to zero. The discrete forms are: $P(X_e) = 0$ and $\dfrac{P(X_e) - P(X_e - 1)}{\Delta X} = 0$.

The film thickness equation is

$$H_i = H_0 + \frac{x_i^2}{2} - \frac{1}{\pi}\sum_{j=1}^{n} K_{ij}P_j \tag{13.12}$$

where K_{ij} is the stiffness coefficient of the elastic deformation.

Load equation

$$\Delta X \sum_{i=1}^{n} \frac{(P_i + P_{i+1})}{2} = \frac{\pi}{2} \tag{13.13}$$

Convergence criterion

$$\frac{\sum |P_i^{k+1} - P_i^k|}{\sum P_i^{k+1}} \le \varepsilon_4 \tag{13.14}$$

where ε_4 is the given error.

13.2.3 Iterative Method

In the iteration, the iterative equation to correct the pressure can be written as

$$\bar{P}_i = \tilde{P}_i + c_1\delta_i \tag{13.15}$$

where c_1 is a relaxation factor, δ_i is the pressure increment and $\tilde{P}_i$ and $\bar{P}_i$ are the pressures before and after iteration.

For the pressure correction, the Gauss–Seidel iteration is commonly used in the low pressure region. However, the method is subject to divergence if the pressure is high, so the Jacobi iteration is used in the high pressure region. The two iteration formulas are as follows.

Gauss–Seidel iteration

$$\delta_i = \left(\frac{\partial L_i}{\partial P_i}\right)^{-1} \bar{\gamma}_i \tag{13.16}$$

Jacobi iteration

$$\delta_i = \left(\frac{\partial L_i}{\partial P_i} - \frac{\partial L_i}{\partial P_{i-1}}\right)^{-1} \tilde{\gamma}_i \tag{13.17}$$

where

$$L_i = \frac{\varepsilon_{i-1/2}P_{i-1} - (\varepsilon_{i-1/2} + \varepsilon_{i+1/2})P_i + \varepsilon_{i+1/2}P_{i+1}}{\Delta X^2} - \frac{\rho_i^* H_i - \rho_{i-1}^* H_{i-1}}{\Delta X} \tag{13.18}$$

$$\bar{\gamma}_i = -\frac{\varepsilon_{i-1/2}\bar{P}_{i-1} - (\varepsilon_{i-1/2} + \varepsilon_{i+1/2})\bar{P}_i + \varepsilon_{i+1/2}\bar{P}_{i+1}}{\Delta X^2} + \frac{\rho_i^* \tilde{H}_i - \rho_{i-1}^* \tilde{H}_{i-1}}{\Delta X} \tag{13.19}$$

$\tilde{\gamma}_i$ is obtained by changing $\bar{P}_{i-1}$ to $\tilde{P}_{i-1}$ in Equation 13.19. $\dfrac{\partial L_i}{\partial P_i}$ is the derivation of L_i to P_i. Because H_i is also the function of pressure, its derivative should be taken into account. For convenience, ε is not considered as a function of P_i. Then, we have

$$\frac{\partial L_i}{\partial P_i} = -\left(\varepsilon_{i-1/2} + \varepsilon_{i+1/2}\right)/\Delta X^2 + \frac{1}{\pi}\left(\rho_i^* K_{ii} - \rho_{i-1}^* K_{i-1,i}\right)/\Delta X \tag{13.20}$$

$\dfrac{\partial L_i}{\partial P_{i-1}}$ can be deduced similarly.

In the Jacobi iteration, we should add δ_i to the pressure of the present node and subtract it from the pressure of the front node, that is

$$\begin{aligned} \bar{P}_i &= \tilde{P}_i + c_2\delta_i \\ \bar{P}_{i-1} &= \tilde{P}_{i-1} - c_2\delta_i \end{aligned} \tag{13.21}$$

where c_2 is the relaxation factor of the Jacobi iteration.

The load balance condition is completed by changing the displacement coefficient H_0. The method is

$$\bar{H}_0 = \tilde{H}_0 + c_3\left[g - \frac{\Delta X}{\pi}\sum_{i=1}^{N-1}\left(P_i + P_{i+1}\right)\right] \tag{13.22}$$

where c_3 is a relaxation factor of load balance and g is a dimensionless load.

13.2.4 Selection of Iterative Methods

During the process of pressure correction, the above two iterations can be used in the different regions for the same problem. This is because an iteration method is only to reduce the local error. If a global error meets the requirement, more iteration methods can be adopted. Then, two different iteration methods will be used in the high and low pressure regions. However, another problem is how to divide the pressure regions. The pressure region can be divided into two parts by the following method.

In Equation 13.20, there are two terms.

One is affected by pressure

$$A_1 = \left(\varepsilon_{i-1/2} + \varepsilon_{i+1/2}\right)/\Delta X^2 \tag{13.23}$$

The other part is affected by film thickness

$$A_2 = \frac{1}{\pi}\left(\rho_i^* K_{ii} - \rho_{i-1}^* K_{i-1,i}\right)/\Delta X \tag{13.24}$$

When A_1 is larger, the Gauss–Seidel method is more effective. When A_2 is larger, the Gauss–Seidel method is easily subject to divergence because the pressure cannot change with film thickness and the film thickness in γ_i has not been corrected. Therefore, the ratio of A_1 and A_2 can be served as a parameter to divide the pressure region. According to the author's practice, if $A_1 \geq 0.1A_2$, the Gauss–Seidel iteration can be adopted. If $A_1 < A_2$, the Jacobi iteration will be more effective.

13.2.5 Relaxation Factors

The choice of relaxation factors often determines the calculation convergence. In the present iteration process, three relaxation factors should be given. They are the relaxation factor c_1 of the Gauss–Seidel iteration, the relaxation factor c_2 of the Jacobi bipolar iteration and the relaxation factor c_3 of load balance. These relaxation factors are experiential. The ranges of the former two relaxation factors are $c_1 = 0.3 \sim 1.0$ and $c_2 = 0.1 \sim 0.6$. It is found that c_2 has a greater impact on convergence. For a heavy load condition, c_2 should be low. As for c_3, there is no definite range. In actuality, an empirical formula can be used effectively and easily to obtain its value, because the relationship between the film thickness and the working conditions can be expressed as

$$h = G^{\alpha} U^{\beta} W^{\gamma} \tag{13.25}$$

where G is shear modulus of elasticity, E is the equivalent modulus of elasticity and α, β and γ are the exponent factors.

If the load is unbalanced, the relationship of the thickness increment and the load increment is

$$dh = \gamma G^{\alpha} U^{\beta} W^{\gamma-1} dW \tag{13.26}$$

Further, because $g - \frac{\Delta X}{\pi}\sum_{i=1}^{N-1}(P_i + P_{i+1})$ is equal to the load increment $-dW$, the increment of the displacement of the rigid body motion is equal to

$$dh = \bar{H}_0 - \tilde{H}_0 = -c_3 dW \tag{13.27}$$

Therefore, the relaxation factor can be determined from the above equation as

$$c_3 \approx -\frac{dh}{dW} = -\gamma G^{\alpha} U^{\beta} W^{\gamma-1} \tag{13.28}$$

13.3 Calculation Program

13.3.1 Calculation Diagram

The calculation diagram for EHL in line contact is shown in Figure 13.1.

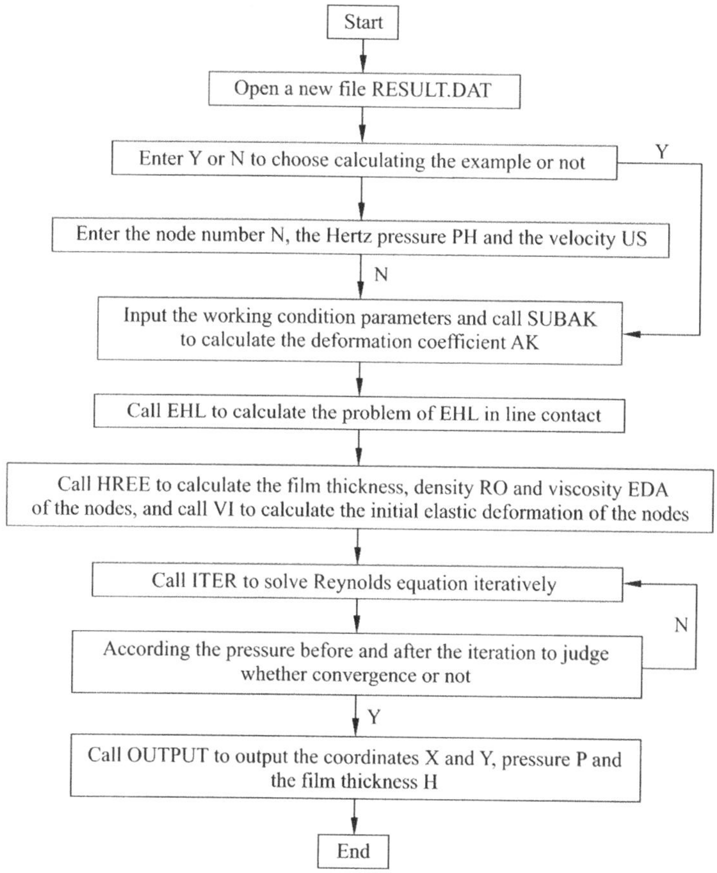

Figure 13.1 Calculation diagram for EHL in line contact.

13.3.2 Calculation Program

Pre-assignment parameters:

Node numbers	N = 129
Dimensionless inlet node coordinate	X0 = −4.0
Dimensionless outlet node coordinate	XE = 1.4
Load	W = 1.E5 N
Equivalent modulus of elasticity	E1 = 2.21E11Pa
Initial viscosity	EDA0 = 0.028 Pa·s
Equivalent radius of the cylinder	R = 0.012183 m
Velocity	US = 0.87 m s^{-1}
Ratio of slide to roll	CU = 0.67
Iteration coefficient	C1 = 0.5

If we need to change the above parameters, we need to recompile and relink the program before executing it.

Input parameters. Input character S. If S = "Y" or "y", calculate the example. If S = "N" or "n", then input the following data:

```
Node number        N=              (N must be 2^n+1, n=5,6, . . . , but N<1100)
Maximum Hertz contact pressure PH=       (Pa)
Velocity                    US=          (m/s)
```

Output parameters. In the file RESULT.DAT, the following data will be output in column:

```
Coordinate array   Pressure array   Film thickness array   Residual array
X(1)               P(1)             H(1)                   R(1)
X(2)               P(2)             H(2)                   R(2)
. . .  . . .       . . .  . . .     . . .  . . .           . . .  . . .
X(N)               P(N)             H(N)                   R(N)
```

```
      PROGRAM LINEEHL
      CHARACTER*1 S,S1,S2
      COMMON /COM1/ENDA,A1,A2,A3,Z,C1,C3,CW/COM2/EDA0/COM4/
X0,XE/COM3/E1,PH,B,U1,U2,R
      DATA PAI,Z,P0/3.14159265,0.68,1.96E8/,S1,S2/1HY,1Hy/
      DATA N,X0,XE,W,E1,EDA0,R,US,CU,C1/129,-4.0,1.4,1.
E5,2.21E11,0.028,0.012183,0.87,0.67,0.5/
      OPEN(8,FILE='RESULT.DAT',STATUS='UNKNOWN')
      WRITE(*,*)'Show the example or not (Y or N)?'
      READ(*,'(A)')S
      IF(S.EQ.S1.OR.S.EQ.S2)THEN
      GOTO 10
      ENDIF
      WRITE(*,*)'N,PH,US='
      READ(*,*)N,PH,US
      W=2.*PAI*R*PH*(PH/E1)
      WRITE(*,*)'W=',W
10    W1=W/(E1*R)
      PH=E1*SQRT(0.5*W1/PAI)
      A1=(ALOG(EDA0)+9.67)
      A2=PH/P0
      A3=0.59/(PH*1.E-9)
      B=4.*R*PH/E1
      ALFA=Z*A1/P0
      G=ALFA*E1
      U=EDA0*US/(2.*E1*R)
      CC1=SQRT(2.*U)
```

```
      AM=2.*PAI*(PH/E1)**2/CC1
      AL=G*SQRT(CC1)
      CW=(PH/E1)*(B/R)
      C3=1.6*(R/B)**2*G**0.6*U**0.7*W1**(-0.13)
      ENDA=3.*(PAI/AM)**2/8.
      U1=0.5*(2.+CU)*U
      U2=0.5*(2.-CU)*U
      CW=-1.13*C3
      WRITE(*,*)N,X0,XE,W,E1,EDA0,R,US
      WRITE(8,*)N,X0,XE,W,E1,EDA0,R,US,B,PH
      WRITE(*,40)
40    FORMAT(2X,'          Wait  Please',//)
      CALL SUBAK(N)
      CALL EHL(N)
      STOP
      END
      SUBROUTINE EHL(N)
      DIMENSION X(1100),P(1100),H(1100),RO(1100),POLD(1100),
     EPS(1100),EDA(1100),R(1100)
      COMMON /COM1/ENDA,A1,A2,A3,Z,C1,C3,CW/COM4/X0,XE
      COMMON /COM3/E1,PH,B,U1,U2,RR
      DATA MK,G0/1,1.570796325/
      NX=N
      DX=(XE-X0)/(N-1.0)
      DO 10 I=1,N
      X(I)=X0+(I-1)*DX
      IF(ABS(X(I)).GE.1.0)P(I)=0.0
      IF(ABS(X(I)).LT.1.0)P(I)=SQRT(1.-X(I)*X(I))
10    CONTINUE
      CALL HREE(N,DX,H00,G0,X,P,H,RO,EPS,EDA)
      CALL FZ(N,P,POLD)
14    KK=19
      CALL ITER(N,KK,DX,H00,G0,X,P,H,RO,EPS,EDA,R)
      MK=MK+1
      CALL ERROP(N,P,POLD,ERP)
      WRITE(*,*)ERP
      IF(MK.EQ.2)THEN
      ENDIF
      IF(ERP.GT.1.E-4.AND.MK.LE.200)THEN
      GOTO 14
      ENDIF
105   IF(MK.GE.200)THEN
      WRITE(*,*)'Pressures are not convergent!!!'
      READ(*,*)
      ENDIF
```

```
      H2=1.E3
      P2=0.0
      DO 106 I=1,N
      IF(H(I).LT.H2)H2=H(I)
      IF(P(I).GT.P2)P2=P(I)
106   CONTINUE
      H3=H2*B*B/RR
      P3=P2*PH
110   FORMAT(6(1X,E12.6))
120   CONTINUE
      WRITE(8,*)'P2,H2,P3,H3=',P2,H2,P3,H3
      CALL OUTHP(N,X,P,H,R)
      RETURN
      END
      SUBROUTINE OUTHP(N,X,P,H,R)
      DIMENSION X(N),P(N),H(N),R(N)
      DX=X(2)-X(1)
      DO 10 I=1,N
      WRITE(8,20)X(I),P(I),H(I),R(I)
10    CONTINUE
20    FORMAT(1X,6(F12.6,1X))
      RETURN
      END
      SUBROUTINE HREE(N,DX,H00,G0,X,P,H,RO,EPS,EDA)
      DIMENSION X(N),P(N),H(N),RO(N),EPS(N),EDA(N)
      DIMENSION W(2200)
      COMMON /COM1/ENDA,A1,A2,A3,Z,C1,C3,CW,K/COM2/EDA0/
COMAK/AK(0:1100)
      DATA KK,NW,PAI1/0,2200,0.318309886/
      IF(KK.NE.0)GOTO 3
      HM0=C3
      H00=0.0
3     W1=0.0
      DO 4 I=1,N
4     W1=W1+P(I)
      C3=(DX*W1)/G0
      DW=1.-C3
      CALL VI(N,DX,P,W)
      HMIN=1.E3
      DO 30 I=1,N
      H0=0.5*X(I)*X(I)+W(I)
      IF(H0.LT.HMIN)HMIN=H0
      H(I)=H0
30    CONTINUE
      IF(KK.NE.0)GOTO 32
```

```
      KK=1
      H00=-HMIN+HM0
32    H0=H00+HMIN
      IF(H0.LE.0.0)GOTO 48
      IF(H0+0.3*CW*DW.GT.0.0)HM0=H0+0.3*CW*DW
      IF(H0+0.3*CW*DW.LE.0.0)HM0=HM0*C3
48    H00=HM0-HMIN
50    DO 60 I=1,N
60    H(I)=H00+H(I)
      DO 100 I=1,N
      EDA(I)=EXP(A1*(-1.+(1.+A2*P(I))**Z))
      RO(I)=(A3+1.34*P(I))/(A3+P(I))
      EPS(I)=RO(I)*H(I)**3/(ENDA*EDA(I))
100   CONTINUE
      RETURN
      END
      SUBROUTINE ITER(N,KK,DX,H00,G0,X,P,H,RO,EPS,EDA,R)
      DIMENSION X(N),P(N),H(N),RO(N),EPS(N),EDA(N),R(N)
      COMMON /COM1/ENDA,A1,A2,A3,Z,C1,C3,CW
      COMMON /COMAK/AK(0:1100)
      DATA KG1,PAI/0,3.14159265/
      IF(KG1.NE.0)GOTO 5
      KG1=1
      DX1=1./DX
      DX2=DX*DX
      DX3=1./DX2
      DX4=DX1/PAI
      DXL=DX*ALOG(DX)
      AK0=DX*AK(0)+DXL
      AK1=DX*AK(1)+DXL
5     DO 100 K=1,KK
      D2=0.5*(EPS(1)+EPS(2))
      D3=0.5*(EPS(2)+EPS(3))
      D5=DX1*(RO(2)*H(2)-RO(1)*H(1))
      D7=DX4*(RO(2)*AK0-RO(1)*AK1)
      PP=0.
      DO 70 I=2,N-1
      D1=D2
      D2=D3
      D4=D5
      D6=D7
      IF(I+2.LE.N)D3=0.5*(EPS(I+1)+EPS(I+2))
      D5=DX1*(RO(I+1)*H(I+1)-RO(I)*H(I))
      D7=DX4*(RO(I+1)*AK0-RO(I)*AK1)
      DD=(D1+D2)*DX3
```

```
      IF(0.05*DD.LT.ABS(D6))GOTO 10
      RI=-DX3*(D1*P(I-1)-(D1+D2)*P(I)+D2*P(I+1))+D4
      R(I)=RI
      DLDP=-DX3*(D1+D2)+D6
      RI=RI/DLDP
      RI=RI/C1
      GOTO 20
10    RI=-DX3*(D1*PP-(D1+D2)*P(I)+D2*P(I+1))+D4
      R(I)=RI
      DLDP=-DX3*(2.*D1+D2)+2.*D6
      RI=RI/DLDP
      RI=0.5*RI
      IF(I.GT.2.AND.P(I-1)-C1*RI.GT.0.0)P(I-1)=P(I-1)-C1*RI
20  PP=P(I)
      P(I)=P(I)+C1*RI
      IF(P(I).LT.0.0)P(I)=0.0
      IF(P(I).LE.0.0)R(I)=0.0
70    CONTINUE
      CALL HREE(N,DX,H00,G0,X,P,H,RO,EPS,EDA)
100  CONTINUE
      RETURN
      END
      SUBROUTINE VI(N,DX,P,V)
      DIMENSION P(N),V(N)
      COMMON /COMAK/AK(0:1100)
      PAI1=0.318309886
      C=ALOG(DX)
      DO 10 I=1,N
      V(I)=0.0
      DO 10 J=1,N
      IJ=IABS(I-J)
10    V(I)=V(I)+(AK(IJ)+C)*DX*P(J)
      DO I=1,N
      V(I)=-PAI1*V(I)
      ENDDO
      RETURN
      END
      SUBROUTINE SUBAK(MM)
      COMMON /COMAK/AK(0:1100)
      DO 10 I=0,MM
10   AK(I)=(I+0.5)*(ALOG(ABS(I+0.5))-1.)-(I-0.5)*(ALOG(ABS(I-
0.5))-1.)
      RETURN
      END
      SUBROUTINE FZ(N,P,POLD)
```

```
      DIMENSION P(N),POLD(N)
      DO 10 I=1,N
10    POLD(I)=P(I)
      RETURN
      END
      SUBROUTINE ERROP(N,P,POLD,ERP)
      DIMENSION P(N),POLD(N)
      SD=0.0
      SUM=0.0
      DO 10 I=1,N
      SD=SD+ABS(P(I)-POLD(I))
      POLD(I)=P(I)
10    SUM=SUM+P(I)
      ERP=SD/SUM
      RETURN
      END
```

13.3.3 Calculation Results

Under the given working conditions of the example in the program, the pressure and film thickness are calculated as shown in Figure 13.2.

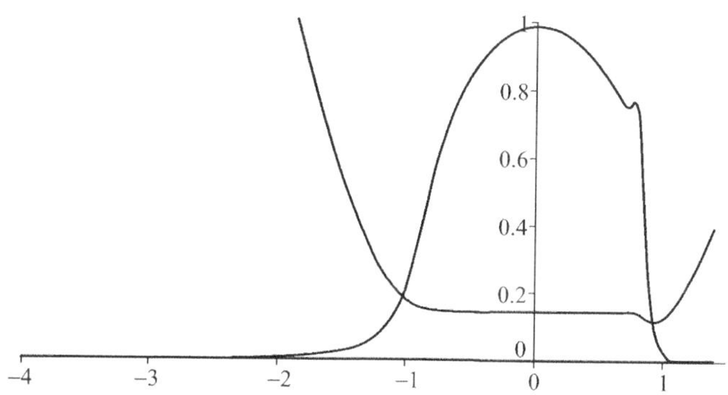

Figure 13.2 Pressure distribution and film thicknesss of EHL in line contact.

14

Numerical Method and Program for EHL in Point Contact

14.1 Basic Equations

Similar to elastohydrodynamic lubrication (EHL) in line contact, the basic equations of EHL in point contact include the Reynolds equation, the film thickness equation, the deformation equation, the viscosity–pressure equation and the density–pressure equation.

Reynolds equation

$$\frac{\partial}{\partial x}\left(\frac{\rho h^3}{\eta}\frac{\partial p}{\partial x}\right)+\frac{\partial}{\partial y}\left(\frac{\rho h^3}{\eta}\frac{\partial p}{\partial y}\right)=-12U\frac{\partial \rho h}{\partial x} \tag{14.1}$$

where ρ is lubricant density, η is lubricant viscosity, U is average velocity [$U=(u_1+u_2)/2$; u_1 and u_2 are the tangential velocities for the two surfaces], p is pressure, h is film thickness, x is the coordinate along the motion and y is the coordinate perpendicular to the motion.

Film thickness equation

$$h(x,y)=h_0+\frac{x^2}{2R_x}+\frac{y^2}{2R_y}+v(x,y) \tag{14.2}$$

where h_0 is a film thickness parameter to be determined in calculation, R_x is the equivalent radius of the surfaces in the x direction $\left(\frac{1}{R_x}=\frac{1}{R_{1x}}+\frac{1}{R_{2x}}\right)$, R_y is the equivalent radius of the surfaces in the y direction $\left(\frac{1}{R_y}=\frac{1}{R_{1y}}+\frac{1}{R_{2y}}\right)$ and v is elastic deformation.

Numerical Calculation of Lubrication: Methods and Programs, First Edition. Ping Huang.

Elastic deformation equation

$$v(x,y) = \frac{2}{\pi E}\iint\limits_{\Omega}\frac{p(s,t)}{\sqrt{(x-s)^2+(y-t)^2}}dsdt \tag{14.3}$$

where E is the equivalent elastic modulus $\left(\frac{1}{E} = \frac{1}{2}\left(\frac{1-\nu_1^2}{E_1}+\frac{1-\nu_2^2}{E_2}\right)\right)$, ν_1 and ν_2 are the Poisson's ratios of the two surfaces and E_1 and E_2 are the elastic modulus of the two surfaces.

Viscosity–pressure equation

$$\eta = \eta_0 \exp\left\{(\ln\eta_0 + 9.67)\left[\left(1+5.1\times 10^{-9}p\right)^{0.68} - 1\right]\right\} \tag{14.4}$$

where η_0 is lubricant viscosity at $p=0$.

Density–pressure equation

$$\rho = \rho_0\left(1+\frac{0.6p}{1+1.7p}\right) \tag{14.5}$$

where ρ_0 is lubricant density at $p=0$; and the unit of pressure p is GPa.

14.2 Numerical Method

14.2.1 Dimensionless Equations

The dimensionless Reynolds equation of isothermal EHL in point contact is

$$\frac{\partial}{\partial X}\left[\varepsilon\frac{\partial P}{\partial X}\right] + \alpha^2\frac{\partial}{\partial Y}\left[\varepsilon\frac{\partial P}{\partial Y}\right] - \frac{\partial(\rho^* H)}{\partial X} = 0 \tag{14.6}$$

where $\varepsilon = \frac{\rho^* H^3}{\eta^*\lambda}$, $\lambda = \frac{12\eta_0 U R_x^2}{a^3 p_H K_{ex}^2}$, η^* is dimensionless lubricant viscosity ($\eta^* = \eta/\eta_0$), ρ^* is dimensionless lubricant density ($\rho^* = \rho/\rho_0$), K_{ex} is the elliptic coefficient of the surface in the x direction, X is the dimensionless coordinate ($X = x/a$; a is the half length of the ellipse contact area in the x direction), Y is the dimensionless coordinate ($Y = y/b$; b is the half length of the ellipse contact area in the y direction), α is a ratio parameter ($\alpha = a/b$), P is dimensionless pressure ($P = p/p_{\mathrm{H}}$), p_{H} is the maximum Hertzian contact stress and H is the dimensionless film thickness $\left(H = \frac{hR_x}{a^2}\right)$.

The boundary conditions of the Reynolds Equation 14.6 are

Inlet boundary condition: $P(X_0, Y) = 0$
Outlet boundary conditions: $P(X_e, Y) = 0$ and $\frac{\partial P(X_e, Y)}{\partial X} = 0$
Side boundary conditions: $P|_{Y=\pm 1} = 0$

where X_0 and X_e are the dimensionless coordinates of the inlet and outlet, X_0 is given but X_e should be determined by the outlet boundary conditions and $Y = \pm 1$ are the two sides of the contact region.

Film thickness equation

$$H(X,Y) = H_0 + \frac{X^2+Y^2}{2} + \frac{2}{\pi^2}\int_{x_0}^{x_e}\int_{y_0}^{y_e}\frac{P(S,T)dSdT}{\sqrt{(X-S)^2+(Y-T)^2}} \tag{14.7}$$

where H_0 is the dimensionless film thickness to be determined in calculation.

Viscosity–pressure equation

$$\eta^* = \exp\left\{[\ln(\eta_0)+9.67]\left[-1+\left(1+5.1\times 10^{-9}P\cdot p_H\right)^{0.68}\right]\right\} \tag{14.8}$$

Density–pressure equation

$$\rho^* = \left(1+\frac{0.6p}{1+1.7p}\right) \tag{14.9}$$

Load equation

$$\int_{x_0}^{x_e}\int_{y_0}^{y_e} P(X,Y)dXdY = \frac{2}{3}\pi \tag{14.10}$$

Compared with the line contact problem, there is an increase in a dimensionless coordinate in the y direction in the above equations. α is the ratio of a and b. K_{ex} is the relative curvature in the x direction. For the equivalent curvature of the point contact, $K_{ex} = 1$. The following discussions are based on $\alpha = 1$ and $K_{ex} = 1$. If $\alpha \neq 1$ or $K_{ex} \neq 1$, we can refer to Wen and Huang, 2011 for the EHL calculation.

14.2.2 Discrete Equations

With the central and forward discrete differential Equations 1.22 and 1.24, the discrete form of the Reynolds Equation 14.6 can be written as

$$\frac{\varepsilon_{i-1/2,j}P_{i-1,j}+\varepsilon_{i+1/2,j}P_{i+1,j}+\varepsilon_{i,j-1/2}P_{i,j-1}+\varepsilon_{i,j+1/2}P_{i,j+1}-\varepsilon_0 P_{ij}}{\Delta X^2} = \frac{\rho^*_{ij}H_{ij}-\rho^*_{i-1,j}H_{i-1,j}}{\Delta X} \tag{14.11}$$

where $\varepsilon_{i\pm1/2,j}=\frac{1}{2}(\varepsilon_{i,j}+\varepsilon_{i\pm1,j})$; $\varepsilon_0=\varepsilon_{i-1/2,j}+\varepsilon_{i+1/2,j}+\varepsilon_{i,j-1/2}+\varepsilon_{i,j+1/2}$, $\Delta X=X_i-X_{i-1}$ and the equidistant grid is used here, that is, $\Delta Y=\Delta X$.

The boundary condition in the inlet: $P(X_0, Y) = 0$; and the boundary of the outlet can be obtained by setting the negative pressure equal to zero. The discrete forms are: $P(X_e, Y) = 0$ and $\frac{P(X_e, Y) - P(X_e - 1, Y)}{\Delta X} = 0$. The side boundary conditions are: $P(X, Y = \pm 1) = 0$.

Film thickness equation

$$H_{ij} = H_0 + \frac{X_i^2 + Y_j^2}{2} + \frac{2}{\pi^2}\sum_{k=1}^{n}\sum_{l=1}^{n} D_{ij}^{kl} P_{kl} \tag{14.12}$$

where D_{ij}^{kl} is the stiffness coefficient of the elastic deformation.

Load equation

$$\Delta X \Delta Y \sum_{i=1}^{n}\sum_{j=1}^{n} P_{ij} = \frac{2\pi}{3} \tag{14.13}$$

Convergence criterion

$$\frac{\sum_i \sum_j |P_{i,j}^{k+1} - P_{i,j}^{k}|}{\sum_i \sum_j P_{i,j}^{k+1}} \leq \varepsilon_4 \tag{14.14}$$

where ε_4 is a given error.

In the above equations, $i = 1, 2 \ldots, n, j = 1, 2 \ldots, n$, where n is the node number.

Similar to a line contact problem, in the solution of a point contact problem, the Gauss–Seidel iteration is used for the low pressure region and the Jacobi iteration is used for the high pressure region. The Gauss–Seidel iteration in a point contact is the same as that in a line contact, but because the pressure of a node is related to the pressure increment of four other adjacent nodes, for the Jacobi iteration method, a matrix equation must be solved to carry out a series of pressure corrections.

14.3 Calculation Program

14.3.1 Calculation Diagram

The calculation diagram for EHL in point contact is shown in Figure 14.1.

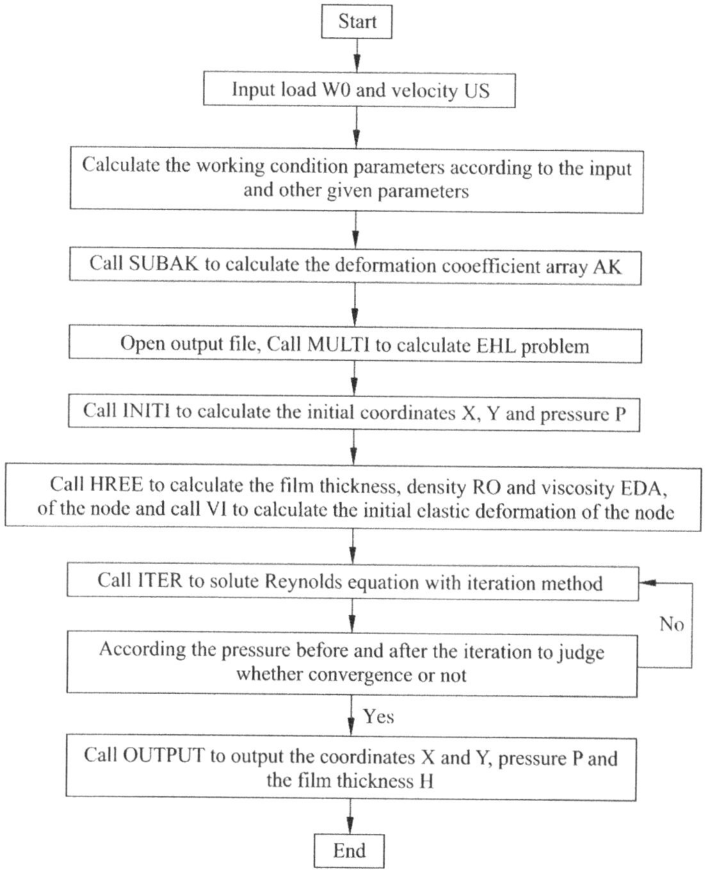

Figure 14.1 Calculation diagram for EHL in point contact.

14.3.2 Calculation Program

14.3.2.1 Pre-assignment Parameters

Node numbers	N = 65 × 65
Dimensionless starting point coordinate in *x* direction	X0 = −2.5
Dimensionless ending point coordinate in *x* direction	XE = 1.5
Dimensionless starting point coordinate in *y* direction	Y0 = −2.0
Dimensionless ending point coordinate in *y* direction	YE = 2.0
Equivalent modulus of elasticity	E1 = 2.21E11 Pa
Initial viscosity	EDA0 = 0.028 Pa s
Radius	RX = RY = 0.05 m

14.3.2.2 Output Parameters

Pressure array P(I, J) will be saved in the file PRESSURE.DAT.

Film thickness array H(I, J) will be saved in the file FILM.DAT.

Other data will be saved in the file OUT.DAT.

If we need to change the above parameters, we need to recompile and relink the program before executing it.

Input parameters are: Load W0 and velocity US

14.3.2.3 Program

```
      PROGRAM POINTEHL
      DIMENSION THETA(15),EALFA(15),EBETA(15)
      COMMON /COM1/ENDA,A1,A2,A3,Z,HM0
      COMMON /COMEK/EK,EAL,EBE
      DATA N,PAI,Z,EAL,EBE,E1,EDA0,RX,RY,X0,XE,W0,US/65,
3.14159265,0.68,1.0,1.0,2.21E11,0.0283,0.05,0.05,-2.5,1.5,
39.24,1.5/
      DATA THETA/10.,20.,30.,35.,40.,45.,50.,55.,60.,65.,70.,
75.,80.,85.,90./
      DATA EALFA/6.612,3.778,2.731,2.397,2.136,1.926,1.754,
1.611,1.486,1.378,1.284,1.202,1.128,1.061,1.0/
      DATA EBETA/0.319,0.408,0.493,0.53,0.567,0.604,0.641,
0.678,0.717,0.759,0.802,0.846,0.893,0.944,1.0/
      EK=RX/RY
      AA=0.5*(1./RX+1./RY)
      BB=0.5*ABS(1./RX-1./RY)
      CC=ACOS(BB/AA)*180.0/PAI
      DO I=1,15
      IF(CC.LT.THETA(I))THEN
      WRITE(*,*)I
      EAL=EALFA(I-1)+(CC-THETA(I-1))*(EALFA(I)-EALFA(I-1))/(THETA(I)-
THETA(I-1))
      EBE=EBETA(I-1)+(CC-THETA(I-1))*(EBETA(I)-EBETA(I-1))/
(THETA(I)-THETA(I-1))

      GOTO 1
      ENDIF
      ENDDO
1     EA=EAL*(1.5*W0/AA/E1)**(1./3.0)
      EB=EBE*(1.5*W0/AA/E1)**(1./3.0)
      PH=1.5*W0/(EA*EB*PAI)
      OPEN(4,FILE='OUT.DAT',STATUS='UNKNOWN')
      OPEN(8,FILE='FILM.DAT',STATUS='UNKNOWN')
      OPEN(10,FILE='PRESSURE.DAT',STATUS='UNKNOWN')
```

```
      WRITE(*,*)N,X0,XE,W0,PH,E1,EDA0,RX,US
      WRITE(4,*)N,X0,XE,W0,PH,E1,EDA0,RX,US
      H00=0.0
      MM=N-1
      U=EDA0*US/(2.*E1*RX)
      A1=ALOG(EDA0)+9.67
      A2=5.1E-9*PH
      A3=0.59/(PH*1.E-9)
      B=PAI*PH*RX/E1
      W=2.*PAI*PH/(3.*E1)*(B/RX)**2
      ALFA=Z*5.1E-9*A1
      G=ALFA*E1
      AHM=1.0-EXP(-0.68*1.03)
      HM0=3.63*(RX/B)**2*G**0.49*U**0.68*W**(-0.073)*AHM
      ENDA=12.*U*(E1/PH)*(RX/B)**3
      UTL=EDA0*US*RX/(B*B*2.E7)
      W0=2.0*PAI*EA*EB*PH/3.0
      WRITE(*,*)'                    Wait please'
      CALL SUBAK(MM)
      CALL MULTI(N,X0,XE,H00)
      STOP
      END
      SUBROUTINE MULTI(N,X0,XE,H00)
      DIMENSION X(65),Y(65),H(4500),RO(4500),EPS(4500),EDA
     (4500),P(4500),POLD(4500)
      COMMON /COMEK/EK,EAL,EBE
      DATA MK,G00/200,2.0943951/
      G0=G00*EAL*EBE
      NX=N
      NY=N
      NN=(N+1)/2
      CALL INITI(N,DX,X0,XE,X,Y,P,POLD)
      CALL HREE(N,DX,H00,G0,X,Y,H,RO,EPS,EDA,P)
      M=0
   14 KK=15
      CALL ITER(N,KK,DX,H00,G0,X,Y,H,RO,EPS,EDA,P)
      M=M+1
      CALL ERP(N,ER,P,POLD)
      WRITE(*,*)'ER=',ER
      IF(M.LT.MK.AND.ER.GT.1.E-5)GOTO 14
      CALL OUTPUT(N,DX,X,Y,H,P)
      RETURN
      END
      SUBROUTINE ERP(N,ER,P,POLD)
      DIMENSION P(N,N),POLD(N,N)
```

```
      ER=0.0
      SUM=0.0
      NN=(N+1)/2
      DO 10 I=1,N
      DO 10 J=1,NN
      ER=ER+ABS(P(I,J)-POLD(I,J))
      SUM=SUM+P(I,J)
10    CONTINUE
      ER=ER/SUM
      DO I=1,N
      DO J=1,N
      POLD(I,J)=P(I,J)
      ENDDO
      ENDDO
      RETURN
      END
      SUBROUTINE INITI(N,DX,X0,XE,X,Y,P,POLD)
      DIMENSION X(N),Y(N),P(N,N),POLD(N,N)
      NN=(N+1)/2
      DX=(XE-X0)/(N-1.)
      Y0=-0.5*(XE-X0)
      DO 5 I=1,N
      X(I)=X0+(I-1)*DX
      Y(I)=Y0+(I-1)*DX
5     CONTINUE
      DO 10 I=1,N
      D=1.-X(I)*X(I)
      DO 10 J=1,NN
      C=D-Y(J)*Y(J)
      IF(C.LE.0.0)P(I,J)=0.0
10    IF(C.GT.0.0)P(I,J)=SQRT(C)
      DO 20 I=1,N
      DO 20 J=NN+1,N
      JJ=N-J+1
20    P(I,J)=P(I,JJ)
      DO I=1,N
      DO J=1,N
      POLD(I,J)=P(I,J)
      ENDDO
      ENDDO
      RETURN
      END
      SUBROUTINE HREE(N,DX,H00,G0,X,Y,H,RO,EPS,EDA,P)
      DIMENSION X(N),Y(N),P(N,N),H(N,N),RO(N,N),EPS(N,N),EDA(N,N)
      DIMENSION W(150,150),P0(150,150)
```

```
      COMMON /COM1/ENDA,A1,A2,A3,Z,HM0/COMAK/AK(0:65,0:65)
      COMMON /COMEK/EK,EAL,EBE
      DATA NW,PAI,PAI1/150,3.14159265,0.2026423/
      NN=(N+1)/2
      CALL VI(NW,N,DX,P,W)
      HMIN=1.E3
      DO 30 I=1,N
      DO 30 J=1,NN
      RAD=X(I)*X(I)+EK*Y(J)*Y(J)
      W1=0.5*RAD
      H0=W1+W(I,J)
      IF(H0.LT.HMIN)HMIN=H0
 30   H(I,J)=H0
      IF(KK.EQ.0)THEN
      KG1=0
      H01=-HMIN+HM0
      DH=0.005*HM0
      H02=-HMIN
      H00=0.5*(H01+H02)
      ENDIF
      W1=0.0
      DO 32 I=1,N
      DO 32 J=1,N
 32   W1=W1+P(I,J)
      W1=DX*DX*W1/G0
      DW=1.-W1
      IF(KK.EQ.0)THEN
      KK=1
      GOTO 50
      ENDIF
      IF(DW.LT.0.0)THEN
      KG1=1
      H00=AMIN1(H01,H00+DH)
      ENDIF
      IF(DW.GT.0.0)THEN
      KG2=2
      H00=AMAX1(H02,H00-DH)
      ENDIF
 50   DO 60 I=1,N
      DO 60 J=1,NN
      H(I,J)=H00+H(I,J)
      IF(P(I,J).LT.0.0)P(I,J)=0.0
      EDA1=EXP(A1*(-1.+(1.+A2*P(I,J))**Z))
      EDA(I,J)=EDA1
 55   RO(I,J)=(A3+1.34*P(I,J))/(A3+P(I,J))
```

```
60   EPS(I,J)=RO(I,J)*H(I,J)**3/(ENDA*EDA1)
     DO 70 J=NN+1,N
     JJ=N-J+1
     DO 70 I=1,N
     H(I,J)=H(I,JJ)
     RO(I,J)=RO(I,JJ)
     EDA(I,J)=EDA(I,JJ)
70   EPS(I,J)=EPS(I,JJ)
     RETURN
     END
     SUBROUTINE ITER(N,KK,DX,H00,G0,X,Y,H,RO,EPS,EDA,P)
     DIMENSION X(N),Y(N),P(N,N),H(N,N),RO(N,N),EPS(N,N),EDA(N,N)
     DIMENSION D(70),A(350),B(210),ID(70)
     COMMON /COM1/ENDA,A1,A2,A3,Z,C3/COMAK/AK(0:65,0:65)
     DATA KG1,PAI1,C1,C2/0,0.2026423,0.31,0.31/
     IF(KG1.NE.0)GOTO 2
     KG1=1
     AK00=AK(0,0)
     AK10=AK(1,0)
     AK20=AK(2,0)
     BK00=AK00-AK10
     BK10=AK10-0.25*(AK00+2.*AK(1,1)+AK(2,0))
     BK20=AK20-0.25*(AK10+2.*AK(2,1)+AK(3,0))
2    NN=(N+1)/2
     MM=N-1
     DX1=1./DX
     DX2=DX*DX
     DX3=1./DX2
     DX4=0.3*DX2
     DO 100 K=1,KK
     PMAX=0.0
     DO 70 J=2,NN
     J0=J-1
     J1=J+1
     JJ=N-J+1
     IA=1
8    MM=N-IA
     IF(P(MM,J0).GT.1.E-6)GOTO 20
     IF(P(MM,J).GT.1.E-6)GOTO 20
     IF(P(MM,J1).GT.1.E-6)GOTO 20
     IA=IA+1
     IF(IA.LT.N)GOTO 8
     GOTO 70
20   IF(MM.LT.N-1)MM=MM+1
     D2=0.5*(EPS(1,J)+EPS(2,J))
```

```
      DO 50 I=2,MM
      I0=I-1
      I1=I+1
      II=5*I0
      D1=D2
      D2=0.5*(EPS(I1,J)+EPS(I,J))
      D4=0.5*(EPS(I,J0)+EPS(I,J))
      D5=0.5*(EPS(I,J1)+EPS(I,J))
      P1=P(I0,JJ)
      P2=P(I1,JJ)
      P3=P(I,JJ)
      P4=P(I,JJ+1)
      P5=P(I,JJ-1)
      D3=D1+D2+D4+D5
      IF(J.EQ.NN.AND.ID(I).EQ.1)P(I,J)=P(I,J)-0.5*C2*D(I)
      IF(H(I,J).LE.0.0)THEN
      ID(I)=2
      A(II+1)=0.0
      A(II+2)=0.0
      A(II+3)=1.0
      A(II+4)=0.0
      A(II+5)=1.0
      A(II-4)=0.0
      GOTO 50
      ENDIF
      IF(D1.GE.DX4)GOTO 30
      IF(D2.GE.DX4)GOTO 30
      IF(D4.GE.DX4)GOTO 30
      IF(D5.GE.DX4)GOTO 30
      ID(I)=1
      IF(J.EQ.NN)P5=P4
      A(II+1)=PAI1*(RO(I0,J)*BK10-RO(I,J)*BK20)
      A(II+2)=DX3*(D1+0.25*D3)+PAI1*(RO(I0,J)*BK00-RO(I,J)*BK10)
      A(II+3)=-1.25*DX3*D3+PAI1*(RO(I0,J)*BK10-RO(I,J)*BK00)
      A(II+4)=DX3*(D2+0.25*D3)+PAI1*(RO(I0,J)*BK20-RO(I,J)*BK10)
      A(II+5)=-DX3*(D1*P1+D2*P2+D4*P4+D5*P5-D3*P3)+DX1*
  (RO(I,J)*H(I,J)-RO(I0,J)*H(I0,J))
      GOTO 50
30    ID(I)=0
      P4=P(I,J0)
      IF(J.EQ.NN)P5=P4
      A(II+1)=PAI1*(RO(I0,J)*AK10-RO(I,J)*AK20)
      A(II+2)=DX3*D1+PAI1*(RO(I0,J)*AK00-RO(I,J)*AK10)
      A(II+3)=-DX3*D3+PAI1*(RO(I0,J)*AK10-RO(I,J)*AK00)
      A(II+4)=DX3*D2+PAI1*(RO(I0,J)*AK20-RO(I,J)*AK10)
```

```
    A(II+5)=-DX3*(D1*P1+D2*P2+D4*P4+D5*P5-D3*P3)+DX1*(RO(I,J)
*H(I,J)-RO(I0,J)*H(I0,J))
50  CONTINUE
    CALL TRA4(MM,D,A,B)
    DO 60 I=2,MM
    IF(ID(I).EQ.2)GOTO 60
    IF(ID(I).EQ.0)GOTO 52
    DD=D(I+1)
    IF(I.EQ.MM)DD=0
    P(I,J)=P(I,J)+C2*(D(I)-0.25*(D(I-1)+DD))
    IF(J0.NE.1)P(I,J0)=P(I,J0)-0.25*C2*D(I)
    IF(P(I,J0).LT.0.)P(I,J0)=0.0
    IF(J1.GE.NN)GOTO 54
    P(I,J1)=P(I,J1)-0.25*C2*D(I)
    GOTO 54
52  P(I,J)=P(I,J)+C1*D(I)
54  IF(P(I,J).LT.0.0)P(I,J)=0.0
    IF(PMAX.LT.P(I,J))PMAX=P(I,J)
60  CONTINUE
70  CONTINUE
    DO 80 J=1,NN
    JJ=N+1-J
    DO 80 I=1,N
80  P(I,JJ)=P(I,J)
    CALL HREE(N,DX,H00,G0,X,Y,H,RO,EPS,EDA,P)
100 CONTINUE
    RETURN
    END
    SUBROUTINE TRA4(N,D,A,B)
    DIMENSION D(N),A(5,N),B(3,N)
    C=1./A(3,N)
    B(1,N)=-A(1,N)*C
    B(2,N)=-A(2,N)*C
    B(3,N)=A(5,N)*C
    DO 10 I=1,N-2
    IN=N-I
    IN1=IN+1
    C=1./(A(3,IN)+A(4,IN)*B(2,IN1))
    B(1,IN)=-A(1,IN)*C
    B(2,IN)=-(A(2,IN)+A(4,IN)*B(1,IN1))*C
10  B(3,IN)=(A(5,IN)-A(4,IN)*B(3,IN1))*C
    D(1)=0.0
    D(2)=B(3,2)
    DO 20 I=3,N
20  D(I)=B(1,I)*D(I-2)+B(2,I)*D(I-1)+B(3,I)
```

```
      RETURN
      END
      SUBROUTINE VI(NW,N,DX,P,V)
      DIMENSION P(N,N),V(NW,NW)
      COMMON /COMAK/AK(0:65,0:65)
      PAI1=0.2026423
      DO 40 I=1,N
      DO 40 J=1,N
      H0=0.0
      DO 30 K=1,N
      IK=IABS(I-K)
      DO 30 L=1,N
      JL=IABS(J-L)
30    H0=H0+AK(IK,JL)*P(K,L)
40    V(I,J)=H0*DX*PAI1
      RETURN
      END
      SUBROUTINE SUBAK(MM)
      COMMON /COMAK/AK(0:65,0:65)
      S(X,Y)=X+SQRT(X**2+Y**2)
      DO 10 I=0,MM
      XP=I+0.5
      XM=I-0.5
      DO 10 J=0,I
      YP=J+0.5
      YM=J-0.5
      A1=S(YP,XP)/S(YM,XP)
      A2=S(XM,YM)/S(XP,YM)
      A3=S(YM,XM)/S(YP,XM)
      A4=S(XP,YP)/S(XM,YP)
      AK(I,J)=XP*ALOG(A1)+YM*ALOG(A2)+XM*ALOG(A3)+YP*ALOG(A4)
10    AK(J,I)=AK(I,J)
      RETURN
      END
      SUBROUTINE OUTPUT(N,DX,X,Y,H,P)
      DIMENSION X(N),Y(N),H(N,N),P(N,N)
      NN=(N+1)/2
      A=0.0
      WRITE(8,110)A,(Y(I),I=1,N)
      O I=1,N
      WRITE(8,110)X(I),(H(I,J),J=1,N)
      ENDDO
      WRITE(10,110)A,(Y(I),I=1,N)
      DO I=1,N
      WRITE(10,110)X(I),(P(I,J),J=1,N)
```

```
      ENDDO
110 FORMAT(66(E12.6,1X))
      RETURN
      END
```

14.3.3 Calculation Results

Under the given working conditions in the program ($w = 36.24$ N, $US = 1.5\,\text{m}\,\text{s}^{-1}$), the pressure distribution and the film thickness are shown in Figure 14.2.

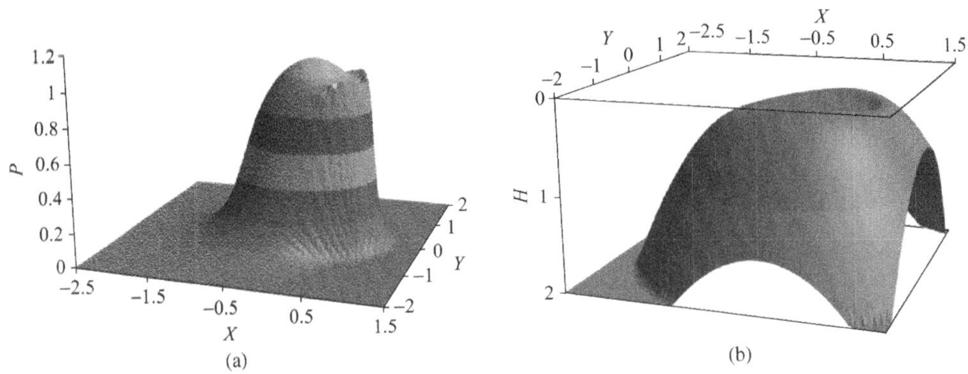

Figure 14.2 Pressure distribution (a) and film thickness (b) of EHL in point contact.

15

Numerical Method and Program for Grease EHL in Line Contact

15.1 Basic Equations

15.1.1 Reynolds Equation

According to Equation 9.5, using the Ostwald model, the Reynolds equation for grease elastohydrodynamic lubrication (GEHL) in line contact is as follows.

$$\frac{n}{2n+1}\cdot\left(\frac{1}{2}\right)^{\frac{n+1}{n}}\left\{\frac{d}{dx}\left[\rho h^{\frac{2n+1}{n}}\left(\frac{1}{\phi}\frac{dp}{dx}\right)^{\frac{1}{n}}\right]\right\} = U\frac{d(\rho h)}{dx} \tag{15.1}$$

where h is film thickness, U is average speed ($U = (u_1 + u_2)/2$, where u_1 and u_2 are the surface tangential velocities of the two surfaces), x is the motion direction, n is rheological parameter of the grease ($n \leq 1$), ϕ is the viscosity of grease and ρ is the density of grease.

15.1.2 Film Thickness Equation

$$h(x) = h_c + \frac{x^2}{2R} - \frac{2}{\pi E}\int_{x_0}^{x_e} p(s)\ln(x-s)^2 ds + c \tag{15.2}$$

where R is the equivalent curvature radius $\left(\frac{1}{R} = \frac{1}{R_1} + \frac{1}{R_2}\right)$, R_1 and R_2 are the radii of the two surfaces respectively, s is additional coordinates in the x direction [the distance between load $p(s)\mathrm{d}s$ and the origin coordinate], $p(s)$ is pressure, x_0 and x_e are the beginning and ending coordinates of pressure $p(x)$, h_c is the central film thickness and c is an integral constant.

Numerical Calculation of Lubrication: Methods and Programs, First Edition. Ping Huang.

15.1.3 Viscosity–Pressure Equation

Because there are no widely recognized viscosity–pressure and density–pressure equations for grease, here we use the equations for oil. If there are corresponding formulas for grease, one can replace the equations in the calculation, which does not affect program execution.

$$\phi = \phi_0 \exp\left\{(\ln \phi_0 + 9.67)\left[\left(1 + 5.1 \times 10^{-9} p\right)^z - 1\right]\right\} \tag{15.3}$$

where z is a constant (here $z = 0.68$) and ϕ_0 is the plastic viscosity of grease at $p = p_0$ and it is equivalent to the viscosity of lubricant.

15.1.4 Density–Pressure Equation

The density of the grease is considered as a constant, that is, $\rho = \text{const}$.

15.2 Numerical Method

15.2.1 Dimensionless Equations

The dimensionless Reynolds equation of GEHL in line contact is

$$\frac{d}{dX}\left[\varepsilon \cdot \left(\frac{dP}{dX}\right)^{1/n}\right] - \frac{dH}{dX} = 0 \tag{15.4}$$

where $\varepsilon = \lambda \dfrac{H^{(2+1/n)}}{\phi^{*1/n}}$, $\lambda = \dfrac{p_H^{1/n} b^{2+1/n}}{2U(2+1/n)R^{(1+1/n)}2^{1/n}\phi_0^{1/n}}$, b is the half width of the contact area, X is a dimensionless coordinate ($X = x/b$), P is dimensionless pressure ($P = p/p_H$), p_H is the maximum Hertz pressure, H is dimensionless film thickness $\left(H = \dfrac{hR}{b^2}\right)$.

The boundary conditions are given as follows.

$$\text{In the inlet:} \quad P(X_0) = 0;$$

$$\text{In the outlet:} \quad P(X_e) = 0 \text{ and } \frac{dP(X_e)}{dX} = 0;$$

where X_0 and X_e are the dimensionless coordinates of the inlet and outlet (X_0 is given, but X_e will be determined by calculation).

Film thickness equation

$$H(X) = H_0 + \frac{X^2}{2} - \frac{1}{\pi}\int_{X_0}^{X_e} \ln|X - X'| p(X') dX' \tag{15.5}$$

where H_0 is dimensionless displacement, including h_c and c from Equation 15.2.

Viscosity–pressure equation

$$\phi^* = \exp\left\{[\ln(\phi_0) + 9.67]\left[-1 + \left(1 + 5.1 \times 10^{-9} P \cdot p_H\right)^{0.68}\right]\right\} \tag{15.6}$$

Load equation

$$W = \int_{X_0}^{X_e} P dx = \frac{\pi}{2} \tag{15.7}$$

15.2.2 Discrete Equations

The dimensionless Reynolds equation can be written in a discrete form as

$$\frac{\varepsilon_{i+1/2}(P_{i+1} - P_i)^{1/n} - \varepsilon_{i-1/2}(P_i - P_{i-1})^{1/n}}{\Delta X^{1+1/n}} - \frac{H_i - H_{i-1}}{\Delta X} = 0 \tag{15.8}$$

where $\varepsilon_{i\pm1/2} = \frac{1}{2}(\varepsilon_i + \varepsilon_{i\pm1})$ and $\Delta X = X_i - X_{i-1}$.

At the inlet, the boundary condition is $P(X_0) = 0$. The boundary of the outlet is determined by setting negative pressure zero.

The dimensionless film thickness equation in a discrete form is

$$H_i = H_0 + \frac{x_i^2}{2} - \frac{1}{\pi}\sum_{j=1}^{n} K_{ij} P_j \tag{15.9}$$

where K_{ij} is the stiffness coefficient of elastic deformation.

The dimensionless load equation in a discrete form is

$$\Delta X \sum_{i=1}^{n} \frac{(P_i + P_{i+1})}{2} = \frac{\pi}{2} \tag{15.10}$$

Convergence criterion

$$\frac{\sum |P_i^{k+1} - P_i^k|}{\sum P_i^{k+1}} \le \varepsilon_4 \tag{15.11}$$

where ε_4 is a given error.

The iterative method of this program is similar to that of EHL in line contact, and so not repeated here.

15.3 Calculation Program

15.3.1 Calculating Diagram

The calculating diagram for GEHL in line contact is shown in Figure 15.1.

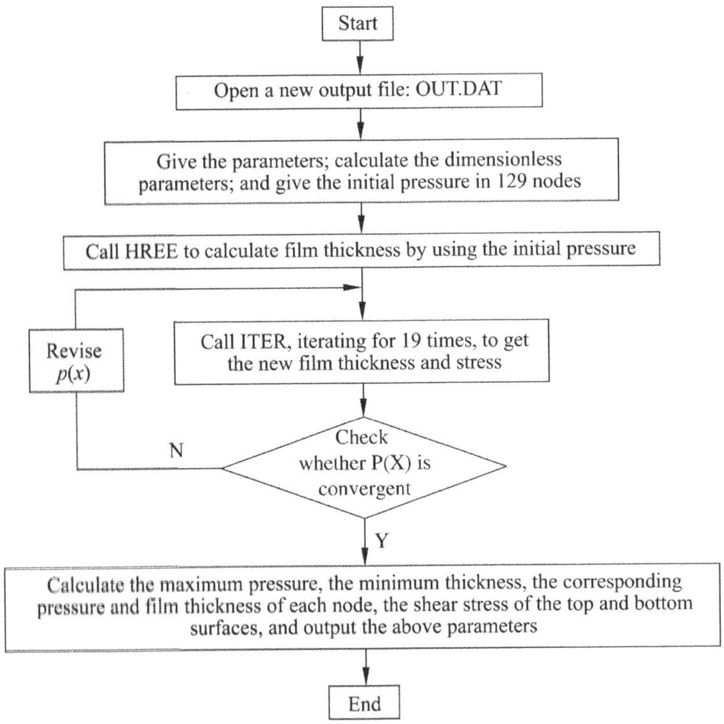

Figure 15.1 Diagram of GEHL in line contact.

15.3.2 Calculation Program

15.3.2.1 Give Parameters

Number of nodes	N = 129
Dimensionless coordinates of inlet	X0 = −4.0
Dimensionless coordinates of outlet	XE = 1.4
Load	W = 3.0E5 (N)
Equivalent elastic modulus	E1 = 2.26E11 (Pa)
Initial viscosity of grease	EDA0 = 0.4 (Pa s)
Equivalent radius	R = 0.012183 (m)
Average velocity	US = 0.87 (m s^{-1})
Ratio of sliding to rolling	CU = 0.67
Iterative coefficient	C1 = 0.5
Rheological parameter	FN = 0.8

15.3.2.2 Input Parameters

When the characteristic coefficient S = "Y" or "y", the program calculates the example and gives the results. If S ≠ "Y" or "y", the program is not an example. One should input the maximum Hertz contact pressure PH.

15.3.2.3 Output Parameters

The data are input in the file OUT.DAT. They are:

```
N, W, E1, EDA0, R,US, B, PH, FF
P2, H2, P3, H3
X(1), p(1), h(1)
X(2), p(2), h(2)
X(N), p(N), h(N)
```

Further, the above given parameters can be changed according to the user's circumstances. However, one needs to recompile and relink it before executing it.

15.3.2.4 Calculation Program

```
PROGRAM GREASELINEEHL
CHARACTER*1 S,S1,S2
      CHARACTER*16 CDATE,CTIME
COMMON /COM1/ENDA,A1,A2,A3,Z,C1,C3,CW,LMAX,FF/COM2/EDA0/COM4/
X0,XE/COM3/E1,PH,B,U1,U2,R
      DATA PAI,Z,P0/3.14159265,0.68,1.96E8/S1,S2/1HY,1Hy/
      DATA N,X0,XE,W,E1,EDA0,R,US,CU,C1,FN/129,-4.,1.4,1.E5,
2.26E11,0.4,0.012183,0.87,0.67,0.5,0.8/CDATE,CTIME/'The date is','The
time is'/
   OPEN(8,FILE='OUT.DAT',STATUS='UNKNOWN')
1    FORMAT(20X,A12,I2.2,':',I2.2,':',I4.4)
2    FORMAT(20X,A12,I2.2,':',I2.2,':',I2.2,'.',I2.2)
     WRITE(*,*)'Show the example or not (Y or N)?'
     READ(*,'(A)')S
     IF(S.EQ.S1.OR.S.EQ.S2)THEN
     GOTO 10
     ENDIF
     WRITE(*,*)'PH='
     READ(*,*)PH
     W=2.*PAI*R*PH*(PH/E1)
     WRITE(*,*)'W=',W
10  CW=N+0.1
  FF=1./FN
```

```
      LMAX=ALOG(CW)/ALOG(2.)
      N=2**LMAX+1
      LMIN=(ALOG(CW)-ALOG(SQRT(CW)))/ALOG(2.)
      LMAX=LMIN
      W1=W/(E1*R)
      PH=E1*SQRT(0.5*W1/PAI)
      A1=(ALOG(EDA0)+9.67)
      A2=PH/P0
      A3=0.59/(PH*1.E-9)
      B=4.*R*PH/E1
      ALFA=Z*A1/P0
      G=ALFA*E1
      U=EDA0*US/(2.*E1*R)
      C3=1.6*(R/B)**2*G**0.6*U**0.7*W1**(-0.13)
      ENDA=B**(2.+FF)*(PH/2/EDA0)**FF/R**(1+FF)/US/(2.+FF)
      U1=0.5*(2.+CU)*U
      U2=0.5*(2.-CU)*U
  WRITE(*,*)'B,PH,G,U=',B,PH,G,U
      CW=-1.13*C3
      WRITE(*,*)N,X0,XE,W,E1,EDA0,R,US,PH
      WRITE(8,*)N,W,E1,EDA0,R,US,B,PH,FF
      WRITE(*,40)
40    FORMAT(2X,'Wait Please',//)
      CALL SUBAK(N)
      CALL MULTI(N)
      STOP
      END
      SUBROUTINE MULTI(N)
      REAL*8
X(1100),P(1100),H(1100),RO(1100),POLD(1100),EPS(1100),EDA
(1100),R(1100),K(1100),E(1100)
      COMMON/COM1/ENDA,A1,A2,A3,Z,C1,C3,CW,LMAX,FF/COM4/X0,XE/
COM3/E1,PH,B,U1,U2,RR
      DATA MK,G0/1,1.570796325/
      NX=N
      DX=(XE-X0)/(N-1.0)
      DO 10 I=1,N
      X(I)=X0+(I-1)*DX
      IF(ABS(X(I)).GE.1.0)P(I)=0.0
      IF(ABS(X(I)).LT.1.0)P(I)=SQRT(1.-X(I)*X(I))
10    CONTINUE
      CALL HREE(N,DX,H00,G0,X,P,H,RO,EPS,EDA)
      CALL FZ(N,P,POLD)
14    KK=19
      CALL ITER(N,KK,DX,H00,G0,X,P,H,RO,EPS,EDA,R)
```

```
      MK=MK+1
      CALL ERROP(N,P,POLD,ERP)
      IF(ERP.GT.1.E-4.AND.MK.LE.800)THEN
      GOTO 14
      ENDIF
      WRITE(*,*)PH,RR,B
105   IF(MK.GE.800)THEN
      WRITE(*,*)'Pressures are not convergent !!!'
      READ(*,*)
      ENDIF
      FM=FRICT(N,DX,X,H,P,EDA)
      H2=1.E3
      P2=0.0
      DO 106 I=1,N
      IF(H(I).LT.H2)H2=H(I)
      IF(P(I).GT.P2)P2=P(I)
106 CONTINUE
   DO 108 I=1,N
   K(I)=P(I)*PH/1.E9
      E(I)=H(I)*B*B*1.E6/RR
108 CONTINUE
      H3=H2*B*B/RR
      P3=P2*PH
110   FORMAT(6(1X,E12.6))
120   CONTINUE
      WRITE(8,*)'P2,H2,P3,H3=',P2,H2,P3,H3
      CALL OUTHP(N,X, K,E)
      RETURN
      END
      SUBROUTINE OUTHP(N,X, K,E)
      REAL*8 X(N), K(N),E(N)
      DX=X(2)-X(1)
      DO 10 I=1,N
      WRITE(8,20)X(I),K(I),E(I)
10  CONTINUE
20  FORMAT(1X,6(F20.6,1X))
      RETURN
      END
      SUBROUTINE HREE(N,DX,H00,G0,X,P,H,RO,EPS,EDA)
  REAL*8 X(N),P(N),H(N),RO(N),EPS(N),EDA(2200)
      REAL*8 W(2200)
COMMON /COM1/ENDA,A1,A2,A3,Z,C1,C3,CW,K,FF/COM2/EDA0/COMAK/AK
(0:1100)
  DATA KK,NW,PAI1/0,2200,0.318309886/
      IF(KK.NE.0)GOTO 3
```

```
      HM0=C3
      H00=0.0
3     W1=0.0
      DO 4 I=1,N
4  W1=W1+P(I)
      C3=(DX*W1)/G0
      DW=1.-C3
      CALL DISP(N,NW,K,DX,P,W)
      HMIN=1.E3
      DO 30 I=1,N
      H0=0.5*X(I)*X(I)-PAI1*W(I)
      IF(H0.LT.HMIN)HMIN=H0
      H(I)=H0
30    CONTINUE
      IF(KK.NE.0)GOTO 32
      KK=1
      H00=-HMIN+HM0
32    H0=H00+HMIN
      IF(H0.LE.0.0)GOTO 48
      IF(H0+0.3*CW*DW.GT.0.0)HM0=H0+0.3*CW*DW
      IF(H0+0.3*CW*DW.LE.0.0)HM0=HM0*C3
48    H00=HM0-HMIN
50    DO 60 I=1,N
60    H(I)=H00+H(I)
      DO 100 I=1,N
      EDA(I)=EXP(A1*(-1.+(1.+A2*P(I))**Z))
      RO(I)=(A3+1.34*P(I))/(A3+P(I))
      EPS(I)=RO(I)*H(I)**(2+FF)*ENDA/EDA(I)**FF
100   CONTINUE
      RETURN
      END
      SUBROUTINE ITER(N,KK,DX,H00,G0,X,P,H,RO,EPS,EDA,R)
      REAL*8 X(N),P(N),H(N),RO(N),EPS(N),EDA(N),R(N)
      COMMON /COM1/ENDA,A1,A2,A3,Z,C1,C3,CW,LMAX,FF/COMAK/AK
      (0:1100)
      DATA KG1,PAI/0,3.14159265/
      IF(KG1.NE.0)GOTO 5
      KG1=1
      DX1=1./DX
      DX2=DX*DX
      DX3=1./DX2
      DX4=DX1/PAI
      DX5=DX1**(1+FF)
      DXL=DX*ALOG(DX)
      AK0=DX*AK(0)+DXL
```

```
      AK1=DX*AK(1)+DXL
5     DO 100 K=1,KK
      D2=0.5*(EPS(1)+EPS(2))
      D3=0.5*(EPS(2)+EPS(3))
      D5=DX1*(RO(2)*H(2)-RO(1)*H(1))
      D7=DX4*(RO(2)*AK0-RO(1)*AK1)
      PP=0.
      DO 70 I=2,N-1
      D1=D2
      D2=D3
      D4=D5
      D6=D7
      IF(I+2.LE.N)D3=0.5*(EPS(I+1)+EPS(I+2))
      D5=DX1*(RO(I+1)*H(I+1)-RO(I)*H(I))
      D7=DX4*(RO(I+1)*AK0-RO(I)*AK1)
      DD=(D1+D2)*DX3
      IF(0.05*DD.LT.ABS(D6))GOTO 10
      RI=-DX5*(D2*SIGN(1.0,(P(I+1)-P(I)))*ABS((P(I+1)-P(I)))**
(FF)-D1*SIGN(1.0,(P(I)-P(I-1)))*ABS((P(I)-P(I-1)))**(FF))+D4
      R(I)=RI
      DLDP=-FF*DX5*(D1*ABS((P(I)-P(I-1)))**(FF-1)+D2*ABS
((P(I+1)-P(I)))**(FF-1))+D6
      RI=RI/DLDP
      RI=RI/C1
      GOTO 20
10    RI=-DX5*(D2*SIGN(1.0,(P(I+1)-P(I)))*ABS((P(I+1)-P(I)))
**(FF)-D1*SIGN(1.0,(P(I)-PP))*ABS((P(I)-PP))**(FF))+D4
      R(I)=RI
      DLDP=-FF*DX5*(2*D1*ABS((P(I)-PP))**(FF-1)+D2*ABS
((P(I+1)-P(I)))**(FF-1))+2.*D6
      RI=RI/DLDP
      RI=0.5*RI
      IF(I.GT.2.AND.P(I-1)-C1*RI.GT.0)P(I-1)=P(I-1)-C1*RI
20  PP=P(I)
      P(I)=P(I)+C1*RI
      IF(P(I).LT.0.0)P(I)=0.0
      IF(P(I).LE.0.0)R(I)=0.0
70    CONTINUE
      CALL HREE(N,DX,H00,G0,X,P,H,RO,EPS,EDA)
100 CONTINUE
      RETURN
      END
      SUBROUTINE DISP(N,NW,KMAX,DX,P1,W)
      REAL*8 P1(N),W(NW),P(2200),AK1(0:50),AK2(0:50)
      COMMON /COMAK/AK(0:1100)
```

```
      DATA NMAX,KMIN/2200,1/
      N2=N
      M=3+2*ALOG(FLOAT(N))
      K1=N+KMAX
      DO 10 I=1,N
10    P(K1+I)=P1(I)
      DO 20 KK=KMIN,KMAX-1
      K=KMAX+KMIN-KK
      N1=(N2+1)/2
      CALL DOWNP(NMAX,N1,N2,K,P)
20    N2=N1
      DX1=DX*2**(KMAX-KMIN)
      CALL WI(NMAX,N1,KMIN,KMAX,DX,DX1,P,W)
      DO 30 K=KMIN+1,KMAX
      N2=2*N1-1
      DX1=DX1/2.
      CALL AKCO(M+5,KMAX,K,AK1)
      CALL AKIN(M+6,AK1,AK2)
      CALL WCOS(NMAX,N1,N2,K,W)
      CALL CORR(NMAX,N2,K,M,1,DX1,P,W,AK1)
      CALL WINT(NMAX,N2,K,W)
      CALL CORR(NMAX,N2,K,M,2,DX1,P,W,AK2)
30    N1=N2
      DO 40 I=1,N
40    W(I)=W(K1+I)
      RETURN
      END
      SUBROUTINE DOWNP(NMAX,N1,N2,K,P)
      REAL*8 P(NMAX)
      K1=N1+K-1
      K2=N2+K-1
      DO 10 I=3,N1-2
      I2=2*I+K2
10    P(K1+I)=(16.*P(I2)+9.*(P(I2-1)+P(I2+1))-(P(I2-3)+P(I2+3)))/32.
      P(K1+2)=0.25*(P(K2+3)+P(K2+5))+0.5*P(K2+4)
      P(K1+N1-1)=0.25*(P(K2+N2-2)+P(K2+N2))+0.5*P(K2+N2-1)
      RETURN
      END
      SUBROUTINE WCOS(NMAX,N1,N2,K,W)
      REAL*8 W(NMAX)
      K1=N1+K-1
      K2=N2+K
      DO 10 I=1,N1
      II=2*I-1
10    W(K2+II)=W(K1+I)
```

```
      RETURN
      END
      SUBROUTINE WINT(NMAX,N,K,W)
      REAL*8 W(NMAX)
      K2=N+K
      DO 10 I=4,N-3,2
      II=K2+I
10    W(II)=(9.*(W(II-1)+W(II+1))-(W(II-3)+W(II+3)))/16.
      I1=K2+2
      I2=K2+N-1
      W(I1)=0.5*(W(I1-1)+W(I1+1))
      W(I2)=0.5*(W(I2-1)+W(I2+1))
      RETURN
      END
      SUBROUTINE CORR(NMAX,N,K,M,I1,DX,P,W,AK)
      REAL*8 P(NMAX),W(NMAX),AK(0:M)
      K1=N+K
      IF(I1.EQ.2)GOTO 20
      DO 10 I=1,N,2
      II=K1+I
      J1=MAX0(1,I-M)
      J2=MIN0(N,I+M)
      DO 10 J=J1,J2
      IJ=IABS(I-J)
10    W(II)=W(II)+AK(IJ)*DX*P(K1+J)
      RETURN
20    DO 30 I=2,N,2
      II=K1+I
      J1=MAX0(1,I-M)
      J2=MIN0(N,I+M)
      DO 30 J=J1,J2
      IJ=IABS(I-J)
30    W(II)=W(II)+AK(IJ)*DX*P(K1+J)
      RETURN
      END
      SUBROUTINE WI(NMAX,N,KMIN,KMAX,DX,DX1,P,W)
      REAL*8 P(NMAX),W(NMAX)
      COMMON /COMAK/AK(0:1100)
      K1=N+1
      K=2**(KMAX-KMIN)
      C=ALOG(DX)
      DO 10 I=1,N
      II=K1+I
      W(II)=0.0
      DO 10 J=1,N
```

```
      IJ=K*IABS(I-J)
10    W(II)=W(II)+(AK(IJ)+C)*DX1*P(K1+J)
      RETURN
      END
      SUBROUTINE AKCO(KA,KMAX,K,AK1)
      REAL*8 AK1(0:KA)
      COMMON /COMAK/AK(0:1100)
      J=2**(KMAX-K)
      DO 10 I=0,KA
      II=J*I
10    AK1(I)=AK(II)
      RETURN
      END
      SUBROUTINE AKIN(KA,AK1,AK2)
      REAL*8 AK1(KA),AK2(KA)
      DO 10 I=4,KA-3
10    AK2(I)=(9.*(AK1(I-1)+AK1(I+1))-(AK1(I-3)+AK1(I+3)))/16.
      AK2(1)=(9.*AK1(2)-AK1(4))/8.
      AK2(2)=(9.*(AK1(1)+AK1(3))-(AK1(3)+AK1(5)))/16.
      AK2(3)=(9.*(AK1(2)+AK1(4))-(AK1(2)+AK1(6)))/16.
      DO 20 I=1,KA
20    AK2(I)=AK1(I)-AK2(I)
      DO 30 I=1,KA-1,2
      I1=I+1
      AK1(I)=0.0
30    AK1(I1)=AK2(I1)
      RETURN
      END
      SUBROUTINE SUBAK(MM)
      COMMON /COMAK/AK(0:1100)
      DO 10 I=0,MM
10    AK(I)=(I+0.5)*(ALOG(ABS(I+0.5))-1.)-(I-0.5)*(ALOG(ABS(I-0.5))-1.)
      RETURN
      END
      FUNCTION FRICT(N,DX,X,H,P,EDA)
      REAL*8 X(N),H(N),P(N),EDA(N)
      COMMON /COM3/E1,PH,B,U1,U2,R
      DATA TAU0,AT/1.98E7,0.078/
      OPEN (10,FILE='TAU.DAT')
      TP=TAU0/PH
      TE=TAU0/E1
      A=AT/TAU0
      BR=B/R
      FRICT=0.0
      DO 10 I=1,N
```

```
      DP=0.0
      A=1.0+AT*P(I)/TAU0
      IF(I.NE.N)DP=(P(I+1)-P(I))/DX
      TAU1=0.5*H(I)*DP*(BR/TP/A)
      TAU2=2.*U1*EDA(I)/(H(I)*BR**2*TE*A)
    FRICT=FRICT+ABS(TAU1)+TAU2
      TAU2=0.05*TAU2
      WRITE(10,5)X(I),TAU1,TAU2,P(I),H(I)
5     FORMAT(5(E12.6,1X))
10  CONTINUE
      FRICT=FRICT*DX*B*TAU0
      RETURN
      END
      SUBROUTINE FZ(N,P,POLD)
      REAL*8 P(N),POLD(N)
      DO 10 I=1,N
10  POLD(I)=P(I)
      RETURN
      END
      SUBROUTINE ERROP(N,P,POLD,ERP)
      REAL*8 P(N),POLD(N)
      SD=0.0
      SUM=0.0
      DO 10 I=1,N
      SD=SD+ABS(P(I)-POLD(I))
      POLD(I)=P(I)
10  SUM=SUM+P(I)
      ERP=SD/SUM
      RETURN
      END
```

15.3.3 Calculation Results

The characteristics of GEHL in line contact can be analyzed according to the operation results. A typical GEHL film is shown in Figure 15.2, with the following main features:

1. The film thickness shape and pressure distribution of GEHL are similar to those of oil lubrication.
2. In the central contact area, the lubrication pressure is close to the Hertz pressure, and the shape of the film thickness is approximately parallel.
3. In the outlet, the pressure appears as an obvious secondary pressure peak, but its width is extremely narrow. Then, the pressure falls sharply to environmental pressure.

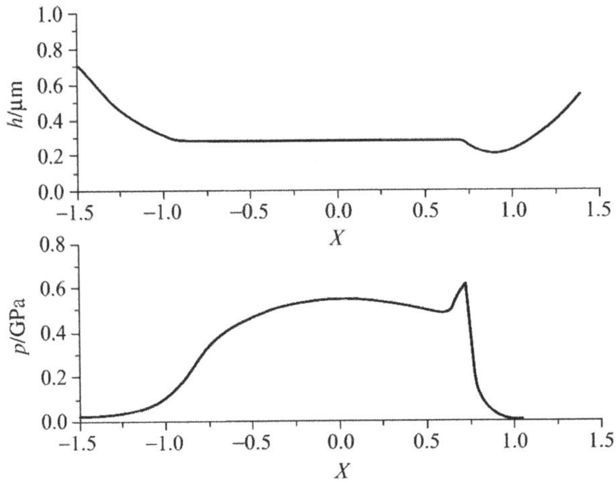

Figure 15.2 Film thickness and pressure distribution of GEHL ($U = 0.87\,\mathrm{m\,s^{-1}}$, $w = 100$ KN, $n = 0.846$).

4. At the second pressure peak, the film begins to shrink and appears the necking. The film thickness at the necking place is the minimum.

Figure 15.3 shows the influence of rheological parameter n on the lubrication film. From the figure, we can see that, as n increases, the pressure peak increases, its position moves towards the inlet and the average film thickness also increases.

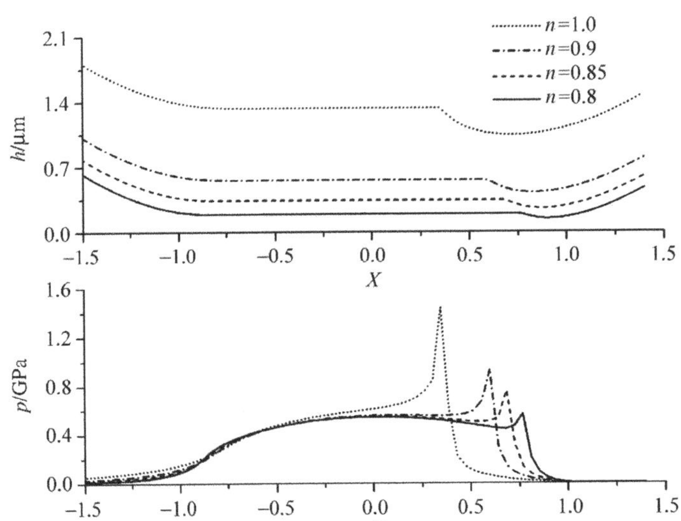

Figure 15.3 Influence of n on lubrication film ($w = 100$ kN, $U = 0.87\,\mathrm{m\,s^{-1}}$).

16

Numerical Method and Program for Grease EHL in Point Contact

16.1 Basic Equations

16.1.1 Reynolds Equation

Based on the Ostwald rheological equation, the Reynolds equation for grease elastohydrodynamic lubrication (GEHL) in a point contact can be written as

$$\frac{n}{2n+1}\left(\frac{1}{2}\right)^{\frac{n+1}{n}}\left\{\frac{\partial}{\partial x}\left[\rho h^{\frac{2n+1}{n}}\left(\frac{1}{\phi}\frac{\partial p}{\partial x}\right)^{\frac{1}{n}}\right]+\frac{\partial}{\partial y}\left[\rho h^{\frac{2n+1}{n}}\left(\frac{1}{\phi}\frac{\partial p}{\partial y}\right)^{\frac{1}{n}}\right]\right\}=U\frac{\partial(\rho h)}{\partial x} \tag{16.1}$$

where p is pressure, h is film thickness, U is average velocity [$U=(u_1+u_2)/2$; u_1 and u_2 are the tangential velocities of the down and up surfaces, respectively], x is the motion direction of the surface, ϕ is the plastic viscosity of grease, ρ is the grease density and n is a rheological parameter ($n \leq 1$).

16.1.2 Film Thickness Equation

$$h(x,y)=h_0+\frac{x^2}{2R_x}+\frac{y^2}{2R_y}+v(x,y) \tag{16.2}$$

where h_0 is a film thickness parameter to be determined in calculation, R_x is the equivalent radius of the surfaces in the x direction $\left(\frac{1}{R_x}=\frac{1}{R_{1x}}+\frac{1}{R_{2x}}\right)$, R_y is the

Numerical Calculation of Lubrication: Methods and Programs, First Edition. Ping Huang.

equivalent radius of the surfaces in the y direction $\left(\frac{1}{R_y} = \frac{1}{R_{1y}} + \frac{1}{R_{2y}}\right)$ and v is the elastic deformation.

16.1.3 *Elastic Deformation Equation*

$$v(x, y) = \frac{2}{\pi E} \iint\limits_{\Omega} \frac{p(s, t)}{\sqrt{(x - s)^2 + (y - t)^2}} ds\, dt \tag{16.3}$$

where E is the equivalent elastic modulus $\frac{1}{E} = \frac{1}{2}\left(\frac{1 - \nu_1^2}{E_1} + \frac{1 - \nu_2^2}{E_2}\right)$; ν_1 and ν_2 are the Poisson's ratios of the two surfaces and E_1 and E_2 are the elastic moduli of the two surfaces.

16.1.4 *Viscosity–Pressure Equation*

$$\phi = \phi_0 \exp\{(\ln \phi_0 + 9.67)\left[-1 + \left(1 + 5.1 \times 10^{-9} p\right)^z\right]\} \tag{16.4}$$

where ϕ_0 is the plastic viscosity of grease at $p = p_0$ (p_0 is atmospheric pressure) and z is a constant ($z = 0.68$).

16.1.5 *Density Equation*

The variation in grease density is so small that it can be assumed that grease is incompressible, that is

$$\rho = \text{const.} \tag{16.5}$$

16.2 Numerical Method

16.2.1 *Dimensionless Equations*

The dimensionless Reynolds equation of grease EHL in a point contact can be expressed as

$$\frac{\partial}{\partial X}\left(\varepsilon' \frac{\partial P}{\partial X}\right) + \frac{\partial}{\partial Y}\left(\varepsilon'' \frac{\partial P}{\partial Y}\right) - \frac{\partial (H)}{\partial X} = 0 \tag{16.6}$$

where $\varepsilon' = \varepsilon \left|\frac{\partial P}{\partial X}\right|^{\frac{1-n}{n}}$, $\varepsilon'' = \varepsilon \left|\frac{\partial P}{\partial Y}\right|^{\frac{1-n}{n}}$, $\varepsilon = \frac{\lambda H^{\frac{2n+1}{n}}}{\phi^{*\frac{1}{n}}}$, $\lambda = \frac{p_H^{\frac{1}{n}} a^{\frac{2n+1}{n}}}{2^{\frac{n+1}{n}} U\left(2 + \frac{1}{n}\right) R_x^{\frac{n+1}{n}} \phi_0^{\frac{1}{n}}}$,

R_x is the equivalent radius of curvature of the surface in the x direction, P is dimensionless pressure ($P = p/p_H$), p_H is the maximum Hertz contact stress, H is dimensionless film thickness $\left(H = \frac{hR_x}{a^2}\right)$, a is the radius of the Hertz contact circle, X is a dimensionless coordinate ($X = x/a$), Y is a dimensionless coordinate ($Y = y/a$) and ϕ^* is the dimensionless plastic viscosity of grease ($\phi^* = \phi/\phi_0$).

The boundary conditions are:

Inlet boundary condition: $P(X_0, Y) = 0$
Outlet boundary conditions: $P(X_e, Y) = 0$ and $\frac{\partial P(X_e, Y)}{\partial X} = 0$
Side boundary conditions: $P|_{Y=\pm 1} = 0$

where X_0 and X_e are the dimensionless coordinates of the inlet and outlet, X_0 is given but X_e should be determined by the outlet boundary conditions and $Y = \pm 1$ are the two sides of the contact region.

Film thickness equation

$$H(X, Y) = H_0 + \frac{X^2 + Y^2}{2} + \frac{2}{\pi^2} \int_{X_0}^{X_e} \int_{Y_0}^{Y_e} \frac{P(S, T)dS\,dT}{\sqrt{(X - S)^2 + (Y - T)^2}} \tag{16.7}$$

where H_0 is the film thickness to be determined by calculation.

Viscosity–pressure equation

$$\phi^* = \exp\left\{[\ln(\phi_0) + 9.67]\left[-1 + \left(1 + 5.1 \times 10^{-9} p\right)^{0.68}\right]\right\} \tag{16.8}$$

Dimensionless density

$$\rho^* = 1 \tag{16.9}$$

Load equation

$$\int_{X_0}^{X_e} \int_{Y_0}^{Y_e} P(X, Y)dX\,dY = \frac{2}{3}\pi \tag{16.10}$$

16.2.2 Discrete Equations

The discrete form of the dimensionless Reynolds equation of grease EHL in a point contact is expressed as

$$\frac{\left(\varepsilon'_{i-1/2,j}P_{i-1,j} + \varepsilon'_{i+1/2,j}P_{i+1,j} + \varepsilon''_{i,j-1/2}P_{i,j-1} + \varepsilon''_{i,j+1/2}P_{i,j+1} - \varepsilon_0 P_{i,j}\right)}{\Delta X^2} - \frac{\left(\rho^*_{i,j}H_{i,j} - \rho^*_{i-1,j}H_{i-1,j}\right)}{\Delta X} = 0 \tag{16.11}$$

where $\varepsilon'_{i-1/2,j} = \frac{1}{2}(\varepsilon_{i,j} + \varepsilon_{i-1,j})\left|\frac{P_{i,j} - P_{i-1,j}}{\Delta X}\right|^{\frac{1-n}{n}}$, $\varepsilon'_{i+1/2,j} = \frac{1}{2}(\varepsilon_{i,j} + \varepsilon_{i+1,j})\left|\frac{P_{i+1,j} - P_{i,j}}{\Delta X}\right|^{\frac{1-n}{n}}$,

$\varepsilon''_{i,j-1/2} = \frac{1}{2}(\varepsilon_{i,j} + \varepsilon_{i,j-1})\left|\frac{P_{i,j} - P_{i,j-1}}{\Delta Y}\right|^{\frac{1-n}{n}}$, $\varepsilon''_{i,j+1/2} = \frac{1}{2}(\varepsilon_{i,j} + \varepsilon_{i,j+1})\left|\frac{P_{i,j} - P_{i,j+1}}{\Delta Y}\right|^{\frac{1-n}{n}}$,

$\varepsilon_0 = \varepsilon'_{i-1/2,j} + \varepsilon'_{i+1/2,j} + \varepsilon''_{i,j-1/2} + \varepsilon''_{i,j+1/2}$.

Since the grids of the mesh are uniform, the dimensionless node spaces are the same, that is, $\Delta Y = \Delta X$.

Film thickness equation

$$H_{ij} = H_0 + \frac{X_i^2 + Y_j^2}{2} + \frac{2}{\pi^2}\sum_{k=1}^{n}\sum_{l=1}^{n} D_{ij}^{kl} P_{kl} \tag{16.12}$$

where D_{ij}^{kl} is the stiffness coefficient of elastic deformation.

Load equation

$$\Delta X \Delta Y \sum_{i=1}^{n}\sum_{j=1}^{n} P_{ij} = 2\pi/3 \tag{16.13}$$

In Chapter 14, when calculating the pressure distribution of oil EHL in a point contact, the Gauss–Seidel iteration method is used for the low pressure region and the Jacobi Bipolar iteration method is used for the high pressure region. In Equation 16.11, the nonlinear part of the pressure gradient is calculated in parameter ε. The Jacobi Bipolar iteration method used in the high pressure region will result easily in iteration divergence and calculation failure. Therefore, in this chapter and Chapter 20, we use the Gauss–Seidel iteration method for the whole region.

Convergence criterion

$$\frac{\sum_i \sum_j |P_{i,j}^{k+1} - P_{i,j}^{k}|}{\sum_i \sum_j P_{i,j}^{k+1}} \le \varepsilon_4 \tag{16.14}$$

where ε_4 is a given error.

16.3 Calculation Program

16.3.1 Calculation Diagram

The calculation diagram of isothermal grease EHL in point contact is shown in Figure 16.1.

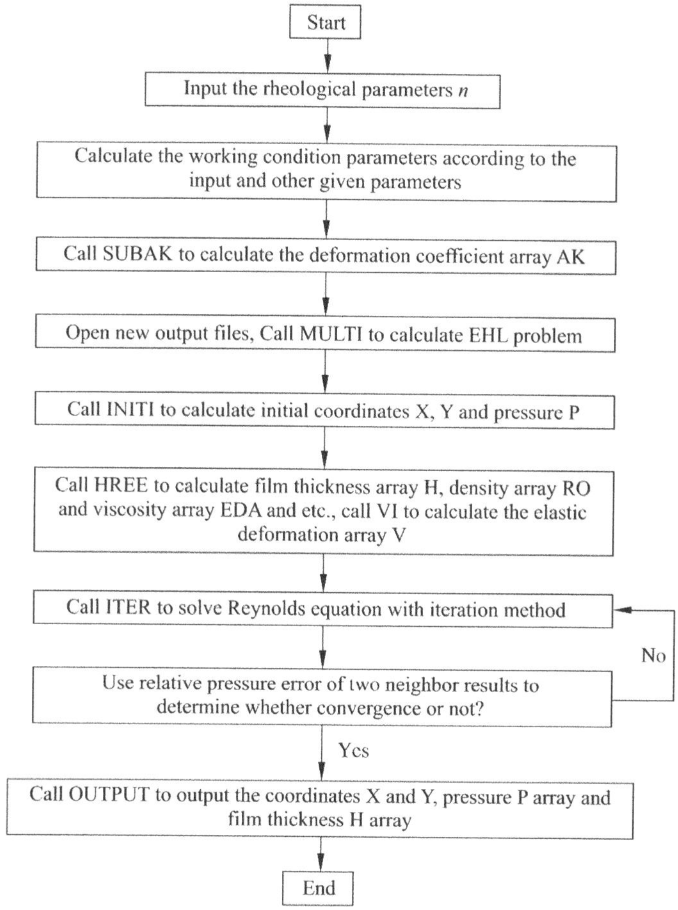

Figure 16.1 Calculation diagram of isothermal grease EHL in point contact.

16.3.2 Calculation Program

In this program, the given parameters are as follows:

Number of nodes	N = 65 × 65
Dimensionless starting coordinate in *x* direction	X0 = −2.5
Dimensionless ending coordinate in *x* direction	XE = 1.5
Dimensionless starting coordinate in *y* direction	Y0 = −2.0
Dimensionless ending coordinate in *y* direction	YE = 2.0
Equivalent modulus of elasticity	E1 = 2.21E11 Pa
Initial plastic viscosity	EDA0 = 0.058 Pa s
Equivalent curvature radius	RX = RY = 0.05 m

Input parameter:

Rheological parameter, n

Output data:

Pressure array P(I, J) will be saved in the file PRESSURE.DAT.
Film thickness array H(I, J) will be saved in the file FILM.DAT.
Other data will be saved in the file OUT.DAT.

```
      PROGRAM GREASEPOINTEHL
      DIMENSION THETA(15),EALFA(15),EBETA(15)
      COMMON /COM1/Z,ENDA,AKC,HM0,HMC,EK,EAL,EBE,AD,AD1,KK1,KK2,
KK3,KK4,FN,FN1,FF
      COMMON /COM2/W0,E1,RX,B,PH,US,U1,U2,EDA0
      COMMON /COM3/A1,A2,A3,LMIN
      DATA PAI,Z,AKC,AD,AD1/3.14159265,0.68,1.0,0.0,0.0/
      DATA T0,EDA0,AK,AK1,AK2,CV,CV1,CV2,RO0,RO1,RO2,S0,D0/
303.,0.058,0.14,46.,46.,2000.,470.,470.,890.,7850.,7850.,
-1.1,-0.00065/
      DATA N,NZ,RX,RY,X0,XE, E1,US,CT,W0/65,5,0.05,0.05,-
2.5,1.5, 2.21E11,1.5,0.31,39.24/
      DATA THETA/10.,20.,30.,35.,40.,45.,50.,55.,60.,65.,70.,
75.,80.,85.,90./
      DATA EALFA/6.612,3.778,2.731,2.397,2.136,1.926,1.754,
1.611,1.486,1.378,1.284,1.202,1.128,1.061,1.0/
      DATA EBETA/0.319,0.408,0.493,0.53,0.567,0.604,0.641,
0.678,0.717,0.759,0.802,0.846,0.893,0.944,1.0/
      DATA KK1,KK2,KK3,KK4/0,0,0,0/
      WRITE(*,*)'n<=1 INPUT n=?'
      READ(*,*)FN
      FN1=1.0/FN
      FF=1.0/FN-1.0
      WRITE(*,*)"FF=",FF
      EK=RX/RY
      AA=0.5*(1./RX+1./RY)
      BB=0.5*ABS(1./RX-1./RY)
      CC=ACOS(BB/AA)*180.0/PAI
      EAL=1.0
      EBE=1.0
      DO I=1,15
      IF(CC.LT.THETA(I))THEN
      WRITE(*,*)I
      EAL=EALFA(I-1)+(CC-THETA(I-1))*(EALFA(I)-EALFA(I-1))/
 (THETA(I)-THETA(I-1))
```

```
      EBE=EBETA(I-1)+(CC-THETA(I-1))*(EBETA(I)-EBETA(I-1))/
     (THETA(I)-THETA(I-1))

      GOTO 10
      ENDIF
      ENDDO
10    EA=EAL*(1.5*W0/AA/E1)**(1./3.0)
      EB=EBE*(1.5*W0/AA/E1)**(1./3.0)
      PH=1.5*W0/(EA*EB*PAI)
      OPEN(8,FILE='FILM.DAT',STATUS='UNKNOWN')
      OPEN(9,FILE='PRESS.DAT',STATUS='UNKNOWN')
      OPEN(10,FILE='OUT.DAT',STATUS='UNKNOWN')
      WRITE(*,*)"N,X0,XE,PH,E1,EDA0,RX,US"
      WRITE(*,*)N,X0,XE,PH,E1,EDA0,RX,US
      WRITE(16,*)"N,X0,XE,PH,E1,EDA0,RX,US"
      WRITE(16,*)N,X0,XE,PH,E1,EDA0,RX,US
      H00=0.0
      MM=N-1
      LMIN=ALOG(N-1.)/ALOG(2.)-1.99
      U=EDA0*(US/2.)**FN/(E1*RX**FN)
      WRITE(*,*)"U=",U
      U1=0.5*(2.+AKC)*U
      U2=0.5*(2.-AKC)*U
      A1=ALOG(EDA0)+9.67
      A2=5.1E-9*PH
      A3=0.59/(PH*1.E-9)
      B=PAI*PH*RX/E1
      W=2.*PAI*PH/(3.*E1)*(B/RX)**2
      ALFA=Z*5.1E-9*A1
      G=ALFA*E1
      AHM=1.0-EXP(-0.68*1.03)
      AHC=1.0-0.61*EXP(-0.73*1.03)
      HM0=3.63*(RX/B)**2*G**0.49*U**0.68*W**(-0.073)*AHM
      HMC=2.69*(RX/B)**2*G**0.53*U**0.67*W**(-0.067)*AHC
      ENDA=2.*U*(3.+FF)*2.0**(1.0+FF)*(E1/PH)**(1.0+FF)*(RX/B)
     **(3.0+FF)
      WRITE(*,*)"ENDA=",ENDA
      UTL=EDA0*US*RX/(B*B*2.E7)
      W0=2.0*PAI*EA*EB*PH/3.0
      T1=PH*B/RX
      T2=EDA0*US*RX/(B*B)
      WRITE(*,*)' Wait please'
      CALL SUBAK(MM)
      CALL MULTI(N,NZ,X0,XE,H00)
      STOP
```

```
      END
      SUBROUTINE MULTI(N,NZ,X0,XE,H00)
      DIMENSION X(65),Y(65),H(4500),RO(4500),EPS(4500),EDA
(4500),P(4500),POLD(4500),T(65,65,5)
      COMMON /COM1/Z,ENDA,AKC,HM0,HMC,EK,EAL,EBE,AD,AD1,KK1,KK2,
KK3,KK4,FN,FN1,FF
      DATA MK,KTK,G00/200,1,2.0943951/
      G0=G00*EAL*EBE
      NX=N
      NY=N
      NN=(N+1)/2
      CALL INITI(N,DX,X0,XE,X,Y,P,POLD)
      CALL HREE(N,DX,H00,G0,X,Y,H,RO,EPS,EDA,P)
      M=0
14    KK=15
      CALL ITER(N,KK,DX,H00,G0,X,Y,H,RO,EPS,EDA,P)
      CALL ERP(N,ER,P,POLD)
      ER=ER/KK
      WRITE(*,*)'ER=',ER
   M=M+1
      IF(M.LT.MK.AND.ER.GT.1.E-4)GOTO 14
      CALL OUPT(N,DX,X,Y,H,P,EDA,TMAX)
      RETURN
      END
      SUBROUTINE INITI(N,DX,X0,XE,X,Y,P,POLD)
      DIMENSION X(N),Y(N),P(N,N),POLD(N,N)
      NN=(N+1)/2
      DX=(XE-X0)/(N-1.)
      Y0=-0.5*(XE-X0)
      DO 5 I=1,N
      X(I)=X0+(I-1)*DX
      Y(I)=Y0+(I-1)*DX
5     CONTINUE
      DO 10 I=1,N
      D=1.-X(I)*X(I)
      DO 10 J=1,NN
      C=D-Y(J)*Y(J)
      IF(C.LE.0.0)P(I,J)=0.0
10    IF(C.GT.0.0)P(I,J)=SQRT(C)
      DO 20 I=1,N
      DO 20 J=NN+1,N
      JJ=N-J+1
20    P(I,J)=P(I,JJ)
      DO I=1,N
      DO J=1,N
```

```
      POLD(I,J)=P(I,J)
      ENDDO
      ENDDO
      RETURN
      END
      SUBROUTINE HREE(N,DX,H00,G0,X,Y,H,RO,EPS,EDA,P)
      DIMENSION X(N),Y(N),P(N,N),H(N,N),RO(N,N),EPS(N,N),EDA(N,N)
      DIMENSION W(150,150),P0(150,150),ROU(65,65)
      COMMON /COM1/Z,ENDA,AKC,HM0,HMC,EK,EAL,EBE,AD,AD1,KK1,KK2,
KK3,KK4,FN,FN1,FF
      COMMON /COM2/W0,E1,RX,B,PH,US,U1,U2,EDA0
      COMMON /COM3/A1,A2,A3,LMIN
      DATA KR,NW,PAI,PAI1,DELTA/0,150,3.14159265,0.2026423,0.0/
      NN=(N+1)/2
      CALL VI(NW,N,DX,P,W)
      HMIN=1.E3
      DO 30 I=1,N
      DO 30 J=1,NN
      RAD=X(I)*X(I)+EK*Y(J)*Y(J)
      W1=0.5*RAD+DELTA
      ZZ=0.5*AD1*AD1+X(I)*ATAN(AD*PAI/180.0)
      IF(W1.LE.ZZ)W1=ZZ
      H0=W1+W(I,J)
      IF(H0.LT.HMIN)HMIN=H0
30    H(I,J)=H0
      IF(KK.EQ.0)THEN
      KG1=0
      H01=-HMIN+HM0
      DH=0.005*HM0
      H02=-HMIN
      H00=0.5*(H01+H02)
      ENDIF
      W1=0.0
      DO 32 I=1,N
      DO 32 J=1,N
32    W1=W1+P(I,J)
      W1=DX*DX*W1/G0
      DW=1.-W1
      IF(KK.EQ.0)THEN
      KK=1
      GOTO 50
      ENDIF
      IF(DW.LT.0.0)THEN
      KG1=1
      H00=AMIN1(H01,H00+DH)
```

```
      ENDIF
      IF(DW.GT.0.0)THEN
      KG2=2
      H00=AMAX1(H02,H00-DH)
      ENDIF
50    DO 60 I=1,N
      DO 60 J=1,NN
      H(I,J)=H00+H(I,J)
      IF(P(I,J).LT.0.0)P(I,J)=0.0
      EDA1=EXP(A1*(-1.+(1.+A2*P(I,J))**Z))
      EDA(I,J)=EDA1
      RO(I,J)=1.
      EPS(I,J)=ENDA*RO(I,J)*H(I,J)**(2.+FN1)/(EDA(I,J)**FN1)
60    CONTINUE
      DO 70 J=NN+1,N
      JJ=N-J+1
      DO 70 I=1,N
      H(I,J)=H(I,JJ)
      RO(I,J)=RO(I,JJ)
      EDA(I,J)=EDA(I,JJ)
70    EPS(I,J)=EPS(I,JJ)
      RETURN
      END
   SUBROUTINE ITER(N,KK,DX,H00,G0,X,Y,H,RO,EPS,EDA,P)
      DIMENSION X(N),Y(N),P(N,N),H(N,N),RO(N,N),EPS(N,N),EDA(N,N)
      DIMENSION D(70),A(350),B(210),ID(70)
      COMMON /COM1/Z,ENDA,AKC,HM0,HMC,EK,EAL,EBE,AD,AD1,KK1,KK2,
 KK3,KK4,FN,FN1,FF
      COMMON /COMAK/AK(0:65,0:65)
      DATA KG1,PAI1,C1,C2/0,0.2026423,0.27,0.27/
      IF(KG1.NE.0)GOTO 2
      KG1=1
      AK00=AK(0,0)
      AK10=AK(1,0)
      AK20=AK(2,0)
      BK00=AK00-AK10
      BK10=AK10-0.25*(AK00+2.*AK(1,1)+AK(2,0))
      BK20=AK20-0.25*(AK10+2.*AK(2,1)+AK(3,0))
  2   NN=(N+1)/2
      MM=N-1
      DX1=1./DX
      DX2=DX*DX
      DX3=1./DX2
      DO 100 K=1,KK
      PMAX=0.0
```

```
      DO 70 J=2,NN
      J0=J-1
      J1=J+1
      IA=1
 8    MM=N-IA
      IF(P(MM,J0).GT.1.E-6)GOTO 20
      IF(P(MM,J).GT.1.E-6)GOTO 20
      IF(P(MM,J1).GT.1.E-6)GOTO 20
      IA=IA+1
      IF(IA.LT.N)GOTO 8
      GOTO 70
20    IF(MM.LT.N-1)MM=MM+1
      DPDX1=ABS((P(2,J)-P(1,J))*DX1)**(FF)
      D2=0.5*(EPS(1,J)+EPS(2,J))*DPDX1
      DO 50 I=2,MM
      I0=I-1
      I1=I+1
      II=5*I0
      DPDX2=ABS((P(I1,J)-P(I,J))*DX1)**(FF)
      DPDY1=ABS((P(I,J)-P(I,J0))*DX1)**(FF)
      DPDY2=ABS((P(I,J1)-P(I,J))*DX1)**(FF)
      D1=D2
      D2=0.5*(EPS(I1,J)+EPS(I,J))*DPDX2
      D4=0.5*(EPS(I,J0)+EPS(I,J))*DPDY1
      D5=0.5*(EPS(I,J1)+EPS(I,J))*DPDY2
      P1=P(I0,J)
      P2=P(I1,J)
      P3=P(I,J)
      P4=P(I,J0)
      P5=P(I,J1)
      D3=D1+D2+D4+D5
      IF(H(I,J).LE.0.0)THEN
      ID(I)=0
      A(II+1)=0.0
      A(II+2)=0.0
      A(II+3)=1.0
      A(II+4)=0.0
      A(II+5)=1.0
      A(II-4)=0.0
      GOTO 50
      ENDIF
      ID(I)=1
      IF(J.EQ.NN)P5=P4
      A(II+1)=PAI1*(RO(I0,J)*AK10-RO(I,J)*AK20)
      A(II+2)=DX3*D1+PAI1*(RO(I0,J)*AK00-RO(I,J)*AK10)
```

```
      A(II+3)=-DX3*D3+PAI1*(RO(I0,J)*AK10-RO(I,J)*AK00)
      A(II+4)=DX3*D2+PAI1*(RO(I0,J)*AK20-RO(I,J)*AK10)
      A(II+5)=-DX3*(D1*P1+D2*P2+D4*P4+D5*P5-D3*P3)+DX1*(RO(I,J)
*H(I,J)-RO(I0,J)*H(I0,J))
50    CONTINUE
      CALL TRA4(MM,D,A,B)
      DO 60 I=2,MM
      IF(ID(I).EQ.1)P(I,J)=P(I,J)+C1*D(I)
      IF(P(I,J).LT.0.0)P(I,J)=0.0
      IF(PMAX.LT.P(I,J))PMAX=P(I,J)
60    CONTINUE
70    CONTINUE
      DO 80 J=1,NN
      JJ=N+1-J
      DO 80 I=1,N
80    P(I,JJ)=P(I,J)
      CALL HREE(N,DX,H00,G0,X,Y,H,RO,EPS,EDA,P)
100   CONTINUE
      RETURN
      END
      SUBROUTINE TRA4(N,D,A,B)
      DIMENSION D(N),A(5,N),B(3,N)
      C=1./A(3,N)
      B(1,N)=-A(1,N)*C
      B(2,N)=-A(2,N)*C
      B(3,N)=A(5,N)*C
      DO 10 I=1,N-2
      IN=N-I
      IN1=IN+1
      C=1./(A(3,IN)+A(4,IN)*B(2,IN1))
      B(1,IN)=-A(1,IN)*C
      B(2,IN)=-(A(2,IN)+A(4,IN)*B(1,IN1))*C
 10   B(3,IN)=(A(5,IN)-A(4,IN)*B(3,IN1))*C
      D(1)=0.0
      D(2)=B(3,2)
      DO 20 I=3,N
 20   D(I)=B(1,I)*D(I-2)+B(2,I)*D(I-1)+B(3,I)
      RETURN
      END
      SUBROUTINE VI(NW,N,DX,P,V)
      DIMENSION P(N,N),V(NW,NW)
      COMMON /COMAK/AK(0:65,0:65)
      PAI1=0.2026423
      DO 40 I=1,N
      DO 40 J=1,N
```

```
      H0=0.0
      DO 30 K=1,N
      IK=IABS(I-K)
      DO 30 L=1,N
      JL=IABS(J-L)
30    H0=H0+AK(IK,JL)*P(K,L)
40    V(I,J)=H0*DX*PAI1
      RETURN
      END
      SUBROUTINE SUBAK(MM)
      COMMON /COMAK/AK(0:65,0:65)
      S(X,Y)=X+SQRT(X**2+Y**2)
      DO 10 I=0,MM
      XP=I+0.5
      XM=I-0.5
      DO 10 J=0,I
      YP=J+0.5
      YM=J-0.5
      A1=S(YP,XP)/S(YM,XP)
      A2=S(XM,YM)/S(XP,YM)
      A3=S(YM,XM)/S(YP,XM)
      A4=S(XP,YP)/S(XM,YP)
      AK(I,J)=XP*ALOG(A1)+YM*ALOG(A2)+XM*ALOG(A3)+YP*ALOG(A4)
10    AK(J,I)=AK(I,J)
      RETURN
      END
      SUBROUTINE ERP(N,ER,P,POLD)
      DIMENSION P(N,N),POLD(N,N)
      ER=0.0
      SUM=0.0
      NN=(N+1)/2
      DO 10 I=1,N
      DO 10 J=1,NN
      ER=ER+ABS(P(I,J)-POLD(I,J))
      SUM=SUM+P(I,J)
10    CONTINUE
      ER=ER/SUM
      DO I=1,N
      DO J=1,N
      POLD(I,J)=P(I,J)
      ENDDO
      ENDDO
      RETURN
      END
      SUBROUTINE OUPT(N,DX,X,Y,H,P,EDA,TMAX)
```

```
      DIMENSION X(N),Y(N),H(N,N),P(N,N),EDA(N,N)
      COMMON /COM1/Z,ENDA,AKC,HM0,HMC,EK,EAL,EBE,AD,AD1,KK1,KK2,
KK3,KK4,FN,FN1,FF
      COMMON /COM2/W0,E1,RX,B,PH,US,U1,U2,EDA0
      A=0.0
      WRITE(8,40)A,(Y(I),I=1,N)
      DO I=1,N
      WRITE(8,40)X(I),(H(I,J),J=1,N)
      ENDDO
      WRITE(9,40)A,(Y(I),I=1,N)
      DO I=1,N
      WRITE(9,40)X(I),(P(I,J),J=1,N)
      ENDDO
40    FORMAT(66(E12.6,1X))
      HMIN=1.E3
      PMAX=0.0
      DO J=1,N
      DO I=2,N
      IF(H(I,J).LT.HMIN)HMIN=H(I,J)
      IF(P(I,J).GT.PMAX)PMAX=P(I,J)
      ENDDO
      ENDDO
      HMIN=HMIN*B*B/RX
      PMAX=PMAX*PH
      WRITE(10,*)'HMIN,PMAX,TMAX',HMIN,PMAX,TMAX
      RETURN
      END
```

16.3.3 Calculation Results

Under the given working conditions in the program, the results of film thickness and pressure distributions are shown in Figure 16.2.

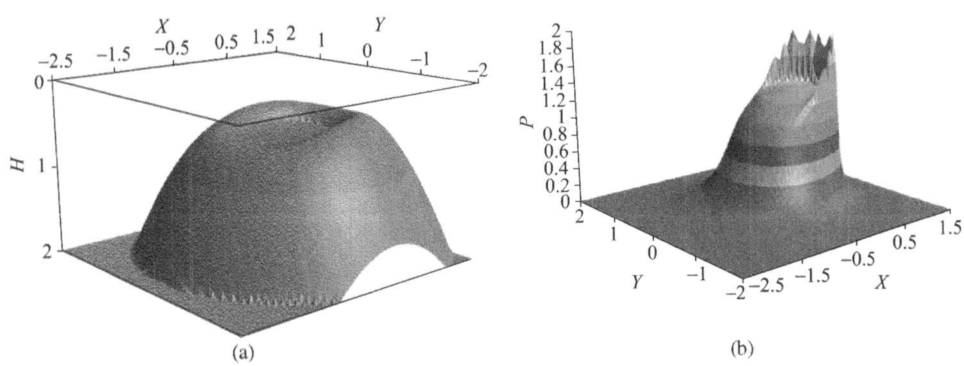

Figure 16.2 Film thickness (a) and pressure distribution (b) of GEHL in point contact ($n = 0.9$).

17

Numerical Method and Program for Thermal EHL in Line Contact

17.1 Basic Equations

The basic equations for thermal elastohydrodynamic lubrication (TEHL) in line contact include all the equations of elastohydrodynamic lubrication (EHL) and a two-dimensional energy equation (in the velocity direction and the film thickness direction). These main equations are as follows.

17.1.1 Reynolds Equation

$$\frac{d}{dx}\left(\frac{\rho h^3}{\eta}\frac{dp}{dx}\right) = 12U\frac{d\rho h}{dx} \tag{17.1}$$

where ρ is the lubricant density, η is the lubricant viscosity, U is the average velocity [$U=(u_1+u_2)/2$; u_1 and u_2 are the tangential velocities of the two surfaces], p is pressure, h is film thickness and x is the coordinate along the motion.

17.1.2 Energy Equation

$$\rho c_p u\frac{\partial T}{\partial x} + \frac{T}{\rho}\frac{\partial \rho}{\partial T}\left(u\frac{\partial p}{\partial x}\right) - K\frac{\partial^2 T}{\partial z^2} = \eta\left(\frac{\partial u}{\partial z}\right)^2 \tag{17.2}$$

where T is temperature, u is flow velocity, c_p is specific heat capacity at constant pressure, K is thermal conductivity and z is the coordinate across the film.

Numerical Calculation of Lubrication: Methods and Programs, First Edition. Ping Huang.

Because $\frac{\partial^2 T}{\partial z^2}$ is of the second derivative, two boundary conditions should be given for solving the energy Equation 17.2. Usually, they can be expressed as follows (see Yang, 1998).

$$\begin{aligned} T(x,0) &= \frac{K}{\sqrt{\pi \rho_1 c_1 K_1 u_1}} \int_{-\infty}^{x} \left.\frac{\partial T}{\partial z}\right|_{x,0} \frac{ds}{\sqrt{x-s}} + T_0 \\ T(x,h) &= \frac{K}{\sqrt{\pi \rho_2 c_2 K_2 u_2}} \int_{-\infty}^{x} \left.\frac{\partial T}{\partial z}\right|_{x,h} \frac{ds}{\sqrt{x-s}} + T_0 \end{aligned} \tag{17.3}$$

where T_0 is ambient temperature, the subscripts 1 and 2 correspond to the down and up surfaces, respectively, ρ_1 and ρ_2 are material densities, c_1 and c_2 are specific heats and K_1 and K_2 are heat transfer coefficients.

17.1.3 Film Thickness Equation

$$h(x) = h_c + \frac{x^2}{2R} + v(x) \tag{17.4}$$

where h_c is the central film thickness to be determined by calculation, R is the equivalent radius of the surfaces $\left(\frac{1}{R} = \frac{1}{R_1} + \frac{1}{R_2}\right)$, and $v(x)$ is the elastic deformation.

17.1.4 Elastic Deformation Equation

In the film thickness Equation 17.4, the deformation $v(x)$ is

$$v(x) = -\frac{2}{\pi E} \int_{x_0}^{x_e} p(s) \ln (s-x)^2 ds + c \tag{17.5}$$

where E is the equivalent elastic modulus $\left[\text{i.e.,} \frac{1}{E} = \frac{1}{2}\left(\frac{1-\nu_1^2}{E_1} + \frac{1-\nu_2^2}{E_2}\right)\right]$, ν_1 and ν_2 are the Poisson's ratios of the two surfaces, E_1 and E_2 are the elastic modulus of the two surfaces, x_0 is the inlet coordinate, x_e is the outlet coordinate and c is an integral constant.

17.1.5 Roelands Viscosity–Pressure–Temperature Equation

$$\eta = \eta_0 e^{-\beta T} \exp\left\{ (\ln \eta_0 + 9.67)\left[(1 + 5.1 \times 10^{-9} p)^{0.68} - 1 \right] \right\} \tag{17.6}$$

where η_0 is the lubricant viscosity at $p = 0$ and $T = T_0$, β is the constant of viscosity and temperature.

17.1.6 Density–Pressure–Temperature Equation

$$\rho = \rho_0\left[1 + \frac{0.6p}{1+1.7p} + D(T - T_0)\right] \tag{17.7}$$

where ρ_0 is the lubricant density at $p=0$ and $T=T_0$, the unit of pressure p is GPa and $D=-0.00065\ \mathrm{K}^{-1}$.

Further, in order to calculate the temperature, not only must the average velocity U of the two surfaces be known, but also both surface velocities u_1 and u_2 should be calculated. If the slip-roll ratio s is known, we can use Equation 17.8 to calculate them.

$$\begin{aligned} u_1 &= 0.5 \times (2+s) \times U \\ u_2 &= 0.5 \times (2-s) \times U \end{aligned} \tag{17.8}$$

17.2 Numerical Method

17.2.1 Dimensionless Equations

The dimensionless Reynolds equation of EHL in line contact is

$$\frac{d}{dX}\left(\varepsilon\frac{dP}{dX}\right) - \frac{d(\rho^* H)}{dX} = 0 \tag{17.9}$$

where $\varepsilon = \frac{\rho^* H^3}{\eta^*\lambda}$, $\lambda = \frac{12\eta_0 U R^2}{b^2 p_H}$, η^* is the dimensionless lubricant viscosity ($\eta^* = \eta/\eta_0$), ρ^* is the dimensionless density, X is the dimensionless coordinate ($X = x/b$; b is the half length of the contact area), P is the dimensionless pressure ($P = p/p_H$; p_H is the maximum Hertzian stress) and H is the dimensionless film thickness $\left(H = \frac{hR}{b^2}\right)$.

The boundary conditions are:

$$\begin{aligned} &\text{At the inlet } P(X_0) = 0 \\ &\text{At the outlet } P(X_e) = 0 \text{ and } \frac{dP(X_e)}{dX} = 0 \end{aligned}$$

where X_0 is the dimensionless inlet coordinate ($X_0 = x_0/b$) and X_e is the dimensionless outlet coordinate ($X_e = x_e/b$).

The dimensionless energy equation is

$$A\rho^* u^* \frac{\partial T}{\partial X} + B\frac{u^* T^*}{\rho^*}\frac{\partial \rho^*}{\partial T^*}\frac{\partial P}{\partial X} + C\frac{\partial^2 T}{\partial Z^2} + D\eta^*\left(\frac{\partial u^*}{\partial Z}\right)^2 = 0 \tag{17.10}$$

where T^* is the dimensionless temperature ($T^* = T/T_0$; T_0 is ambient temperature), $A = \rho_0 c_p U T_0/b$, $B = U p_H/b$, $C = -K T_0 R/b^2$, $D = -\eta_0 \left(\frac{UR}{b^2}\right)^2$, u^* is the dimensionless flow velocity ($u^* = u/U$) and Z is the dimensionless coordinate across the film thickness ($Z = z/h$).

The dimensionless film thickness equation is

$$H(X) = H_0 + \frac{X^2}{2} - \frac{1}{\pi}\int_{X_0}^{X_e} \ln|X - X'|P(X')dX' \tag{17.11}$$

where H_0 is the dimensionless film thickness of $h_0 = h_c + c$, which will be determined in calculation.

The dimensionless viscosity–pressure–temperature relationship is

$$\eta^* = \exp\left[-0.05\left(T - T_0\right)\right]\exp\left\{[\ln(\eta_0) + 9.67]\left[-1 + \left(1 + 5.1 \times 10^{-9} P \cdot p_H\right)^{0.68}\right]\right\} \tag{17.12}$$

where η^* is the dimensionless viscosity of lubricant ($\eta^* = \eta/\eta_0$).

The dimensionless density–pressure–temperature relationship is

$$\rho^* = \left(1 + \frac{0.6p}{1 + 1.7p} + D(T - T_0)\right) \tag{17.13}$$

where ρ^* is the dimensionless viscosity of lubricant ($\rho^* = \rho/\rho_0$).

The dimensionless load equation is

$$W = \int_{X_0}^{X_e} P dx = \frac{\pi}{2} \tag{17.14}$$

17.2.2 Discrete Equations

By using the central and forward differential discretization equations (Equations 1.21 and 1.22), we can obtain the discrete differential Reynolds equation in line contact. The discrete forms of the energy equation, the film thickness equation and the load equation can also be obtained by the numerical integration method. These equations can be written as

$$\frac{\varepsilon_{i-1/2}P_{i-1} - (\varepsilon_{i-1/2} + \varepsilon_{i+1/2})P_i + \varepsilon_{i+1/2}P_{i+1}}{\Delta X^2} = \frac{\rho_i^* H_i - \rho_{i-1}^* H_{i-1}}{\Delta X} \tag{17.15}$$

where $\varepsilon_{i\pm1/2} = \frac{1}{2}(\varepsilon_i + \varepsilon_{i\pm1})$ and $\Delta X = X_i - X_{i-1}$, where we use the equidistant grid, that is, ΔX is constant.

The inlet boundary condition is $P|_{X=0} = 0$; and the outlet boundary condition will be achieved by setting the negative pressure to zero.

The discrete form of the energy equation can be written as

$$A\rho^* u^* \left(\frac{T_{i,k} - T_{i-1,k}}{\Delta X}\right) + B\frac{T^*}{\rho^*}\frac{\partial \rho^*}{\partial T^*}\left(u^* \frac{P_i - P_{i-1}}{\Delta X}\right)$$

$$+ C\frac{T_{i,k+1} - 2T_{i,k} + T_{i,k-1}}{\Delta Z^2} + D\eta^* \left(\frac{u^*_{i,k+1} - u^*_{i,k}}{\Delta Z}\right)^2 = 0 \qquad (17.16)$$

where the subscript k indicates the node number in the thickness direction. ρ^*, η^* and $\frac{\partial \rho^*}{\partial T^*}$ can be obtained by the expressions shown before, so we do not need to discretize them. The film thickness is divided into k layers and ΔZ is the dimensionless distance.

The discrete film thickness equation is

$$H_i = H_0 + \frac{X_i^2}{2} - \frac{1}{\pi}\sum_{j=1}^{n} K_{ij} P_j \qquad (17.17)$$

where K_{ij} is the stiffness coefficient of the elastic deformation.

The discrete load equation is

$$\Delta X \sum_{i=1}^{n} \frac{(P_i + P_{i+1})}{2} = \frac{\pi}{2} \qquad (17.18)$$

The following convergent criterion of pressure is used to determine the convergency.

$$\frac{\sum |P_i^{k+1} - P_i^k|}{\sum P_i^{k+1}} \le \varepsilon_4 \qquad (17.19)$$

where ε_4 is a given error.

Because the temperature in the energy Equation 17.16, the viscosity in the viscous–pressure–temperature Equation 17.12 and the elastic deformation in the film thickness Equation 17.17 vary with pressure, for the general solving method, we should first give an initial pressure distribution (such as the Hertz contact pressure distribution) and an initial temperature distribution (such as the uniform temperature distribution). Then, we can use Equations 17.17 and 17.12 to calculate a new thickness and a new viscosity. After that, we substitute them into the Reynolds Equation 17.15 to obtain a new pressure distribution with the previous pressure distribution by the iterative method. Then, we substitute the new pressure distribution into the energy Equation 17.16 to calculate a new temperature distribution. We use the new temperature distribution to correct the viscosity iteratively by Equation 17.12 and to iterate another new pressure distribution by Equation 17.15 again. We repeat the above steps until the pressure difference between two neighboring iterations is less than the given error, as shown in Equation 17.19. We finally obtain the temperature distribution, the pressure distribution and the film thickness of thermal EHL in line contact.

17.3 Calculation Program

In the following calculation program of TEHL in line contact, beside the methods for solving the Reynolds equation and the elastic deformation, the multi-grid method and the temperature equation solving method are used. The following block diagrams highlight the two solving methods.

17.3.1 Calculation Diagram of Multigrid Method

The calculation diagram of the multigrid method is shown in Figure 17.1. In this diagram, KG is a switch variable, G(k) is the dimensionless load of the *k*th grid, H is

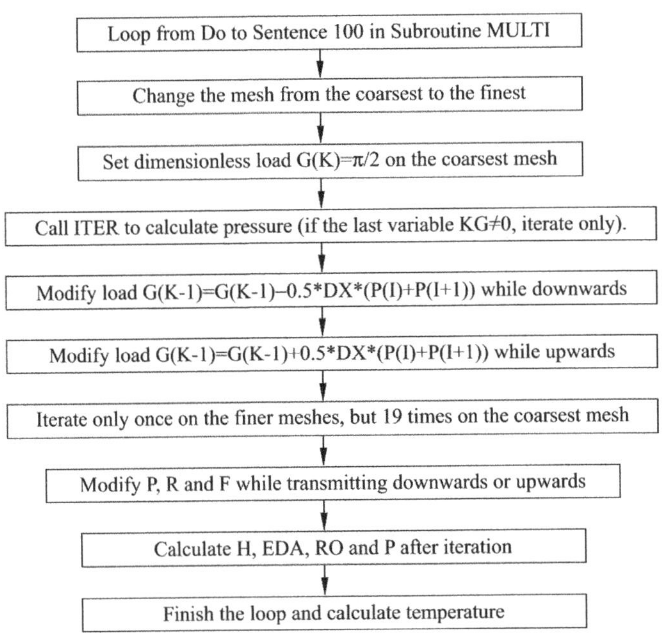

Figure 17.1 Calculation diagram of multigrid method.

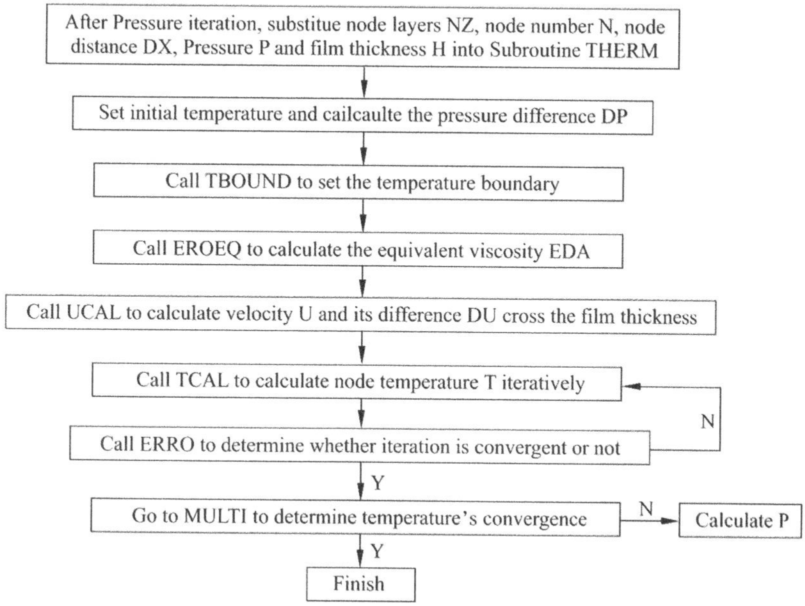

Figure 17.2 Calculation diagram of temperature.

the dimensionless film thickness, EDA is the dimensionless viscosity, RO is the dimensionless density, P is the dimensionless pressure, R is the dimensionless error and F is the dimensionless residual of the discrete Reynolds equation.

17.3.2 Calculation Diagram of Temperature

After the THERM subroutine, the program will calculate the temperature of the film thickness. It consists mainly of the master program THERM for temperature calculating, the temperature boundary condition subroutine TBOUD, the equivalent viscosity subroutine EROEQ, the velocity distribution calculation subroutine UCAL, the iteratively calculating temperature subroutine TCAL and the error calculating subroutine ERRO. The calculation block diagram is shown in Figure 17.2.

In the diagram, NZ is the number of layers across the film thickness, N is the number of nodes in the x direction, DP is the maximum difference of two iteration pressures, DU is the velocity difference between two neighboring layers and T is the dimensionless temperature.

The total calculation diagram to calculate TEHL in line contact is shown in Figure 17.3. In the diagram, AK is the coefficient array of the elastic deformation.

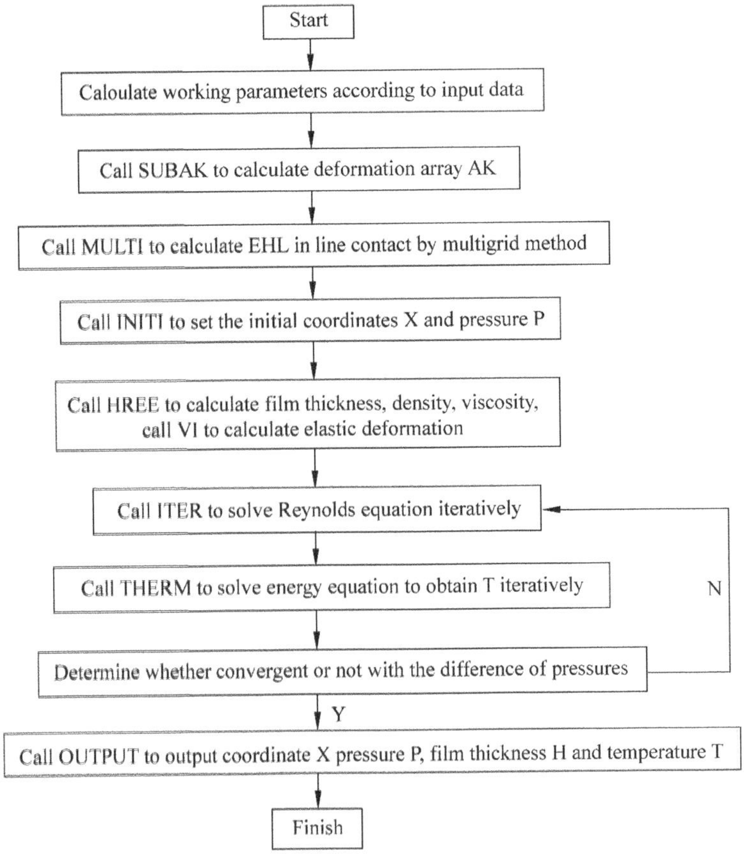

Figure 17.3 Calculation diagram of TEHL in line contact.

17.3.3 Calculation Program

Preassignment parameters:

Node number	N = 129
Dimensionless inlet node coordinate	X0 = −4.0
Dimensionless outlet node coordinate	XE = 1.4
Load	W = 1.768E5 N
Equivalent modulus of elasticity	E1 = 2.21E11Pa
Initial viscosity	EDA0 = 0.03 Pa s
Equivalent radius of cylinder	R = 0.012183 m
Velocity	US = 1.77 m s^{-1}
Layer number cross the film thickness	NZ = 5

Ratio of slide to roll	CU = 0.25
Iteration coefficient	C1 = 0.37
Iteration coefficient	C2 = 0.37
Temperature iteration coefficient	CT = 0.35

In addition, in BLOCK DATA the following parameters are assigned.

Initial temperature	T0 = 303 K
Thermal conductivity of lubricant	AK0 = 0.14 W m^{-1} K
Thermal conductivity of down surface	AK1 = 46 W m^{-1} K
Thermal conductivity of up surface	AK2 = 46 W m^{-1} K
Specific heat of lubricant	CV = 2000 J kg^{-1} K^{-1}
Specific heats of down surface	CV1 = 470 J kg^{-1} K^{-1}
Specific heats of up surface	CV2 = 470 J kg^{-1} K^{-1}
Initial density of lubricant	RO0 = 890 kg m^{-3}
Densities of down surface	RO1 = 7850 kg m^{-3}
Densities of down surface	RO2 = 7850 kg m^{-3}
Temperature–viscosity coefficient	S0 = −1.1
Temperature–density coefficient of lubricant	D0 = −0.00065.

If we need to change the above parameters, we need to recompile and relink the program before executing it.

Input parameters:

Input Character S. If S = "Y" or "y", calculate the example. If S = "N" or "n", then input the following data:

```
                    Input: KT= If KT=2 consider temperature or not
consider temperature
Input: N,X0,XE,W,E1,EDA0,R,US=
If consider temperature, that is, KT=2, then input: NZ,CU=
                    If change the iteration factors, input: Y, and
then input: C1,C2=
Velocity                                              US=
```

Output parameters:

In the file OUT.DAT, the following data will be output in column:

```
Coordinate array        Pressure array              Film thickness array
X(1)                    P(1)                  H(1)
X(2)                    P(2)                  H(2)
......                  ......                ......
X(N)                    P(N)                  H(N)
```

In the file OUPT.DAT, the temperature array will be output in columns:

```
Coordinate array  Temperature in layer
X(1)                 T(1,1)          T(1,2)          T(1,3)
        T(1,4)            T(1,5)
X(2)                 T(2,1)          T(2,2)          T(2,3)
        T(2,4)            T(2,5)
......                ......          ......                ......
X(N)                 T(N,1)          T(N,2)          T(N,3)
        T(N,4)            T(N,5)
```

```
PROGRAM LINEEHLT
CHARACTER*1 S,S1,S2
      CHARACTER*16 FILEO,CDATE,CTIME
      COMMON /COM1/ENDA,A1,A2,A3,Z,C1,C2,C3,CW
      COMMON /COM2/T0,EDA0,AK0,AK1,AK2,CV,CV1,CV2,RO0,RO1,RO2,D0
      COMMON/COM3/E1,PH,B,U1,U2,R,CT/COM4/X0,XE/COM5/H2,P2,T2,
ROM,HM,FM
      DATA PAI,Z,P0/3.14159265,0.68,1.96E8/,KT,S1,S2/0,1HY,1Hy/
      DATA N,X0,XE,W,E1,EDA0,R,US,C1,C2,NZ,CU,CT/129,-4.,1.4,
1.768E5,2.21E11,0.03,0.02,1.77,0.37,0.37,5,0.25,0.35/
      OPEN(8,FILE='OUT.DAT',STATUS='UNKNOWN')
      WRITE(*,*)'Show the example or not (Y or N)?'
      READ(*,'(A)')S
      IF(S.EQ.S1.OR.S.EQ.S2)THEN
      KT=2
      GOTO 10
      ELSE
      WRITE(*,*)' Temperature is considered or not (Y or N) ?'
      READ(*,'(A)')S
      IF(S.EQ.S1.OR.S.EQ.S2)KT=2
      ENDIF
      WRITE(*,*)'N,X0,XE,W,E,EDA0,R,US='
      READ(*,*)N,X0,XE,W,E1,EDA0,R,US
      IF(KT.EQ.2)THEN
      WRITE(*,*)'NZ,CU='
      READ(*,*)NZ,CU
      ENDIF
      WRITE(*,*)' Change iteration factors or not (Y or N) ?'
      READ(*,'(A)')S
      IF(S.EQ.S1.OR.S.EQ.S2)THEN
      WRITE(*,*)'C1,C2='
      READ(*,*)C1,C2
      ENDIF
 10   CW=N+0.1
      LMAX=ALOG(CW)/ALOG(2.)
      N=2**LMAX+1
      LMIN=(ALOG(CW)-ALOG(SQRT(CW)))/ALOG(2.)
      LMAX=LMIN
      H00=0.0
```

```
      W1=W/(E1*R)
      PH=E1*SQRT(0.5*W1/PAI)
      A1=(ALOG(EDA0)+9.67)
      A2=PH/P0
      A3=0.59/(PH*1.E-9)
      T2=0.0
      B=4.*R*PH/E1
      ALFA=Z*A1/P0
      G=ALFA*E1
      U=EDA0*US/(2.*E1*R)
      CC1=SQRT(2.*U)
      AM=2.*PAI*(PH/E1)**2/CC1
      AL=G*SQRT(CC1)
      CW=(PH/E1)*(B/R)
      C3=1.6*(R/B)**2*G**0.6*U**0.7*W1**(-0.13)
      ENDA=3.*(PAI/AM)**2/8.
      U1=0.5*(2.+CU)*U
      U2=0.5*(2.-CU)*U
      CW=-1.13*C3
      WRITE(*,40)
40 FORMAT(2X,'                    Wait       Please',//)
      CALL SUBAK(N)
      CALL MULTI(N,NZ,KT,LMIN,LMAX,H00)
      STOP
      END
      SUBROUTINE MULTI(N,NZ,KT,LMIN,LMAX,H00)
      DIMENSION X(1100),P(1100),H(1100),RO(1100),POLD(1100),EPS(1100), EDA(1100),
P0(2200),F(1100),F0(2200),R(1100),R0(2200),
G(10),T(22000)
      COMMON /COM1/ENDA,A1,A2,A3,Z,C1,C2,C3,CW
      COMMON/COMK/K/COMT/LT,T1(1100)/COM3/E1,PH,B,U1,U2,RR,CT/COM5/
H2,P2,T2,RM,HM,FM
      DATA MK,IT,KH,NMAX,PAI,G0/0,0,0,1100,3.14159265,1.570796325/
      LT=LMAX
      NX=N
      K=LMIN
      N0=(N-1)/2**(LMIN-1)
      CALL KNDX(K,N,N0,N1,NMAX,DX,X)
      DO 10 I=1,N
      T1(I)=1.0
      IF(ABS(X(I)).GE.1.0)P(I)=0.0
10    IF(ABS(X(I)).LT.1.0)P(I)=SQRT(1.-X(I)*X(I))
12    CALL HREE(N,DX,H00,G0,X,P,H,RO,EPS,EDA,F,0)
      IF(KH.NE.0)GOTO 14
      KH=1
      GOTO 12
14    CALL FZ(N,P,POLD)
      DO 100 L=LMIN,LMAX
      K=L
```

```
      G(K)=PAI/2.
      DO 18 I=1,N
      R(I)=0.0
      F(I)=0.0
      RO(N1+I)=0.0
18    FO(N1+I)=0.0
20    KK=2
      CALL ITER(N,KK,DX,H00,G0,X,P,H,RO,EPS,EDA,F,R,0)
      KK=1
      CALL ITER(N,KK,DX,H00,G0,X,P,H,RO,EPS,EDA,F,R,1)
      G(K-1)=G(K)
      DO 24 I=1,N
      IF(I.LT.N)G(K-1)=G(K-1)-0.5*DX*(P(I)+P(I+1))
24    PO(N1+I)=P(I)
      N2=N
      K=K-1
      CALL KNDX(K,N,N0,N1,NMAX,DX,X)
      CALL TRANS(N,N2,P,H,RO,EPS,EDA,R)
      CALL ITER(N,KK,DX,H00,G0,X,P,H,RO,EPS,EDA,F,R,2)
      DO 26 I=1,N
      IF(I.LT.N)G(K)=G(K)+0.5*DX*(P(I)+P(I+1))
26    F(I)=H(I)
      G0=G(K)
      CALL HREE(N,DX,H00,G0,X,P,H,RO,EPS,EDA,F,1)
      DO 28 I=1,N
      RO(N1+I)=R(I)
28    FO(N1+I)=F(I)
      IF(K.NE.1)GOTO 20
      KK=19
      CALL ITER(N,KK,DX,H00,G0,X,P,H,RO,EPS,EDA,F,R,0)
40    DO 42 I=1,N
42    PO(N1+I)=P(I)
      N2=N1
      K=K+1
      CALL KNDX(K,N,N0,N1,NMAX,DX,X)
      G0=G(K)
      DO 50 I=2,N,2
      I1=N1+I
      I2=N2+I/2
      P(I-1)=PO(I2)
      P(I)=PO(I1)+0.5*(PO(I2)+PO(I2+1)-PO(I1-1)-PO(I1+1))
50    IF(P(I).LT.0.0)P(I)=0.
      DO 52 I=1,N
      R(I)=RO(N1+I)
52    F(I)=FO(N1+I)
      CALL HREE(N,DX,H00,G0,X,P,H,RO,EPS,EDA,F,0)
      KK=1
      CALL ITER(N,KK,DX,H00,G0,X,P,H,RO,EPS,EDA,F,R,0)
      IF(K.LT.L)GOTO 40
```

```
100   CONTINUE
      MK=MK+1
      CALL ERROP(N,P,POLD,ERP)
      IF(ERP.GT.0.01*C2.AND.MK.LE.12)GOTO 14
      MK=8
      IF(KT.NE.2)GOTO 105
      CALL THERM(NX,NZ,DX,T,P,H)
      CALL ERROM(NX,NZ,T1,T,KT)
      IT=IT+1
      IF(KT.EQ.2.AND.IT.LT.10)GOTO 14
      KT=2
      IF(IT.GE.10)THEN
      WRITE(*,*)'Temperature is not convergent !!!'
      READ(*,*)
      ENDIF
105   IF(MK.GE.10)THEN
      WRITE(*,*)'Pressures are not convergent !!!'
      READ(*,*)
      ENDIF
      FM=FRICT(N,DX,H,P,EDA)
      DO I=1,N
      WRITE(8,110)X(I),P(I),H(I)
      H(I)=H(I)*B*B/RR
      P(I)=P(I)*PH
      ENDDO
110   FORMAT(1X,6(E12.6,1X))
      DO I=2,N-1
      IF(P(I).GE.P(I-1).AND.P(I).GE.P(I+1))THEN
      HM=H(I)
      RM=RO(I)
      GOTO 120
      ENDIF
      ENDDO
120   DO I=1,N
      H(I)=H(I)*1.E6
      P(I)=P(I)*1.E-9
      ENDDO
      IF(KT.EQ.2)THEN
      CALL OUPT(NX,NZ,X,T)
      ENDIF
      RETURN
      END
      SUBROUTINE HREE(N,DX,H00,G0,X,P,H,RO,EPS,EDA,F0,KG)
      DIMENSION X(N),P(N),H(N),RO(N),EPS(N),EDA(N),F0(N)
      DIMENSION W(2200)
      COMMON /COM1/ENDA,A1,A2,A3,Z,C1,C2,C3,CW/COMK/K/COMT/LT,T1(1100)/COM2/
      T0,EDA0,AK0,AK1,AK2,CV,CV1,CV2,RO0,RO1,RO2,D0/COMAK/AK(0:1100)
      DATA KK,MK1,MK2,NW,PAI1/0,3,0,2200,0.318309886/
      IF(KK.NE.0)GOTO 3
```

```
      HM0=C3
3     W1=0.0
      DO 4 I=1,N
4     W1=W1+P(I)
      C3=(DX*W1)/G0
      DW=1.-C3
      IF(K.EQ.1)GOTO 6
      CALL VI(N,DX,P,W)
      GOTO 10
6     WX=-PAI1*W1*DX*ALOG(DX)
      DO 8 I=1,N
      W(I)=WX
      DO 8 J=1,N
      IJ=IABS(I-J)
8     W(I)=W(I)-PAI1*AK(IJ)*P(J)*DX
10    HMIN=1.E3
      DO 30 I=1,N
      H0=0.5*X(I)*X(I)+W(I)
      IF(KG.EQ.1)GOTO 20
      IF(H0+F0(I).LT.HMIN)HMIN=H0+F0(I)
      H(I)=H0
      GOTO 30
20    F0(I)=F0(I)-H00-H0
30    CONTINUE
      IF(KG.EQ.1)RETURN
      H0=H00+HMIN
      IF(KK.NE.0)GOTO 32
      KK=1
      H00=-H0+HM0
32    IF(H0.LE.0.0)GOTO 48
      IF(K.NE.1)GOTO 50
40    MK=MK+1
      IF(MK.LE.MK1)GOTO 50
      IF(MK.GE.MK2)MK=0
      IF(H0+CW*DW.GT.0.0)HM0=H0+CW*DW
      IF(H0+CW*DW.LE.0.0)HM0=HM0*C3
48    H00=HM0-HMIN
50    DO 60 I=1,N
60    H(I)=H00+H(I)+F0(I)
      IT=2**(LT-K)
      DO 100 I=1,N
      II=IT*(I-1)+1
      CT1=-0.05*T0*(T1(II)-1.0)
      CT2=D0*T0*(T1(II)-1.)
      EDA(I)=EXP(CT1)*EXP(A1*(-1.+(1.+A2*P(I))**Z))
      RO(I)=(A3+1.34*P(I))/(A3+P(I))+CT2
      EPS(I)=RO(I)*H(I)**3/(ENDA*EDA(I))
100   CONTINUE
      RETURN
```

```
      END
      SUBROUTINE ITER(N,KK,DX,H00,G0,X,P,H,RO,EPS,EDA,F0,R0,KG)
      DIMENSION X(N),P(N),H(N),RO(N),EPS(N),EDA(N),F0(N),R0(N)
      COMMON /COM1/ENDA,A1,A2,A3,Z,C1,C2,C3/COMAK/AK(0:1100)
      DATA PAI/3.14159265/
      DX1=1./DX
      DX2=DX*DX
      DX3=1./DX2
      DX4=DX1/PAI
      DXL=DX*ALOG(DX)
      AK0=DX*AK(0)+DXL
      AK1=DX*AK(1)+DXL
      DO 100 K=1,KK
      RMAX=0.0
      D2=0.5*(EPS(1)+EPS(2))
      D3=0.5*(EPS(2)+EPS(3))
      D5=DX1*(RO(2)*H(2)-RO(1)*H(1))
      D7=DX4*(RO(2)*AK0-RO(1)*AK1)
      PP=0.
      DO 70 I=2,N-1
      D1=D2
      D2=D3
      D4=D5
      D6=D7
      IF(I+2.LE.N)D3=0.5*(EPS(I+1)+EPS(I+2))
      D5=DX1*(RO(I+1)*H(I+1)-RO(I)*H(I))
      D7=DX4*(RO(I+1)*AK0-RO(I)*AK1)
      IF(KG.NE.0)GOTO 30
      DD=(D1+D2)*DX3
      IF(DD.LT.0.1*ABS(D6))GOTO 10
      RI=-DX3*(D1*P(I-1)-(D1+D2)*P(I)+D2*P(I+1))+D4+R0(I)
      DLDP=-DX3*(D1+D2)+D6
      RI=C1*RI/DLDP
      GOTO 20
10    RI=-DX3*(D1*PP-(D1+D2)*P(I)+D2*P(I+1))+D4+R0(I)
      DLDP=-DX3*(2.*D1+D2)+2.*D6
      RI=C2*RI/DLDP
      IF(I.GT.2.AND.P(I-1)-RI.GT.0.0)P(I-1)=P(I-1)-RI
20    PP=P(I)
      P(I)=P(I)+RI
      IF(P(I).LT.0.0)P(I)=0.0
      IF(K.NE.KK)GOTO 70
      IF(RMAX.LT.ABS(RI).AND.P(I).GT.0.0)RMAX=ABS(RI)
      GOTO 70
30    IF(KG.EQ.2)GOTO 40
      R0(I)=-DX3*(D1*P(I-1)-(D1+D2)*P(I)+D2*P(I+1))+D4+R0(I)
      GOTO 70
40    R0(I)=DX3*(D1*P(I-1)-(D1+D2)*P(I)+D2*P(I+1))-D4+R0(I)
70    CONTINUE
```

```
      IF(KG.NE.0)GOTO 100
      CALL HREE(N,DX,H00,G0,X,P,H,RO,EPS,EDA,F0,0)
100   CONTINUE
      RETURN
      END
      SUBROUTINE VI(N,DX,P,V)
      DIMENSION P(N),V(N)
      COMMON /COMAK/AK(0:1100)
      PAI1=0.318309886
      C=ALOG(DX)
      DO 10 I=1,N
      V(I)=0.0
      DO 10 J=1,N
      IJ=IABS(I-J)
10    V(I)=V(I)+(AK(IJ)+C)*DX*P(J)
      DO I=1,N
      V(I)=-PAI1*V(I)
      ENDDO
      RETURN
      END
      SUBROUTINE SUBAK(MM)
      COMMON /COMAK/AK(0:1100)
      DO 10 I=0,MM
10    AK(I)=(I+0.5)*(ALOG(ABS(I+0.5))-1.)-(I-0.5)*(ALOG(ABS(I-0.5))-1.)
      RETURN
      END
      FUNCTION FRICT(N,DX,H,P,EDA)
      DIMENSION H(N),P(N),EDA(N)
      COMMON /COM3/E1,PH,B,U1,U2,R,CT
      DATA TAU0/4.E7/
      TP=TAU0/PH
      TE=TAU0/E1
      BR=B/R
      FRICT=0.0
      DO I=1,N
      DP=0.0
      IF(I.NE.N)DP=(P(I+1)-P(I))/DX
      TAU=0.5*H(I)*ABS(DP)*(BR/TP)+2.*ABS(U1-U2)*EDA(I)/(H(I)
      *BR**2*TE)
      FRICT=FRICT+TAU
      ENDDO
      FRICT=FRICT*DX*B*TAU0
      RETURN
      END
      SUBROUTINE FZ(N,P,POLD)
      DIMENSION P(N),POLD(N)
      DO 10 I=1,N
10    POLD(I)=P(I)
      RETURN
```

```
      END
      SUBROUTINE ERROP(N,P,POLD,ERP)
      DIMENSION P(N),POLD(N)
      SD=0.0
      SUM=0.0
      DO 10 I=1,N
      SD=SD+ABS(P(I)-POLD(I))
10    SUM=SUM+P(I)
      ERP=SD/SUM
      RETURN
      END
      SUBROUTINE KNDX(K,N,N0,N1,NMAX,DX,X)
      DIMENSION X(NMAX)
      COMMON /COM4/X0,XE
      N=2**(K-1)*N0
      DX=(XE-X0)/N
      N=N+1
      N1=N+K
      DO 10 I=1,N
10    X(I)=X0+(I-1)*DX
      RETURN
      END
      SUBROUTINE TRANS(N1,N2,P,H,RO,EPS,EDA,R)
      DIMENSION P(N2),H(N2),RO(N2),EPS(N2),EDA(N2),R(N2)
      DO 10 I=1,N1
      II=2*I-1
      P(I)=P(II)
      H(I)=H(II)
      R(I)=R(II)
      RO(I)=RO(II)
      EPS(I)=EPS(II)
10    EDA(I)=EDA(II)
      RETURN
      END
      SUBROUTINE OUPT(NX,NZ,X,T)
      DIMENSION X(NX),T(NX,NZ)
      DO I=1,NX
      DO K=1,NZ
      T(I,K)=303.*(T(I,K)-1.0)
      END DO
      END DO
      DO I=1,NX
      WRITE(8,30)X(I),(T(I,K),K=1,NZ)
      ENDDO
30    FORMAT(6(1X,E12.6))
      RETURN
      END
      SUBROUTINE ERROM(NX,NZ,T1,T,KT)
      DIMENSION T(NX,NZ),T1(NX)
```

```
      KT=2
      ERM=0.
      C1=1./FLOAT(NZ)
      DO 20 I=1,NX
      TT=0.
      DO 10 K=1,NZ
10    TT=TT+T(I,K)
      TT=C1*TT
      ER=ABS((TT-T1(I))/TT)
      IF(ER.GT.ERM)ERM=ER
20    T1(I)=TT
      IF(ERM.LT.0.003)KT=1
      RETURN
      END
      SUBROUTINE THERM(NX,NZ,DX,T,P,H)
      DIMENSION T(NX,NZ),P(NX),H(NX),T1(21),TI(21),TF(21),U(21),DU
      (21),W(21),EDA(21),RO(21),EDA1(21),EDA2(21),ROR(21),UU(21)
      DATA KK/0/
      IF(KK.NE.0)GOTO 4
      DO 2 K=1,NZ
2     T(1,K)=1.0
4     DO 30 I=2,NX
      KG=0
      DO 8 K=1,NZ
      TF(K)=T(I-1,K)
      IF(KK.NE.0)GOTO 6
      T1(K)=T(I-1,K)
      GOTO 8
6     T1(K)=T(I,K)
8     TI(K)=T1(K)
      P1=P(I)
      H1=H(I)
      DP=(P(I)-P(I-1))/DX
      CALL TBOUD(NX,NZ,I,CC1,CC2,T)
10    CALL EROEQ(NZ,T1,P1,EDA,RO,EDA1,EDA2,KG)
      CALL UCAL(NZ,DX,H1,EDA,RO,ROR,EDA1,EDA2,U,UU,DU,W,DP)
      CALL TCAL(NZ,DX,CC1,CC2,T1,TF,U,W,DU,H1,DP,EDA,RO)
      CALL ERRO(NZ,TI,T1,ETS)
      KG=KG+3
      IF(ETS.GT.1.E-4.AND.KG.LE.50)GOTO 10
      DO 20 K=1,NZ
      ROR(K)=RO(K)
      UU(K)=U(K)
20    T(I,K)=T1(K)
30    CONTINUE
      KK=1
      RETURN
      END
      SUBROUTINE TBOUD(NX,NZ,I,CC1,CC2,T)
```

```
      DIMENSION T(NX,NZ)
      CC1=0.
      CC2=0.
      DO 10 L=1,I-1
      DS=1./SQRT(FLOAT(I-L))
      IF(L.EQ.I-1)DS=1.1666667
      CC1=CC1+DS*(T(L,2)-T(L,1))
10    CC2=CC2+DS*(T(L,NZ)-T(L,NZ-1))
      RETURN
      END
      SUBROUTINE ERRO(NZ,T0,T,ETS)
      DIMENSION T0(NZ),T(NZ)
      ETS=0.0
      DO 10 K=1,NZ
      IF(T(K).LT.1.E-5)ETS0=1.
      IF(T(K).GE.1.E-5)ETS0=ABS((T(K)-T0(K))/T(K))
      IF(ETS0.GT.ETS)ETS=ETS0
10    T0(K)=T(K)
      RETURN
      END
      SUBROUTINE EROEQ(NZ,T,P,EDA,RO,EDA1,EDA2,KG)
      DIMENSION T(NZ),EDA(NZ),RO(NZ),EDA1(NZ),EDA2(NZ)
      COMMON /COM1/ENDA,A1,A2,A3,Z,O1,O2,O3/COM2/T0,EDA0,AK0,AK1,AK2,
      CV,CV1,CV2,RO0,RO1,RO2,D0/COM3/E1,PH,B,U1,U2,R,CC
      DATA A4,A5/0.455445545,0.544554455/
      IF(KG.NE.0)GOTO 20
      B1=(1.+A2*P)**Z
      B2=(A3+1.34*P)/(A3+P)
      B3=-0.05*T0
20    DO 30 K=1,NZ
      EDA(K)=EXP(A1*(-1.+B1))*EXP(B3*(T(K)-1.0))
30    RO(K)=B2+D0*T0*(T(K)-1.)
      CC1=0.5/(NZ-1.)
      CC2=1./(NZ-1.)
      C1=0.
      C2=0.
      DO 40 K=1,NZ
      IF(K.EQ.1)GOTO 32
      C1=C1+0.5/EDA(K)+0.5/EDA(K-1)
      C2=C2+CC1*((K-1.)/EDA(K)+(K-2.)/EDA(K-1))
32    EDA1(K)=C1*CC2
40    EDA2(K)=C2*CC2
      IF(KG.NE.2)RETURN
      C1=0.
      C2=0.
      C3=0.
      DO 50 K=1,NZ
      IF(K.EQ.1)GOTO 50
      C1=C1+0.5*(RO(K)+RO(K-1))
```

```
      C2=C2+0.5*(RO(K)*EDA1(K)+RO(K)*EDA1(K-1))
      C3=C3+0.5*(RO(K)*EDA2(K)+RO(K)*EDA2(K-1))
50    CONTINUE
      B1=12.*CC2*(C1*EDA2(NZ)/EDA1(NZ)-C2)
60    B2=2.*CC2/(U1+U2)*(C1*(U1-U2)/EDA1(NZ)+C3*U1)
      RETURN
      END
      SUBROUTINE UCAL(NZ,DX,H,EDA,RO,ROR,EDA1,EDA2,U,UU,DU,W,DP)
      DIMENSION U(NZ),DU(NZ),W(NZ),ROR(NZ),UU(NZ),EDA(NZ),RO(NZ),EDA1
      (NZ),EDA2(NZ)
      COMMON /COM2/T0,EDA0,AK0,AK1,AK2,CV,CV1,CV2,RO0,RO1,RO2,D0/
      COM3/E1,PH,B,U1,U2,R,CC
      DATA KK/0/
      IF(KK.NE.0)GOTO 20
      A1=U1
      A2=PH*(B/R)**3/E1
      A3=U2-U1
20    CC1=A2*DP*H
      CC2=CC1*H
      CC3=A3/H
      CC4=1./EDA1(NZ)
      DO 30 K=1,NZ
      U(K)=A1+CC2*(EDA2(K)-CC4*EDA2(NZ)*EDA1(K))+A3*CC4*EDA1(K)
      IF(U(K).LT.0.0)U(K)=0.
30    DU(K)=CC1/EDA(K)*((K-1.)/(NZ-1.)-CC4*EDA2(NZ))+CC3*CC4/EDA(K)
      A4=B/((NZ-1)*R*DX)
      C1=A4*H
      IF(KK.EQ.0)GOTO 50
      DO 40 K=2,NZ-1
      W(K)=(RO(K-1)*W(K-1)+C1*(RO(K)*U(K)-ROR(K)*UU(K)))/RO(K)
40    CONTINUE
50    KK=1
      RETURN
      END
      SUBROUTINE TCAL(NZ,DX,CC1,CC2,T,TF,U,W,DU,H,DP,EDA,RO)
      DIMENSION T(NZ),TF(NZ),U(NZ),DU(NZ),W(NZ),EDA(NZ),RO(NZ),A
      (4,21),D(21),AA(2,21)
      COMMON /COM2/T0,EDA0,AK0,AK1,AK2,CV,CV1,CV2,RO0,RO1,RO2,D0/
      COM3/E1,PH,B,U1,U2,R,CC
      DATA KK,CC5,PAI,TAU0/0,0.6666667,3.14159265,4.E7/
      IF(KK.NE.0)GOTO 5
      KK=1
      TAU=TAU0*B*B/(E1*R*R)
      A2=-CV*RO0*E1*B**3/(EDA0*AK0*R)
      A3=-E1*PH*B**3*D0/(AK0*EDA0*T0*R)
      A4=-(E1*R)**2/(AK0*EDA0*T0)
      A5=0.5*R/B*A2
      A6=AK0*SQRT(EDA0*R/(PAI*RO1*CV1*U1*E1*AK1*B**3))
      A7=AK0*SQRT(EDA0*R/(PAI*RO2*CV2*U2*E1*AK2*B**3))
```

```
5       CC3=A6*SQRT(DX)
        CC4=A7*SQRT(DX)
        DZ=H/(NZ-1.)
        DZ1=1./DZ
        DZ2=DZ1*DZ1
        CC6=A3*DP
        DO 10 K=2,NZ-1
        A(1,K)=DZ2+DZ1*A5*RO(K)*W(K)
        A(2,K)=-2.*DZ2+A2*RO(K)*U(K)/DX+CC6*U(K)/RO(K)
        A(3,K)=DZ2-DZ1*A5*RO(K)*W(K)
        AE=ABS(EDA(K)*DU(K))
10      A(4,K)=A4*ABS(DU(K))*AE+A2*RO(K)*U(K)*TF(K)/DX
        A(1,1)=0.
        A(2,1)=1.+2.*DZ1*CC3*CC5
        A(3,1)=-2.*DZ1*CC3*CC5
        A(1,NZ)=-2.*DZ1*CC4*CC5
        A(2,NZ)=1.+2.*DZ1*CC4*CC5
        A(3,NZ)=0.
        A(4,1)=1.+CC1*CC3*DZ1
        A(4,NZ)=1.-CC2*CC4*DZ1
        CALL TRA3(NZ,D,A,AA)
        DO 20 K=1,NZ
20      T(K)=(1.-CC)*T(K)+CC*D(K)
30      CONTINUE
        RETURN
        END
        SUBROUTINE TRA3(N,D,A,B)
        DIMENSION D(N),A(4,N),B(2,N)
        C=1./A(2,N)
        B(1,N)=-A(1,N)*C
        B(2,N)=A(4,N)*C
        DO 10 I=1,N-1
        IN=N-I
        IN1=IN+1
        C=1./(A(2,IN)+A(3,IN)*B(1,IN1))
        B(1,IN)=-A(1,IN)*C
10      B(2,IN)=(A(4,IN)-A(3,IN)*B(2,IN1))*C
        D(1)=B(2,1)
        DO 20 I=2,N
20      D(I)=B(1,I)*D(I-1)+B(2,I)
        RETURN
        END
        BLOCK DATA
        COMMON /COM2/T0,EDA0,AK0,AK1,AK2,CV,CV1,CV2,RO0,RO1,RO2,D0
        DATA T0,AK0,AK1,AK2,CV,CV1,CV2,RO0,RO1,RO2,D0/303.,0.14,46.,46.,2000.,
        470.,470.,890.,7850.,7850.,-0.00065/
        END
```

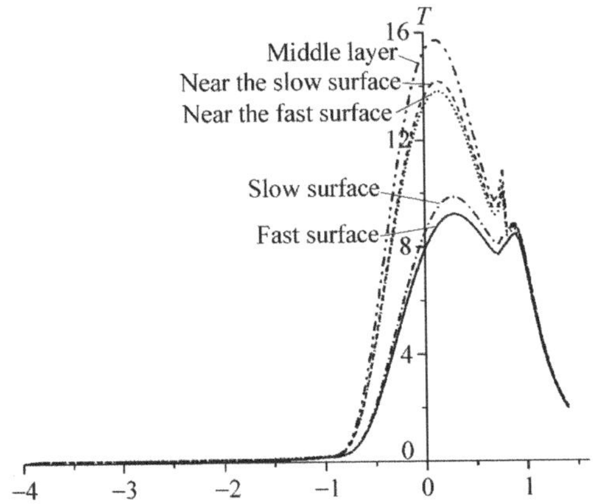

Figure 17.4 Temperature distribution of each layer for TEHL in line contact.

17.3.4 Calculation Results

With the given working conditions and dividing the film thickness into five layers across the thickness direction, the calculated temperature distribution curves are shown in Figure 17.4. We can see that the temperature is the highest on the middle layer and the lowest on the fast moving surface.

18

Numerical Method and Program for Thermal EHL in Point Contact

18.1 Basic Equations

The basic equations for thermal elastohydrodynamic lubrication (TEHL) in point contact include all the equations of EHL in point contact and the three-dimensional energy equation (in two velocity directions and the film thickness direction). These main equations are as follows.

18.1.1 Reynolds Equation

$$\frac{\partial}{\partial x}\left(\frac{\rho h^3}{\eta}\frac{\partial p}{\partial x}\right)+\frac{\partial}{\partial y}\left(\frac{\rho h^3}{\eta}\frac{\partial p}{\partial y}\right)=-12U\frac{\partial \rho h}{\partial x} \tag{18.1}$$

where ρ is the lubricant density, η is the lubricant viscosity, U is the average velocity $[U=(u_1+u_2)/2]$, u_1 and u_2 are the tangential velocities for the two surfaces, p is the pressure, h is the film thickness, x is the coordinate along the motion and y is the coordinate perpendicular to the motion.

18.1.2 Energy Equation

$$\rho c_\rho\left(u\frac{\partial T}{\partial x}+v\frac{\partial T}{\partial y}\right)=k\frac{\partial^2 T}{\partial z^2}-\frac{T}{\rho}\frac{\partial \rho}{\partial T}\left(u\frac{\partial p}{\partial x}+v\frac{\partial p}{\partial y}\right)+\eta\left[\left(\frac{\partial u}{\partial z}\right)^2+\left(\frac{\partial v}{\partial z}\right)^2\right] \tag{18.2}$$

where T is the temperature, u is the flow velocity, c_p is the specific heat capacity at the constant pressure, K is the thermal conductivity and z is the coordinate across the film.

Numerical Calculation of Lubrication: Methods and Programs, First Edition. Ping Huang.

Because $\frac{\partial^2 T}{\partial z^2}$ is of the second derivative, two boundary conditions should be given for solving the energy Equation 18.2. Usually, they can be expressed as follows (see Yang, 1998).

$$T(x,y,0) = \frac{K}{\sqrt{\pi\rho_1 c_1 K_1 u_1}} \int_{-\infty}^{x} \left.\frac{\partial T}{\partial z}\right|_{x,y,0} \frac{ds}{\sqrt{x-s}} + T_0$$
$$T(x,y,h) = \frac{K}{\sqrt{\pi\rho_2 c_2 K_2 u_2}} \int_{-\infty}^{x} \left.\frac{\partial T}{\partial z}\right|_{x,y,h} \frac{ds}{\sqrt{x-s}} + T_0 \tag{18.3}$$

where T_0 is the ambient temperature; the subscripts 1 and 2 correspond, respectively, to the down and up surfaces, ρ_1 and ρ_2 are the material densities, $c_1$and c_2 are the specific heats, K_1 and K_2 are the heat transfer coefficients and u_1 and u_2 are the tangential velocities.

In order to calculate the temperature, not only must we know the average velocity U of the two surfaces, but also both the surface velocities u_1 and u_2 should be calculated. If the slip-roll ratio s is known, we can use Equation 18.4 to calculate them.

$$\begin{aligned} u_1 &= 0.5 \times (2+s) \times U \\ u_2 &= 0.5 \times (2-s) \times U \end{aligned} \tag{18.4}$$

18.1.3 Film Thickness Equation

$$h(x,y) = h_0 + \frac{x^2}{2R_x} + \frac{y^2}{2R_y} + v(x,y) \tag{18.5}$$

where h_0 is a film thickness parameter to be determined by calculation, R_x is the equivalent radius of the surfaces in the x direction $\left(\frac{1}{R_x} = \frac{1}{R_{1x}} + \frac{1}{R_{2x}}\right)$, R_y is the equivalent radius of the surfaces in the y direction $\left(\frac{1}{R_y} = \frac{1}{R_{1y}} + \frac{1}{R_{2y}}\right)$ and v is the elastic deformation.

18.1.4 Elastic Deformation Equation

In the film thickness Equation 18.4, the deformation $v(x, y)$ is

$$v(x,y) = \frac{2}{\pi E} \iint_{\Omega} \frac{p(s,t)}{\sqrt{(x-s)^2 + (y-t)^2}} ds dt \tag{18.6}$$

where E is the equivalent elastic modulus $\left[\frac{1}{E}=\frac{1}{2}\left(\frac{1-\nu_1^2}{E_1}+\frac{1-\nu_2^2}{E_2}\right)\right]$, ν_1 and ν_2 are the Poisson's ratios of the two surfaces and E_1 and E_2 are the elastic modulus of the two surfaces.

18.1.5 Roelands Viscosity–Pressure–Temperature Equation

$$\eta=\eta_0\exp\left\{(\ln\eta_0+9.67)\left[(1+5.1\times10^{-9}p)^{0.68}\times\left(\frac{T-138}{T_0-138}\right)^{-1.1}-1\right]\right\} \tag{18.7}$$

where η_0 is the viscosity of lubricant at $p=0$ and $T=T_0$.

18.1.6 Density–Pressure–Temperature Equation

$$\rho=\rho_0\left[1+\frac{0.6p}{1+1.7p}+D(T-T_0)\right] \tag{18.8}$$

where ρ_0 is the density of lubricant at $p=0$ and $T=T_0$, the unit of pressure p is GPa and $D=-0.00065\,\mathrm{K}^{-1}$.

18.2 Numerical Method

18.2.1 Dimensionless Equations

The dimensionless Reynolds equation of EHL in point contact is

$$\frac{\partial}{\partial X}\left[\varepsilon\frac{\partial P}{\partial X}\right]+\alpha^2\frac{\partial}{\partial Y}\left[\varepsilon\frac{\partial P}{\partial Y}\right]-\frac{\partial(\rho^*H)}{\partial X}=0 \tag{18.9}$$

where $\varepsilon=\frac{\rho^*H^3}{\eta^*\lambda}$, $\lambda=\frac{12\eta_0UR_x^2}{a^3p_HK_{ex}^2}$, η^* is the dimensionless viscosity of lubricant ($\eta^*=\eta/\eta_0$), ρ^* is the dimensionless density of lubricant ($\rho^*=\rho/\rho_0$), K_{ex} is the elliptic coefficient of the surface in the x direction, X is the dimensionless coordinate ($X=x/a$), a is the half length of the ellipse contact area in the x direction, Y is the dimensionless coordinate ($Y=y/b$), b is the half length of the ellipse contact area in the y direction, α is a ratio parameter ($\alpha=a/b$), P is the dimensionless pressure ($P=p/p_{\mathrm{H}}$), p_{H} is the maximum Hertzian contact stress and H is the dimensionless film thickness $\left(H=\frac{hR_x}{a^2}\right)$.

The boundary conditions are:

$$\text{At the inlet} \quad P(X_0, Y) = 0$$
$$\text{At the outlet} \quad P(X_e, Y) = 0 \quad \frac{\partial P(X_e, Y)}{\partial X} = 0$$
$$\text{At the two sides} \quad P|_{Y=\pm 1} = 0$$

where X_0 is the dimensionless inlet coordinate ($X_0 = x_0/b$), X_e is the dimensionless outlet coordinate ($X_e = x_e/b$) and the boundaries of $Y = \pm 1$ are the two sides perpendicular to the velocity U.

The dimensionless energy equation is

$$A\rho^* u^* \left(\frac{\partial T}{\partial X} + \alpha \frac{\partial T}{\partial Y}\right) + B\frac{T^*}{\rho^*}\frac{\partial \rho^*}{\partial T^*}\left(u^*\frac{\partial P}{\partial X} + \alpha v^* \frac{\partial P}{\partial Y}\right) + C\frac{\partial^2 T}{\partial Z^2} + D\eta^*\left(\frac{\partial u^*}{\partial Z}\right)^2 = 0 \tag{18.10}$$

where T^* is the dimensionless temperature ($T^* = T/T_0$), T_0 is the ambient temperature, $A = \rho_0 c_p U T_0/b, B = U p_H/b, C = -K T_0 R/b^2, D = -\eta_0 \left(\frac{UR}{b^2}\right)^2$, u^* is the dimensionless flow velocity in the x direction ($u^* = u/U$), v^* is the dimensionless flow velocity in the y direction ($v^* = v/U$) and Z is the dimensionless coordinate across the film thickness ($Z = z/h$).

The dimensionless film thickness equation is

$$H(X, Y) = H_0 + \frac{X^2 + Y^2}{2} + \frac{2}{\pi^2}\int_{X_0}^{X_e}\int_{Y_0}^{Y_e} \frac{P(S, T)dSdT}{\sqrt{(X - S)^2 + (Y - T)^2}} \tag{18.11}$$

where H_0 is the dimensionless film thickness to be determined by calculation.

The dimensionless viscosity–pressure–temperature equation is

$$\eta^* = \exp\left\{(\ln \eta_0 + 9.67)\left[(1 + 5.1 \times 10^{-9} p)^{0.68} \times \left(\frac{T^* - 0.455445545}{0.544554455}\right)^{-1.1} - 1\right]\right\} \tag{18.12}$$

where η^* is the dimensionless viscosity of lubricant, $\eta^* = \eta/\eta_0$, and $T_0 = 303$ K.

The dimensionless density–pressure–temperature equation is

$$\rho* = 1 + \frac{0.6p}{1 + 1.7p} + D(T - T_0) \tag{18.13}$$

where ρ^* is the dimensionless viscosity of lubricant, $\rho^* = \rho/\rho_0$.

The dimensionless load equation is

$$W = \int_{X_0}^{X_e} \int_{Y_0}^{Y_e} P(X, Y)dXdY = \frac{2}{3}\pi \tag{18.14}$$

18.2.2 Discrete Equations

By using the central and the forward differential discretization equations, we can obtain the discrete differential Reynolds equation in point contact as Equation 14.11. The energy equation, the film thickness equation and the load equation of the discrete form can also be obtained by the numerical integration method. These equations can be written as follows.

Discrete dimensionless Reynolds equation

$$\frac{\varepsilon_{i-1/2,j}P_{i-1,j} + \varepsilon_{i+1/2,j}P_{i+1,j} + \varepsilon_{i,j-1/2}P_{i,j-1} + \varepsilon_{i,j+1/2}P_{i,j+1} - \varepsilon_0 P_{ij}}{\Delta X^2} = \frac{\rho^*_{ij}H_{ij} - \rho^*_{i-1,j}H_{i-1,j}}{\Delta X} \tag{18.15}$$

where $\varepsilon_{i\pm 1/2,j} = \frac{1}{2}(\varepsilon_{i,j} + \varepsilon_{i\pm 1,j})$, $\varepsilon_0 = \varepsilon_{i-1/2,j} + \varepsilon_{i+1/2,j} + \varepsilon_{i,j-1/2} + \varepsilon_{i,j+1/2}$, $\Delta X = X_i - X_{i-1}$ and the equidistant grid is used here, that is, $\Delta Y = \Delta X$.

The boundary condition at the inlet: $P(X_0, Y) = 0$; and the boundary of the outlet can be obtained by set the negative pressure to be equal to zero. The discrete forms are: $P(X_e, Y) = 0$ and $\frac{P(X_e, Y) - P(X_e - 1, Y)}{\Delta X} = 0$. The side boundary conditions are: $P(X, Y = \pm 1) = 0$.

The discrete form of the energy equation can be written as

$$\begin{aligned} & A\rho^* u^* \left(\frac{T_{i,j,k} - T_{i-1,j,k}}{\Delta X} + \alpha \frac{T_{i,j,k} - T_{i,j-1,k}}{\Delta Y} \right) \\ & + B \frac{T^*}{\rho^*} \frac{\partial \rho^*}{\partial T^*} \left(u^* \frac{P_{i,j} - P_{i-1,j}}{\Delta X} + \alpha v^* \frac{P_{i,j} - P_{i,j-1}}{\Delta Y} \right) \\ & + C \frac{T_{i,j,k+1} - 2T_{i,j,k} + T_{i,j,k-1}}{\Delta Z^2} + D\eta^* \left(\frac{u^*_{i,j,k+1} - u^*_{i,j,k}}{\Delta Z} \right)^2 = 0 \end{aligned} \tag{18.16}$$

where the subscript k is indicates the node number in the thickness direction, ρ^*, η^* and $\frac{\partial \rho^*}{\partial T^*}$ can be obtained by the expressions before so we do not need to discretize them, the film thickness is divided into k layers and ΔZ is the dimensionless distance.

The discrete film thickness equation is

$$H_{ij} = H_0 + \frac{X_i^2 + Y_j^2}{2} + \frac{2}{\pi^2} \sum_{k=1}^{n} \sum_{l=1}^{n} D_{ij}^{kl} P_{kl} \tag{18.17}$$

where D_{ij}^{kl} is the stiffness coefficient of the elastic deformation.

The discrete load equation is

$$\Delta X \Delta Y \sum_{i=1}^{n} \sum_{j=1}^{n} P_{ij} = \frac{2\pi}{3} \tag{18.18}$$

The following convergent criterion of pressure is used to determine the convergency.

$$\frac{\sum_{i} \sum_{j} \left| P_{i,j}^{k+1} - P_{i,j}^{k} \right|}{\sum_i \sum_j P_{i,j}^{k+1}} \leq \varepsilon_4 \tag{18.19}$$

where ε_4 is a given error.

In the above equations, $i = 1, 2 \ldots, n, j = 1, 2 \ldots, n$, where n is the node number.

Because the temperature in the energy Equation 18.16, the viscosity in the viscous–pressure–temperature Equation 18.12 and the elastic deformation in the film thickness Equation 18.17 vary with pressure, we should first give an initial pressure distribution (such as a Hertz contact pressure distribution) and a temperature distribution (such as a uniform temperature distribution). Then, we use Equations 18.17 and 18.12 to calculate the new thickness and the new viscosity. We substitute them into the Reynolds Equation 18.15 to solve a new pressure distribution iteratively and then substitute it into the energy Equation 18.16 to correct the temperature further. Then, we use the new temperature to correct the viscosity by Equation 18.12. In addition, we use Equation 18.15 to iterate the pressure again. We repeat these steps until the pressure difference between two neighboring iterations is less than the given error, as shown in Equation 18.19. Therefore, we can finally obtain the final pressure distribution, the film thickness and the temperature distribution of the thermal elastohydrodynamic lubrication in the point contact.

18.3 Calculation Program

18.3.1 Calculation Diagram

The block diagram of calculation is shown in Figure 18.1.

18.3.2 Calculation Program

The basic steps of TEHL calculation in point contact are similar to the thermal elastohydrodynamic program in line contact. Besides, the ellipticity calculation is added in the main program. The data pre-assigned by the data block and the interpolation is used to obtain the needed value. The calculation result of an example of TEHL in the elliptical contact is given in the following program.

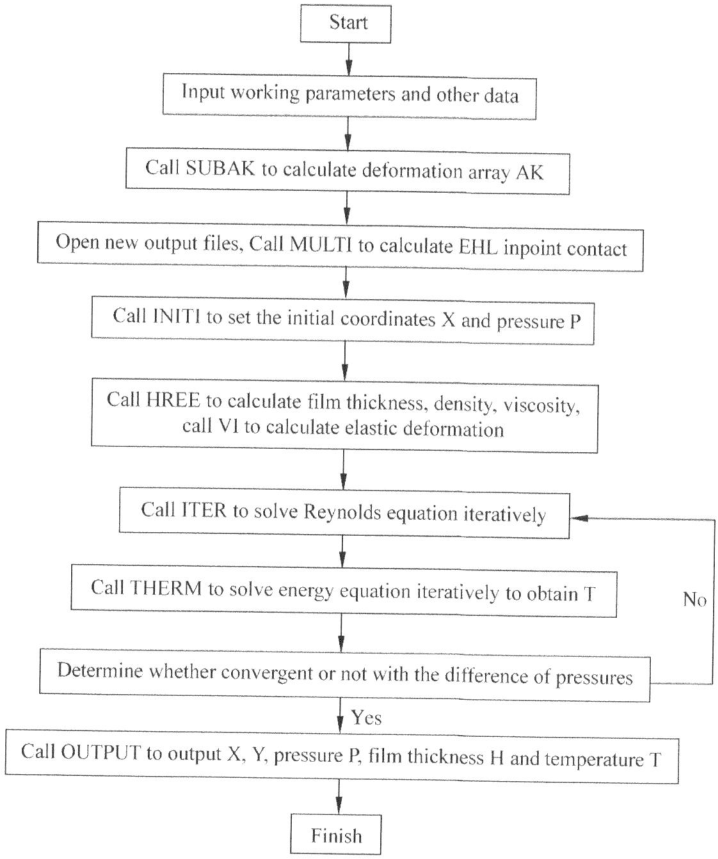

Figure 18.1 Calculation block diagram of TEHL in point contact.

Viscosity–pressure coefficient	$Z = 0.68$
Number of nodes	$N = 65 \times 65$
Load	$W0 = 39.24$ N
Equivalent modulus of elasticity	$E1 = 2.21E11$ Pa
Initial plastic viscosity	$EDA0 = 0.03$ Pa·s
Equivalent curvature radius	$RX = 0.01$ m, $RY = 0.04$ m
Average velocity	$US = 1.5$ m s^{-1}
Dimensionless starting coordinate in x direction	$X0 = -2.5$
Dimensionless ending coordinate in x direction	$XE = 1.5$
Dimensionless starting coordinate in y direction	$Y0 = -2.0$
Dimensionless ending coordinate in y direction	$YE = 2.0$
Number of layers across the film thickness direction	$NZ = 5$
Iteration coefficient of temperature	$CT = 0.31$
Sliding–rolling ratio	$AKC = 1.0$.

In the BLOCK DATA, the following parameters are given:

Initial temperature	T0 = 303 K
Thermal conductivity of lubricant	AK0 = 0.14 W·m^{-1}·K
Thermal conductivity of down surface	AK1 = 46 W·m^{-1}·K
Thermal conductivity of up surface	AK2 = 46 W·m^{-1}·K;
Specific heat of lubricant	CV = 2000 J·kg^{-1}·K^{-1}
Specific heat of down surface	CV1 = 470 J·kg^{-1}·K^{-1}
Specific heat of up surface	CV2 = 470 J·kg^{-1}·K^{-1}
Initial density of lubricant	RO0 = 890 kg·m^{-3}
Density of down surface	RO1 = 7850 kg·m^{-3}
Density of up surface	RO2 = 7850 kg·m^{-3}
Viscosity–temperature coefficient	S0 = −1.1
Density–temperature coefficient	D0 = −0.00065.

Furthermore, in the DATA sentences, the data for θ, e_α and e_β are given as follows for interpolating the coefficients of ellipticity.

```
    DATA THETA/10.,20.,30.,35.,40.,45.,50.,55.,60.,65.,70.,75.,
80.,85.,90./
    DATA EALFA/6.612,3.778,2.731,2.397,2.136,1.926,1.754,1.611,1.486,
1.378,1.284,1.202,1.128,1.061,1.0/
    DATA EBETA/0.319,0.408,0.493,0.53,0.567,0.604,0.641,0.678,
0.717,0.759,0.802,0.846,0.893,0.944,1.0/
```

According to the working conditions, the user can change the above parameters' values. However, the program must be re-compiled and re-linked before executing.

Further, during execution of the program, if input KT = 2, it will carry out the thermal calculation; otherwise we do not consider the temperature.

Output parameters: the basic data is in the file OUT.DAT, the pressure array P(I, J) will be saved in the file PRESSURE.DAT, the film thickness array H(I, J) will be saved in the file FILM.DAT, the average temperature array T1(I, J) will be saved in the file TEM.DAT.

```
      PROGRAM POINTEHLT
      DIMENSION THETA(15),EALFA(15),EBETA(15)
      COMMON /COM1/ENDA,A1,A2,A3,Z,HM0/COM3/E1,PH1,B1,U1,U2,RE,
CT/COMK/LMIN,AKC
      COMMON /COM2/T0,AK0,AK1,AK2,CV,CV1,CV2,RO0,RO1,RO2,S0,D0
      COMMON /COMW/W0,T1,T2,RX,B,PH/COMC/KT,NF/COMD/AD,AD1,KK1,
KK2,KK3,KK4/COME/US,EDA0/COMH/HMC
      COMMON /COMEK/EK,EAL,EBE
      DATA PAI,Z/3.14159265,0.68/,N,W0,E1,EDA0,RX,RY,US,X0,XE/
65,39.24,2.21E11,0.03,0.01,0.03,1.5,-2.5,1.5/
```

```
      DATA NZ,CT,AKC/5,0.31,1.0/
      DATA THETA/10.,20.,30.,35.,40.,45.,50.,55.,60.,65.,70.,
75.,80.,85.,90./
      DATA EALFA/6.612,3.778,2.731,2.397,2.136,1.926,1.754,
1.611,1.486,1.378,
1.284,1.202,1.128,1.061,1.0/
      DATA EBETA/0.319,0.408,0.493,0.53,0.567,0.604,0.641,
0.678,0.717,0.759,0.802,0.846,0.893,0.944,1.0/
      DATA KK1,KK2,KK3,KK4,NF,AD,AD1,EAL,EBE/0,0,0,0,0,0.0,
0.0,1.0,1.0/
      EK=RX/RY
      WRITE(*,*)'KT='
      READ(*,*)KT
      AA=0.5*(1./RX+1./RY)
      BB=0.5*ABS(1./RX-1./RY)
      CC=ACOS(BB/AA)*180.0/PAI
      DO I=1,15
      IF(CC.LT.THETA(I))THEN
      WRITE(*,*)I
      EAL=EALFA(I-1)+(CC-THETA(I-1))*(EALFA(I)-EALFA(I-1))/
(THETA(I)-THETA(I-1))
      EBE=EBETA(I-1)+(CC-THETA(I-1))*(EBETA(I)-EBETA(I-1))/
(THETA(I)-THETA(I-1))
      GOTO 1
      ENDIF
      ENDDO
1     EA=EAL*(1.5*W0/AA/E1)**(1./3.0)
      EB=EBE*(1.5*W0/AA/E1)**(1./3.0)
      PH=1.5*W0/(EA*EB*PAI)
      OPEN(8,FILE='FILM.DAT',STATUS='UNKNOWN')
      OPEN(9,FILE='PRESS.DAT',STATUS='UNKNOWN')
      OPEN(10,FILE='OUT.DAT',STATUS='UNKNOWN')
      WRITE(*,*)N,X0,XE,PH,E1,EDA0,RX,US
      H00=0.0
      MM=N-1
      LMIN=ALOG(N-1.)/ALOG(2.)-1.99
      U=EDA0*US/(2.*E1*RX)
      U1=0.5*(2.+AKC)*U
      U2=0.5*(2.-AKC)*U
      A1=ALOG(EDA0)+9.67
      A2=5.1E-9*PH
      A3=0.59/(PH*1.E-9)
      B=PAI*PH*RX/E1
      PH1=PH
      B1=B
```

```
      RE=RX
      W=2.*PAI*PH/(3.*E1)*(B/RX)**2
      ALFA=Z*5.1E-9*A1
      G=ALFA*E1
      AHM=1.0-EXP(-0.68*1.03)
      AHC=1.0-0.61*EXP(-0.73*1.03)
      HM0=3.63*(RX/B)**2*G**0.49*U**0.68*W**(-0.073)*AHM
      HMC=2.69*(RX/B)**2*G**0.53*U**0.67*W**(-0.067)*AHC
      ENDA=12.*U*(E1/PH)*(RX/B)**3
      UTL=EDA0*US*RX/(B*B*2.E7)
      W0=2.0*PAI*EA*EB*PH/3.0
      T1=PH*B/RX
      T2=EDA0*US*RX/(B*B)
      WRITE(*,*)' Wait please'
      CALL SUBAK(MM)
      CALL MULTI(N,NZ,X0,XE,H00)
      STOP
      END
      SUBROUTINE MULTI(N,NZ,X0,XE,H00)
      DIMENSION X(65),Y(65),H(4500),RO(4500),EPS(4500),EDA(4500),
     P(4500),POLD(4500),T(65,65,5)
      COMMON /COMT/T1(65,65)/COMC/KT,NF
      COMMON /COMEK/EK,EAL,EBE
      DATA MK,KTK,G00/200,1,2.0943951/
      G0=G00*EAL*EBE
      NX=N
      NY=N
      NN=(N+1)/2
      DO I=1,N
      DO J=1,N
      T1(I,J)=1.0
      DO K=1,5
      T(I,J,K)=1.0
      ENDDO
      ENDDO
      ENDDO
      CALL INITI(N,DX,X0,XE,X,Y,P,POLD)
      CALL HREE(N,DX,H00,G0,X,Y,H,RO,EPS,EDA,P)
      M=0
      KTK=0
 14   KK=15
 15   CALL ITER(N,KK,DX,H00,G0,X,Y,H,RO,EPS,EDA,P)
      M=M+1
      CALL ERP(N,ER,P,POLD)
      ER=ER/KK
```

```
      WRITE(*,*)'ER=',ER
      IF(KT.NE.0)GOTO 17
      IF(M.LT.MK.AND.ER.GT.1.E-7)GOTO 14
      GOTO 120
 17   KT1=0
 18   CALL THERM(NX,NY,NZ,DX,P,H,T)
      CALL ERROM(NX,NY,NZ,T,ERM)
      IF(ER.LT.1.0E-5)GOTO 120
      IF(KT1.LT.1)THEN
      KT1=KT1+1
      GOTO 18
      ENDIF
      IF(KTK.LT.MK)THEN
      KTK=KTK+1
      GOTO 14
      ENDIF
 120  CONTINUE
      OPEN(11,FILE='TEM.DAT',STATUS='UNKNOWN')
      WRITE(11,110)X0,(Y(I),I=1,N)
      TMAX=0.0
      DO I=1,N
      WRITE(11,110)X(I),(273.0*(T1(I,JJ)-1.),JJ=1,N)
      DO J=1,N
      IF(TMAX.LT.273.0*(T1(I,J)-1.))TMAX=273.*(T1(I,J)-1.)
      ENDDO
      ENDDO
 110  FORMAT(66(E12.6,1X))
 130  CALL OUPT(N,DX,X,Y,H,P,EDA,TMAX)
      RETURN
      END
      SUBROUTINE INITI(N,DX,X0,XE,X,Y,P,POLD)
      DIMENSION X(N),Y(N),P(N,N),POLD(N,N)
      NN=(N+1)/2
      DX=(XE-X0)/(N-1.)
      Y0=-0.5*(XE-X0)
      DO 5 I=1,N
      X(I)=X0+(I-1)*DX
      Y(I)=Y0+(I-1)*DX
 5    CONTINUE
      DO 10 I=1,N
      D=1.-X(I)*X(I)
      DO 10 J=1,NN
      C=D-Y(J)*Y(J)
      IF(C.LE.0.0)P(I,J)=0.0
 10   IF(C.GT.0.0)P(I,J)=SQRT(C)
```

```
      DO 20 I=1,N
      DO 20 J=NN+1,N
      JJ=N-J+1
20    P(I,J)=P(I,JJ)
      DO I=1,N
      DO J=1,N
      POLD(I,J)=P(I,J)
      ENDDO
      ENDDO
      RETURN
      END
      SUBROUTINE HREE(N,DX,H00,G0,X,Y,H,RO,EPS,EDA,P)
      DIMENSION X(N),Y(N),P(N,N),H(N,N),RO(N,N),EPS(N,N),EDA(N,N)
      DIMENSION W(150,150),P0(150,150),ROU(65,65)
      COMMON /COM1/ENDA,A1,A2,A3,Z,HM0/COMAK/AK(0:65,0:65)
      COMMON /COM2/T0,EAK,EAK1,EAK2,CV,CV1,CV2,RO0,RO1,RO2,S0,D0
      COMMON /COMT/T1(65,65)/COMK/LMIN,AKC/COMD/AD,AD1,KK,KK2,
 KK3,KK4/COMC/KT,NF
      COMMON /COMEK/EK,EAL,EBE
      DATA KR,NW,pai,PAI1,delta/0,150,3.14159265,0.2026423,0.0/
      NN=(N+1)/2
      CALL VI(NW,N,DX,P,W)
      HMIN=1.E3
      IF(KR.EQ.0)THEN
      OPEN(12,FILE='ROUGH2.DAT',STATUS='UNKNOWN')
      DO I=1,N
      DO J=NN+1,N
      ROU(I,J)=ROU(I,N+1-J)
      ENDDO
 100  FORMAT(33(1X,F10.6))
      ENDDO
      CLOSE(12)
      KR=1
      ENDIF
      DO 30 I=1,N
      DO 30 J=1,NN
      RAD=X(I)*X(I)+EK*Y(J)*Y(J)
      W1=0.5*RAD+DELTA
      ZZ=0.5*AD1*AD1+X(I)*ATAN(AD*PAI/180.0)
      IF(W1.LE.ZZ)W1=ZZ
      H0=W1+W(I,J)
      IF(H0.LT.HMIN)HMIN=H0
 30   H(I,J)=H0
      IF(KK.EQ.0)THEN
      KG1=0
```

```
      H01=-HMIN+HM0
      DH=0.005*HM0
      H02=-HMIN
      H00=0.5*(H01+H02)
      ENDIF
      W1=0.0
      DO 32 I=1,N
      DO 32 J=1,N
32    W1=W1+P(I,J)
      W1=DX*DX*W1/G0
      DW=1.-W1
      IF(KK.EQ.0)THEN
      KK=1
      GOTO 50
      ENDIF
      IF(DW.LT.0.0)THEN
      KG1=1
      H00=AMIN1(H01,H00+DH)
      ENDIF
      IF(DW.GT.0.0)THEN
      KG2=2
      H00=AMAX1(H02,H00-DH)
      ENDIF
50    DO 60 I=1,N
      DO 60 J=1,NN
      H(I,J)=H00+H(I,J)
      CT1=((T1(I,J)-0.455445545)/0.544554455)**S0
      CT2=D0*T0*(T1(I,J)-1.)
      IF(P(I,J).LT.0.0)P(I,J)=0.0
      EDA1=EXP(A1*(-1.+(1.+A2*P(I,J))**Z*CT1))
      EDA(I,J)=EDA1
      IF(NF.EQ.0)GOTO 55
      IF(I.NE.1.AND.J.NE.1)THEN
      DPDX=(P(I,J)-P(I-1,J))/DX
      DPDY=(P(I,J)-P(I,J-1))/DX
      EDA(I,J)=EQEDA(DPDX,DPDY,P(I,J),H(I,J),EDA1)
      ENDIF
      EDA1=EDA(I,J)
55    RO(I,J)=(A3+1.34*P(I,J))/(A3+P(I,J))+CT2
60    EPS(I,J)=RO(I,J)*H(I,J)**3/(ENDA*EDA1)
      DO 70 J=NN+1,N
      JJ=N-J+1
      DO 70 I=1,N
      H(I,J)=H(I,JJ)
      RO(I,J)=RO(I,JJ)
```

```
      EDA(I,J)=EDA(I,JJ)
70    EPS(I,J)=EPS(I,JJ)
      RETURN
      END
      SUBROUTINE ITER(N,KK,DX,H00,G0,X,Y,H,RO,EPS,EDA,P)
      DIMENSION X(N),Y(N),P(N,N),H(N,N),RO(N,N),EPS(N,N),EDA(N,N)
      DIMENSION D(70),A(350),B(210),ID(70)
      COMMON /COM1/ENDA,A1,A2,A3,Z,C3/COMAK/AK(0:65,0:65)
      DATA KG1,PAI1,C1,C2/0,0.2026423,0.31,0.31/
      IF(KG1.NE.0)GOTO 2
      KG1=1
      AK00=AK(0,0)
      AK10=AK(1,0)
      AK20=AK(2,0)
      BK00=AK00-AK10
      BK10=AK10-0.25*(AK00+2.*AK(1,1)+AK(2,0))
      BK20=AK20-0.25*(AK10+2.*AK(2,1)+AK(3,0))
2     NN=(N+1)/2
      MM=N-1
      DX1=1./DX
      DX2=DX*DX
      DX3=1./DX2
      DX4=0.3*DX2
      DO 100 K=1,KK
      PMAX=0.0
      DO 70 J=2,NN
      J0=J-1
      J1=J+1
      JJ=N-J+1
      IA=1
8     MM=N-IA
      IF(P(MM,J0).GT.1.E-6)GOTO 20
      IF(P(MM,J).GT.1.E-6)GOTO 20
      IF(P(MM,J1).GT.1.E-6)GOTO 20
      IA=IA+1
      IF(IA.LT.N)GOTO 8
      GOTO 70
20    IF(MM.LT.N-1)MM=MM+1] ]>
      D2=0.5*(EPS(1,J)+EPS(2,J))
      DO 50 I=2,MM
      I0=I-1
      I1=I+1
      II=5*I0
      D1=D2
      D2=0.5*(EPS(I1,J)+EPS(I,J))
```

```
      D4=0.5*(EPS(I,J0)+EPS(I,J))
      D5=0.5*(EPS(I,J1)+EPS(I,J))
      P1=P(I0,JJ)
      P2=P(I1,JJ)
      P3=P(I,JJ)
      P4=P(I,JJ+1)
      P5=P(I,JJ-1)
      D3=D1+D2+D4+D5
      IF(J.EQ.NN.AND.ID(I).EQ.1)P(I,J)=P(I,J)-0.5*C2*D(I)
      IF(H(I,J).LE.0.0)THEN
      ID(I)=2
      A(II+1)=0.0
      A(II+2)=0.0
      A(II+3)=1.0
      A(II+4)=0.0
      A(II+5)=1.0
      A(II-4)=0.0
      GOTO 50
      ENDIF
      IF(D1.GE.DX4)GOTO 30
      IF(D2.GE.DX4)GOTO 30
      IF(D4.GE.DX4)GOTO 30
      IF(D5.GE.DX4)GOTO 30
      ID(I)=1
      IF(J.EQ.NN)P5=P4
      A(II+1)=PAI1*(RO(I0,J)*BK10-RO(I,J)*BK20)
      A(II+2)=DX3*(D1+0.25*D3)+PAI1*(RO(I0,J)*BK00-RO(I,J)*BK10)
      A(II+3)=-1.25*DX3*D3+PAI1*(RO(I0,J)*BK10-RO(I,J)*BK00)
      A(II+4)=DX3*(D2+0.25*D3)+PAI1*(RO(I0,J)*BK20-RO(I,J)*BK10)
      A(II+5)=-DX3*(D1*P1+D2*P2+D4*P4+D5*P5-D3*P3)+DX1*(RO(I,
J)*H(I,J)-RO(I0,J)*H(I0,J))
      GOTO 50
30    ID(I)=0
      P4=P(I,J0)
      IF(J.EQ.NN)P5=P4
      A(II+1)=PAI1*(RO(I0,J)*AK10-RO(I,J)*AK20)
      A(II+2)=DX3*D1+PAI1*(RO(I0,J)*AK00-RO(I,J)*AK10)
      A(II+3)=-DX3*D3+PAI1*(RO(I0,J)*AK10-RO(I,J)*AK00)
      A(II+4)=DX3*D2+PAI1*(RO(I0,J)*AK20-RO(I,J)*AK10)
      A(II+5)=-DX3*(D1*P1+D2*P2+D4*P4+D5*P5-D3*P3)+DX1*(RO(I,
J)*H(I,J)-RO(I0,J)*H(I0,J))
50    CONTINUE
      CALL TRA4(MM,D,A,B)
      DO 60 I=2,MM
      IF(ID(I).EQ.2)GOTO 60
```

```
      IF(ID(I).EQ.0)GOTO 52
      DD=D(I+1)
      IF(I.EQ.MM)DD=0
      P(I,J)=P(I,J)+C2*(D(I)-0.25*(D(I-1)+DD))
      IF(J0.NE.1)P(I,J0)=P(I,J0)-0.25*C2*D(I)
      IF(P(I,J0).LT.0.)P(I,J0)=0.0
      IF(J1.GE.NN)GOTO 54
      P(I,J1)=P(I,J1)-0.25*C2*D(I)
      GOTO 54
52    P(I,J)=P(I,J)+C1*D(I)
54    IF(P(I,J).LT.0.0)P(I,J)=0.0
      IF(PMAX.LT.P(I,J))PMAX=P(I,J)
60    CONTINUE
70    CONTINUE
      DO 80 J=1,NN
      JJ=N+1-J
      DO 80 I=1,N
80    P(I,JJ)=P(I,J)
      CALL HREE(N,DX,H00,G0,X,Y,H,RO,EPS,EDA,P)
100   CONTINUE
      RETURN
      END
      SUBROUTINE TRA4(N,D,A,B)
      DIMENSION D(N),A(5,N),B(3,N)
      C=1./A(3,N)
      B(1,N)=-A(1,N)*C
      B(2,N)=-A(2,N)*C
      B(3,N)=A(5,N)*C
      DO 10 I=1,N-2
      IN=N-I
      IN1=IN+1
      C=1./(A(3,IN)+A(4,IN)*B(2,IN1))
      B(1,IN)=-A(1,IN)*C
      B(2,IN)=-(A(2,IN)+A(4,IN)*B(1,IN1))*C
10    B(3,IN)=(A(5,IN)-A(4,IN)*B(3,IN1))*C
      D(1)=0.0
      D(2)=B(3,2)
      DO 20 I=3,N
20    D(I)=B(1,I)*D(I-2)+B(2,I)*D(I-1)+B(3,I)
      RETURN
      END
      SUBROUTINE ERP(N,ER,P,POLD)
      DIMENSION P(N,N),POLD(N,N)
      ER=0.0
      SUM=0.0
```

```
      NN=(N+1)/2
      DO 10 I=1,N
      DO 10 J=1,NN
      ER=ER+ABS(P(I,J)-POLD(I,J))
      SUM=SUM+P(I,J)
10    CONTINUE
      ER=ER/SUM
      DO I=1,N
      DO J=1,N
      POLD(I,J)=P(I,J)
      ENDDO
      ENDDO
      RETURN
      END
      SUBROUTINE VI(NW,N,DX,P,V)
      DIMENSION P(N,N),V(NW,NW)
      COMMON /COMAK/AK(0:65,0:65)
      PAI1=0.2026423
      DO 40 I=1,N
      DO 40 J=1,N
      H0=0.0
      DO 30 K=1,N
      IK=IABS(I-K)
      DO 30 L=1,N
      JL=IABS(J-L)
30    H0=H0+AK(IK,JL)*P(K,L)
40    V(I,J)=H0*DX*PAI1
      RETURN
      END
      SUBROUTINE SUBAK(MM)
      COMMON /COMAK/AK(0:65,0:65)
      S(X,Y)=X+SQRT(X**2+Y**2)
      DO 10 I=0,MM
      XP=I+0.5
      XM=I-0.5
      DO 10 J=0,I
      YP=J+0.5
      YM=J-0.5
      A1=S(YP,XP)/S(YM,XP)
      A2=S(XM,YM)/S(XP,YM)
      A3=S(YM,XM)/S(YP,XM)
      A4=S(XP,YP)/S(XM,YP)
      AK(I,J)=XP*ALOG(A1)+YM*ALOG(A2)+XM*ALOG(A3)+YP*ALOG(A4)
10    AK(J,I)=AK(I,J)
      RETURN
```

```
      END
      SUBROUTINE THERM(NX,NY,NZ,DX,P,H,T)
      DIMENSION T(NX,NY,NZ),T1(21),TI(21),U(21),DU(21),UU(21),
V(21),DV(21),VV(21),W(21),EDA(21),RO(21),EDA1(21),EDA2(21),
ROR(21),P(NX,NX),H(NX,NX),TFX(21),TFY(21)
      COMMON /COMD/AD,AD1,KK1,KK,KK3,KK4
      IF(KK.NE.0)GOTO 4
      DO 2 K=1,NZ
      DO 1 J=1,NY
1     T(1,J,K)=1.0
      DO 2 I=1,NX
2     T(I,1,K)=1.0
4     DO 30 I=2,NX
      DO 30 J=2,NY
      KG=0
      DO 6 K=1,NZ
      TFX(K)=T(I-1,J,K)
      TFY(K)=T(I,J-1,K)
      IF(KK.NE.0)GOTO 5
      T1(K)=T(I-1,J,K)
      GOTO 6
5     T1(K)=T(I,J,K)
6     TI(K)=T1(K)
      P1=P(I,J)
      H1=H(I,J)
      DPX=(P(I,J)-P(I-1,J))/DX
      DPY=(P(I,J)-P(I,J-1))/DX
      CALL TBOUD(NX,NY,NZ,I,J,CC1,CC2,T)
10    CALL EROEQ(NZ,T1,P1,H1,DPX,DPY,EDA,RO,EDA1,EDA2,KG)
      CALL UCAL(NZ,DX,H1,EDA,RO,ROR,EDA1,EDA2,U,UU,DU,V,VV,DV,
W,DPX,DPY)
      CALL TCAL(NZ,DX,CC1,CC2,T1,TFX,TFY,U,V,W,DU,DV,H1,DPX,DPY,
EDA,RO)
      CALL ERRO(NZ,TI,T1,ETS)
      KG=KG+3
      IF(ETS.GT.1.E-4.AND.KG.LE.50)GOTO 10
      DO 20 K=1,NZ
      ROR(K)=RO(K)
      UU(K)=U(K)
      VV(K)=V(K)
20    T(I,J,K)=T1(K)
30    CONTINUE
      KK=1
      RETURN
      END
```

```
      SUBROUTINE TBOUD(NX,NY,NZ,I,J,CC1,CC2,T)
      DIMENSION T(NX,NY,NZ)
      CC1=0.
      CC2=0.
      DO 10 L=1,I-1
      DS=1./SQRT(FLOAT(I-L))
      IF(L.EQ.I-1)DS=1.1666667
      CC1=CC1+DS*(T(L,J,2)-T(L,J,1))
10    CC2=CC2+DS*(T(L,J,NZ)-T(L,J,NZ-1))
      RETURN
      END
      SUBROUTINE ERRO(NZ,T0,T,ETS)
      DIMENSION T0(NZ),T(NZ)
      ETS=0.0
      DO 10 K=1,NZ
      IF(T(K).LT.1.E-5)ETS0=1.
      IF(T(K).GE.1.E-5)ETS0=ABS((T(K)-T0(K))/T(K))
      IF(ETS0.GT.ETS)ETS=ETS0
10    T0(K)=T(K)
      RETURN
      END
      SUBROUTINE EROEQ(NZ,T,P,H,DPX,DPY,EDA,RO,EDA1,EDA2,KG)
      DIMENSION T(NZ),EDA(NZ),RO(NZ),EDA1(NZ),EDA2(NZ)
      COMMON /COM1/ENDA,A1,A2,A3,Z,C3/COM2/T0,AK0,AK1,AK2,CV,
CV1,CV2,RO0,RO1,RO2,S0,D0/COM3/E1,PH,B,U1,U2,R,CC/COMC/KT,NF
      DATA A4,A5/0.455445545,0.544554455/
      IF(KG.NE.0)GOTO 20
      B1=(1.+A2*P)**Z
      B2=(A3+1.34*P)/(A3+P)
20    DO 30 K=1,NZ
      EDA3=EXP(A1*(-1.+B1*((T(K)-A4)/A5)**S0))
      EDA(K)=EDA3
      IF(NF.NE.0)EDA(K)=EQEDA(DPX,DPY,P,H,EDA3)
30    RO(K)=B2+D0*T0*(T(K)-1.)
      CC1=0.5/(NZ-1.)
      CC2=1./(NZ-1.)
      C1=0.
      C2=0.
      DO 40 K=1,NZ
      IF(K.EQ.1)GOTO 32
      C1=C1+0.5/EDA(K)+0.5/EDA(K-1)
      C2=C2+CC1*((K-1.)/EDA(K)+(K-2.)/EDA(K-1))
32    EDA1(K)=C1*CC2
40    EDA2(K)=C2*CC2
      RETURN
```

```
      END
      SUBROUTINE UCAL(NZ,DX,H,EDA,RO,ROR,EDA1,EDA2,U,UU,DU,V,
VV,DV,W,DPX,DPY)
      DIMENSION U(NZ),UU(NZ),DU(NZ),V(NZ),VV(NZ),DV(NZ),W(NZ),
ROR(NZ),EDA(NZ),RO(NZ),EDA1(NZ),EDA2(NZ)
      COMMON /COM2/T0,AK0,AK1,AK2,CV,CV1,CV2,RO0,RO1,RO2,S0,D0/
COM3/E1,PH,B,U1,U2,R,CC
      COMMON /COMD/AD,AD1,KK1,KK2,KK,KK4
      IF(KK.NE.0)GOTO 20
      A1=U1
      A2=PH*(B/R)**3/E1
      A3=U2-U1
20    CUA=A2*DPX*H
      CUB=CUA*H
      CVA=A2*DPY*H
      CVB=CVA*H
      CC3=A3/H
      CC4=1./EDA1(NZ)
      DO 30 K=1,NZ
      U(K)=A1+CUB*(EDA2(K)-CC4*EDA2(NZ)*EDA1(K))+A3*CC4*EDA1(K)
      V(K)=CVB*(EDA2(K)-CC4*EDA2(NZ)*EDA1(K))
      DU(K)=CUA/EDA(K)*((K-1.)/(NZ-1.)-CC4*EDA2(NZ))+CC3*CC4/EDA(K)
30    DV(K)=CVA/EDA(K)*((K-1.)/(NZ-1.)-CC4*EDA2(NZ))
      A4=B/((NZ-1)*R*DX)
      C1=A4*H
      IF(KK.EQ.0)GOTO 50
      DO 40 K=2,NZ-1
      W(K)=(RO(K-1)*W(K-1)+C1*(RO(K)*(U(K)+V(K))-ROR(K)*(UU(K)+
VV(K))))/RO(K)
40    CONTINUE
50    KK=1
      RETURN
      END
      SUBROUTINE TCAL(NZ,DX,CC1,CC2,T,TFX,TFY,U,V,W,DU,DV,H,
DPX,DPY,EDA,RO)
      DIMENSION T(NZ),U(NZ),DU(NZ),V(NZ),DV(NZ),W(NZ),EDA(NZ),
RO(NZ),A(4,21),D(21),AA(2,21),TFX(NZ),TFY(NZ)
      COMMON /COM2/T0,AK0,AK1,AK2,CV,CV1,CV2,RO0,RO1,RO2,S0,
D0/COM3/E1,PH,B,U1,U2,R,CC
      COMMON /COMD/AD,AD1,KK1,KK2,KK3,KK/COME/US,EDA0
      DATA CC5,PAI/0.6666667,3.14159265/
      IF(KK.NE.0)GOTO 5
      KK=1
      A2=-CV*RO0*E1*B**3/(EDA0*AK0*R)
      A3=-E1*PH*B**3*D0/(AK0*EDA0*T0*R)
```

```
      A4=-(E1*R)**2/(AK0*EDA0*T0)
      A5=0.5*R/B*A2
      A6=AK0*SQRT(EDA0*R/(PAI*RO1*CV1*U1*E1*AK1*B**3))
      A7=AK0*SQRT(EDA0*R/(PAI*RO2*CV2*U2*E1*AK2*B**3))
5     CC3=A6*SQRT(DX)
      CC4=A7*SQRT(DX)
      DZ=H/(NZ-1.)
      DZ1=1./DZ
      DZ2=DZ1*DZ1
      CC6=A3*DPX
      CC7=A3*DPY
      DO 10 K=2,NZ-1
      A(1,K)=DZ2+DZ1*A5*RO(K)*W(K)
      A(2,K)=-2.*DZ2+A2*RO(K)*(U(K)+V(K))/DX+(CC6*U(K)+CC7*V
(K))/RO(K)
      A(3,K)=DZ2-DZ1*A5*RO(K)*W(K)
10    A(4,K)=A4*EDA(K)*(DU(K)**2+DV(K)**2)+A2*RO(K)*(U(K)*TFX
(K)+V(K)*TFY(K))/DX
      A(1,1)=0.
      A(2,1)=1.+2.*DZ1*CC3*CC5
      A(3,1)=-2.*DZ1*CC3*CC5
      A(1,NZ)=-2.*DZ1*CC4*CC5
      A(2,NZ)=1.+2.*DZ1*CC4*CC5
      A(3,NZ)=0.
      A(4,1)=1.+CC1*CC3*DZ1
      A(4,NZ)=1.-CC2*CC4*DZ1
      CALL TRA3(NZ,D,A,AA)
      DO 20 K=1,NZ
      T(K)=(1.-CC)*T(K)+CC*D(K)
20    IF(T(K).LT.1.)T(K)=1.
30    CONTINUE
      RETURN
      END
      SUBROUTINE TRA3(N,D,A,B)
      DIMENSION D(N),A(4,N),B(2,N)
      C=1./A(2,N)
      B(1,N)=-A(1,N)*C
      B(2,N)=A(4,N)*C
      DO 10 I=1,N-1
      IN=N-I
      IN1=IN+1
      C=1./(A(2,IN)+A(3,IN)*B(1,IN1))
      B(1,IN)=-A(1,IN)*C
10    B(2,IN)=(A(4,IN)-A(3,IN)*B(2,IN1))*C
      D(1)=B(2,1)
```

```
      DO 20 I=2,N
20    D(I)=B(1,I)*D(I-1)+B(2,I)
      RETURN
      END
      SUBROUTINE ERROM(NX,NY,NZ,T,ERM)
      DIMENSION T(NX,NY,NZ)
      COMMON /COMT/T1(65,65)
      ERM=0.
      C1=1./FLOAT(NZ)
      DO 20 I=2,NX
      DO 20 J=2,NY
      TT=0.
      DO 10 K=1,NZ
10    TT=TT+T(I,J,K)
      TT=C1*TT
      ER=ABS((TT-T1(I,J))/TT)
      IF(ER.GT.ERM)ERM=ER
20    T1(I,J)=TT
      RETURN
      END
      SUBROUTINE OUPT(N,DX,X,Y,H,P,EDA,TMAX)
      DIMENSION X(N),Y(N),H(N,N),P(N,N),EDA(N,N)
      COMMON /COM1/ENDA,A1,A2,A3,Z,HM0/COMH/HMC
      COMMON /COMW/W0,T1,T2,RX,B,PH/COMK/LMIN,AKC/COMD/AD,AD1,
 KK1,KK2,KK3,KK4
      NN=(N+1)/2
      A=0.0
      WRITE(8,110)A,(Y(I),I=1,N)
      DO I=1,N
      WRITE(8,110)X(I),(H(I,J),J=1,N)
      ENDDO
      WRITE(9,110)A,(Y(I),I=1,N)
      DO I=1,N
      WRITE(9,110)X(I),(P(I,J),J=1,N)
      ENDDO
110   FORMAT(66(E12.6,1X))
      F=0.0
      HMIN=H(1,1)
      PMAX=0.0
      HM=0.0
      NCOUN=0
      NPA=0
      DO I=2,N
      DO J=1,N
      IF(X(I).LE.0.0.AND.X(I+1).GE.0.0)THEN
```

```
      IF(Y(J).LE.0.0.AND.Y(I+1).GE.0.0)HC=H(I,J)
      ENDIF
      DPDX=(P(I,J)-P(I-1,J))/DX
      TAU=T1*DPDX*H(I,J)+0.5*AKC*T2*EDA(I,J)/H(I,J)
      F=F+TAU
      IF(H(I,J).LT.HMIN)HMIN=H(I,J)
      IF(P(I,J).GT.PMAX)PMAX=P(I,J)
      RAD=SQRT(X(I)*X(I)+Y(J)*Y(J))
      IF(RAD.LE.0.5)THEN
      NCOUN=NCOUN+1
      HM=HM+H(I,J)
      ENDIF
      IF(P(I,J).GT.1.E-6)THEN
      PA=PA+P(I,J)
      NPA=NPA+1
      ENDIF
      ENDDO
      ENDDO
      PA=PA/FLOAT(NPA)
      HM=HM/FLOAT(NCOUN)
      F=B*B*F*DX*DX/W0
      HMIN=HMIN*B*B/RX
      HM=HM*B*B/RX
      HC=HC*B*B/RX
      PMAX=PMAX*PH
      HDM=HM0*B*B/RX
      HDC=HMC*B*B/RX
      WRITE(10,*)'W0,F,HMIN,HC,HDM,HDC,PMAX,HM,TMAX,PA'
      WRITE(10,120)W0,F,HMIN,HC,HDM,HDC,PMAX,HM,TMAX,PA
120   FORMAT(10(1X,E12.6))
      RETURN
      END
      FUNCTION EQEDA(DPDX,DPDY,P,H,EDA)
      COMMON /COME/U0,EDA0/COMW/W0,T1,T2,R,B,PH/COMK/LMIN,AKC
      DATA TAU0/2.E7/
      DPDX1=DPDX*PH/B
      DPDY1=DPDY*PH/B
      P1=P*PH
      H1=H*B*B/R
      EDA1=EDA*EDA0
      TAUL=TAU0+0.036*P1
      C1=-0.5*EDA1*AKC*U0/H1-0.5*H1*DPDX1
      TAU1=DPDX1*H1+C1
      TAU2=C1
      TAUY=0.5*DPDY1*H1
```

```
      TAUX=AMAX1(ABS(TAU1),ABS(TAU2))
      TAU=SQRT(TAUX**2+TAUY**2)
      X=TAUL/TAU
      EQEDA=EDA
      IF(X.LT.1)THEN
      EQEDA=EDA*X
      ENDIF
      IF(EQEDA.LT.1)EQEDA=1.
      RETURN
      END
      BLOCK DATA
      COMMON /COM2/T0,AK0,AK1,AK2,CV,CV1,CV2,RO0,RO1,RO2,S0,D0
      DATA T0,AK0,AK1,AK2,CV,CV1,CV2,RO0,RO1,RO2,S0,D0/303.,0.14,
 46.,46.,2000.,470.,470.,890.,7850.,7850.,-1.1,-0.00065/
      END
```

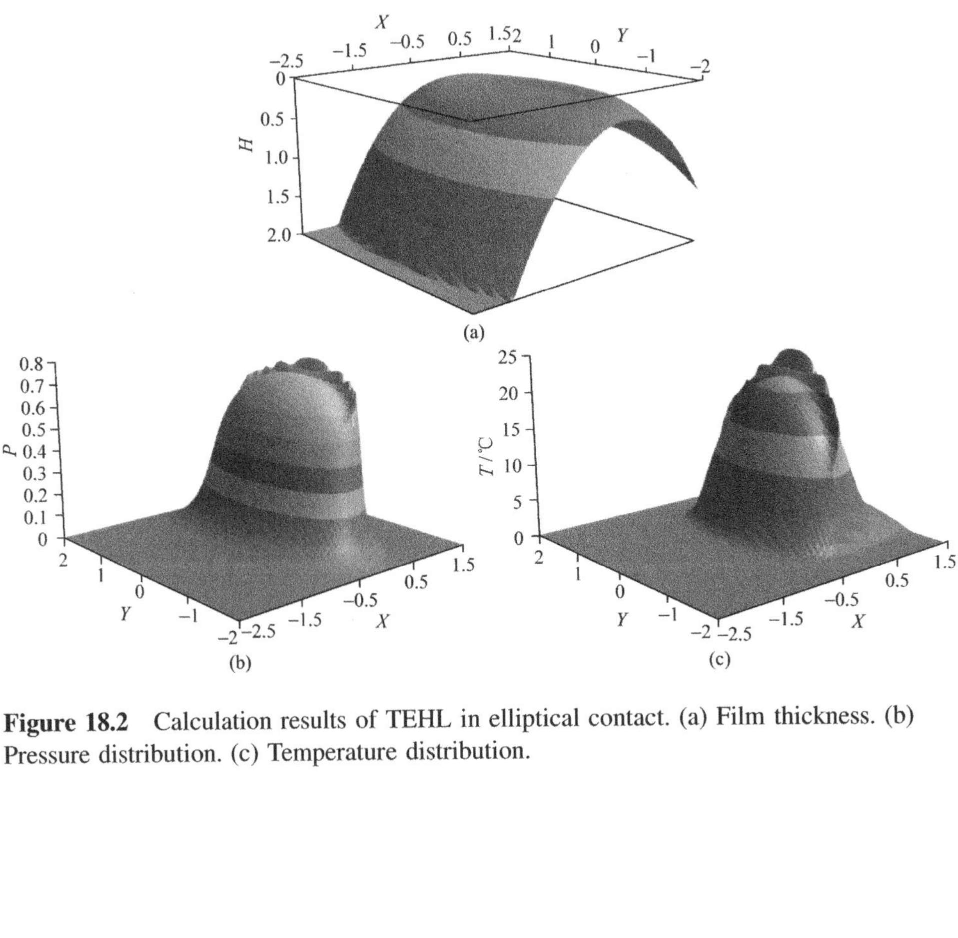

Figure 18.2 Calculation results of TEHL in elliptical contact. (a) Film thickness. (b) Pressure distribution. (c) Temperature distribution.

18.3.3 Calculation Results

Using the conditions given in the program, the film thickness *H*, the pressure distribution *P* and the average temperature distribution *T* (°C) are shown in Figure 18.2. Because the curvature radii in the *x* and *y* directions are different, this is an elliptical contact problem. From the results we can see that, in the contact area, the film thickness, the pressure distribution and the temperature distribution are elliptical.

19

Numerical Method and Program for Thermal Grease EHL in Line Contact

19.1 Basic Equations

The Reynolds equation and film thickness equation for thermal grease EHL in line contact are the same as those for isothermal grease EHL in line contact. Compared to the calculation for isothermal grease EHL in line contact, the calculation for thermal grease EHL in line contact adds the energy equation. Therefore, the calculation for thermal grease EHL in line contact needs the following equations.

19.1.1 Reynolds Equation

According to Equation 9.5, the Reynolds equation for grease based on the Ostwald model becomes

$$\frac{n}{2n+1}\cdot\left(\frac{1}{2}\right)^{\frac{n+1}{n}}\left\{\frac{d}{dx}\left[\rho h^{\frac{2n+1}{n}}\left(\frac{1}{\phi}\frac{dp}{dx}\right)^{\frac{1}{n}}\right]\right\}=U\frac{d(\rho h)}{dx} \tag{19.1}$$

where n is the rheological parameter ($n \le 1$), ϕ is the plastic viscosity of grease, x is the coordinate in the motion direction, ρ is the density of grease, p is the pressure, h is the film thickness, $\bar{h}$ is the film thickness when $\dfrac{dp}{dx}=0$, U is the average velocity [$U=(u_1+u_2)/2$] and u_1 and u_2 are the tangential velocities of the two surfaces.

Numerical Calculation of Lubrication: Methods and Programs, First Edition. Ping Huang.

19.1.2 Energy Equation

The two-dimensional energy equation including the film thickness direction can be written as

$$\rho c_p u \frac{\partial T}{\partial x} = K \frac{\partial^2 T}{\partial z^2} - \frac{T}{\rho} \cdot \frac{\partial \rho}{\partial T} u \frac{\partial p}{\partial x} + \phi \left(\frac{\partial u}{\partial z} \right)^2 \tag{19.2}$$

where T is lubricant temperature, u is the grease flow velocity, c_p is the specific heat at constant pressure, K is the coefficient of heat conduction and z is the coordinate across the film.

Because $\frac{\partial^2 T}{\partial z^2}$ is of the second derivative, two boundary conditions should be given for solving the energy Equation 19.2. They are

$$\begin{aligned} T(x,0) &= \frac{K}{\sqrt{\pi \rho_1 c_1 K_1 u_1}} \int_{-\infty}^{x} \left. \frac{\partial T}{\partial z} \right|_{x,0} \frac{ds}{\sqrt{x-s}} + T_0 \\ T(x,h) &= \frac{K}{\sqrt{\pi \rho_2 c_2 K_2 u_2}} \int_{-\infty}^{x} \left. \frac{\partial T}{\partial z} \right|_{x,h} \frac{ds}{\sqrt{x-s}} + T_0 \end{aligned} \tag{19.3}$$

where T_0 is the initial temperature, the subscripts 1 and 2 correspond, respectively, to the down and up surfaces, ρ_1 and ρ_2 are the material densities, $c_1$and c_2 are the material specific heats and K_1 and K_2 are the material heat transfer coefficients.

In order to calculate the temperature, not only must we know the average velocity U of the two surfaces, but also the both surface velocities u_1 and u_2 should be calculated. If the slip–roll ratio s is known, we can use the following equation to calculate them.

$$\begin{cases} u_1 = 0.5 \times (2 + s) \times U \\ u_2 = 0.5 \times (2 - s) \times U \end{cases} \tag{19.4}$$

19.1.3 Film Thickness Equation

$$h(x) = h_c + \frac{x^2}{2R} + v(x) \tag{19.5}$$

where h_c is the central film thickness to be determined by calculation, R is the equivalent radius of the surfaces $\left(\frac{1}{R} = \frac{1}{R_1} + \frac{1}{R_2} \right)$ and $v(x)$ is the elastic deformation.

19.1.4 Elastic Deformation Equation

In the film thickness Equation 17.4, the deformation $v(x)$ is

$$v(x) = -\frac{2}{\pi E}\int_{x_0}^{x_e} p(s)\ln(s-x)^2 ds + c \tag{19.6}$$

where E is the equivalent elastic modulus $\left[\text{i.e.}, \frac{1}{E} = \frac{1}{2}\left(\frac{1-\nu_1{}^2}{E_1} + \frac{1-\nu_2^2}{E_2}\right)\right]$ ν_1 and ν_2 are the Poisson's ratios of the two surfaces, E_1 and E_2 are the elastic modulus of the two surfaces, x_0 is the inlet coordinate, x_e is the outlet coordinate and c is the integral constant.

19.1.5 Viscosity–Pressure–Temperature Equation

$$\phi = \phi_0 \exp\left\{(\ln\phi_0 + 9.67)\left[(1 + 5.1\times 10^{-9}p)^{0.68} \times \left(\frac{T-138}{T_0-138}\right)^{-1.1} - 1\right]\right\} \tag{19.7}$$

where ϕ_0 is the viscosity of grease at $p=0$ and $T=T_0$, $p=0$ means the pressure is equal to the ambient pressure and T_0 is the ambient temperature.

19.1.6 Density–Pressure–Temperature Equation

$$\rho = \rho_0\left[1 + \frac{0.6p}{1+1.7p} + D(T-T_0)\right] \tag{19.8}$$

where ρ_0 is the density of grease at $p=0$ and $T=T_0$, the unit of pressure p is GPa and $D = -0.00\,065\ \text{K}^{-1}$.

19.2 Numerical Method

19.2.1 Dimensionless Equations

The dimensionless Reynolds equation for grease EHL in line contact is given as

$$\frac{d}{dX}\left[\varepsilon\left(\frac{dP}{dX}\right)^{1/n}\right] - \frac{d(\rho^* H)}{dX} = 0 \tag{19.9}$$

where $\varepsilon = \lambda\frac{\rho^* H^{(2+1/n)}}{\phi^{*1/n}}$, $\lambda = \frac{p_H^{1/n} b^{2+1/n}}{2U(2+1/n)R^{(1+1/n)}2^{1/n}\phi_0{}^{1/n}}$, ϕ^* is the dimensionless viscosity of grease ($\phi^* = \phi/\phi_0$), ρ^* is the dimensionless density of grease, X is the dimensionless coordinate ($X = x/b$), b is the half width of the contact area, P is the dimensionless pressure ($P = p/p_H$; p_H is the maximum Hertzian pressure) and H is the dimensionless film thickness ($H = \frac{hR}{b^2}$).

The boundary conditions are

$$\text{At the inlet}: \quad P(X_0) = 0$$

$$\text{At the outlet}: \quad P(X_e) = 0 \text{ and } \frac{dP(X_e)}{dX} = 0$$

where X_0 is the dimensionless inlet coordinate ($X_0 = x_0/b$) and X_e is the dimensionless outlet coordinate ($X_e = x_e/b$).

The dimensionless energy equation can be written as

$$A\rho^* u^* \frac{\partial T}{\partial X} + B\frac{u^* T^*}{\rho^*}\frac{\partial \rho^*}{\partial T^*}\frac{\partial P}{\partial X} + C\frac{\partial^2 T}{\partial Z^2} + D\phi^*\left(\frac{\partial u^*}{\partial Z}\right)^2 = 0 \tag{19.10}$$

where T^* is the dimensionless temperature ($T^* = T/T_0$), T_0 is the ambient temperature, $A = \rho_0 c_p U T_0/b$, $B = U p_H/b$, $C = -K T_0 R/b^2$, $D = -\phi_0\left(\frac{UR}{b^2}\right)^2$, u^* is the dimensionless flow velocity ($u^* = u/U$) and Z is the dimensionless coordinates cross the film thickness ($Z = z/h$).

The dimensionless film thickness equation is

$$H(X) = H_0 + \frac{X^2}{2} - \frac{1}{\pi}\int_{X_0}^{X_e} \ln|X - X'|P(X')dX' \tag{19.11}$$

where H_0 is the dimensionless film thickness of $h_0 = h_c + c$, which will be determined by calculation.

The viscosity–pressure and viscosity–temperature equations in dimensionless form are

$$\phi^* = \exp\left\{(\ln\phi_0 + 9.67)\left[(1 + 5.1\times 10^{-9}p)^{0.68} \times \left(\frac{T - 138}{T_0 - 138}\right)^{-1.1} - 1\right]\right\} \tag{19.12}$$

where ϕ^* is the dimensionless viscosity of lubricant, $\phi^* = \phi/\phi_0$.

The dimensionless density–temperature–pressure equation is

$$\rho^* = \left(1 + \frac{0.6p}{1 + 1.7p} + D(T - T_0)\right) \tag{19.13}$$

where ρ^* is the dimensionless viscosity of lubricant, $\rho^* = \rho/\rho_0$.

The dimensionless load equation is

$$\int_{X_0}^{X_e} P dx = \frac{\pi}{2} \tag{19.14}$$

19.2.2 Discrete Equations

The discrete form of the dimensionless Reynolds equation is

$$\frac{\varepsilon_{i+1/2}(P_{i+1} - P_i)^{1/n} - \varepsilon_{i-1/2}(P_i - P_{i-1})^{1/n}}{\Delta X^{1+1/n}} - \frac{\rho_i^* H_i - \rho_{i-1}^* H_{i-1}}{\Delta X} = 0 \tag{19.15}$$

where $\varepsilon_{i\pm1/2} = \frac{1}{2}(\varepsilon_i + \varepsilon_{i\pm1})$ and $\Delta X = X_i - X_{i-1}$.

In the inlet area, the boundary condition is $P(X_0) = 0$; The boundary for the outlet is determined by setting the negative pressure zero.

The dimensionless energy equation in a discrete form is

$$\begin{aligned} &A\rho^* u^* \left(\frac{T_{i,k} - T_{i-1,k}}{\Delta X}\right) + B\frac{T^*}{\rho^*}\frac{\partial \rho^*}{\partial T^*}\left(u^* \frac{P_{i,j} - P_{i-1,j}}{\Delta X}\right) \\ &+ C\frac{T_{i,k+1} - 2T_{i,k} + T_{i,k-1}}{\Delta Z^2} + D\phi^* \left(\frac{u^*_{i,k+1} - u^*_{i,k}}{\Delta Z}\right)^2 = 0 \end{aligned} \tag{19.16}$$

where the subscript k indicates the node number in the thickness direction. ρ^*, ϕ^*, and $\frac{\partial \rho^*}{\partial T^*}$ can be obtained from earlier expressions so that they do not need to be discretized, the film thickness is divided into k layers and ΔZ is the dimensionless distance.

The discrete film thickness equation is

$$H_i = H_0 + \frac{x_i^2}{2} - \frac{1}{\pi}\sum_{j=1}^{n} K_{ij} P_j \tag{19.17}$$

where K_{ij} is the stiffness coefficient of the elastic deformation.

The discrete dimensionless load equation is

$$\Delta X \sum_{i=1}^{n} \frac{(P_i + P_{i+1})}{2} = \frac{\pi}{2} \tag{19.18}$$

The following convergent criterion of pressure is used to determine the convergency.

$$\frac{\sum \left|P_i^{k+1} - P_i^k\right|}{\sum P_i^{k+1}} \le \varepsilon_4 \tag{19.19}$$

where ε_4 is a given error.

Similar to the EHL of oil, and because the temperature in the energy equation, the viscosity in the viscosity–pressure and viscosity–temperature equations and the elastic deformation in the film thickness equation all change with pressure, we should first give an initial pressure distribution (such as a Hertz contact pressure) and a temperature distribution (uniform temperature field) and then calculate the pressure and temperature with an iteration method.

19.3 Calculation Program

19.3.1 Calculation Diagram

The calculation program for thermal grease EHL in line contact contains a main program and many subroutines. The role of the main program is to pre-assign initial parameters, calculate the dimensionless parameters and set the output parameters. Then, we call the subroutines to solve the Reynolds equation and to calculate the lubrication film thickness. Subroutine THERM is used to calculate the temperature. It gets the pressure, film thickness, viscosity and so on from the above subroutines. After we substitute them into the energy equation and with the thermal boundary conditions, the temperature can be calculated. Subroutine ERRO is to calculate the error. If the error meets the accuracy requirements, the calculation is finished. The program diagram is shown in Figure 19.1.

19.3.2 Calculation Program

Preassignment parameters:

Node numbers	N = 129
Dimensionless inlet node coordinate	X0 = −4.0
Dimensionless outlet node coordinate	XE = 1.4
Load	W = 1.768E5 N
Equivalent modulus of elasticity	E1 = 2.21E11 Pa
Equivalent radius of cylinder	R = 0.023 m
Velocity	US = 1.77 m s^{-1}
Layer number across the film thickness	NZ = 5
Ratio of slide to roll	CU = 0.25
Iteration coefficient 1	C1 = 0.37
Iteration coefficient 2	C2 = 0.37
Temperature iteration coefficient	CT = 0.35
Difference of temperature	T2 = 0.0 K
Rheology parameter of grease	FF = 0.85

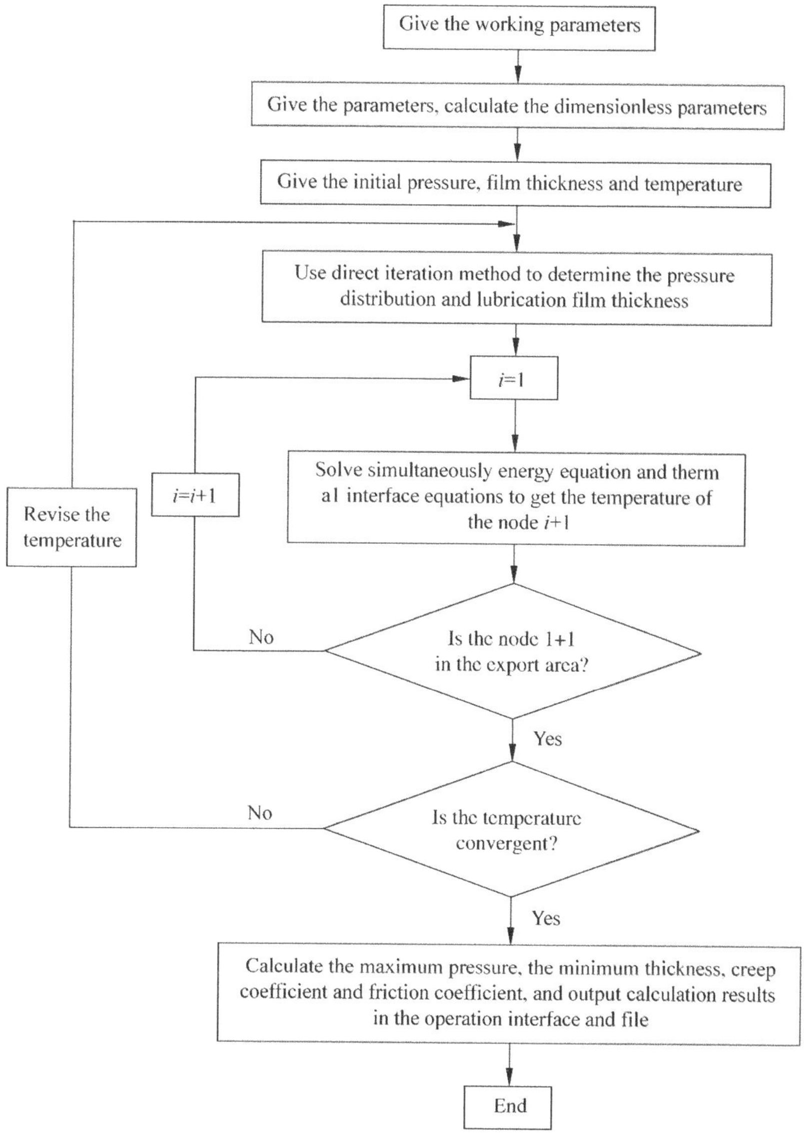

Figure 19.1 Diagram of calculation program.

In addition, in BLOCK DATA the following parameters are assigned.

Ambient temperature	T0 = 303 K
Initial viscosity of grease	EDA0 = 0.08 Pa·s
Thermal conductivity of lubricant	AK0 = 0.14 W·m^{-1}·K

Thermal conductivity of down surface	$AK1 = 46\ W{\cdot}m^{-1}{\cdot}K$
Thermal conductivity of up surface	$AK2 = 46\ W{\cdot}m^{-1}{\cdot}K$
Specific heat of lubricant	$CV = 2000\ J{\cdot}kg^{-1}{\cdot}K^{-1}$
Specific heat of down surface	$CV1 = 470\ J{\cdot}kg^{-1}{\cdot}K^{-1}$
Specific heat of up surface	$CV2 = 470\ J{\cdot}kg^{-1}{\cdot}K^{-1}$
Initial density of lubricant	$RO0 = 890\ kg{\cdot}m^{-3}$
Density 1 of down surface	$RO1 = 7850\ kg{\cdot}m^{-3}$
Density 2 of down surface	$RO2 = 7850\ kg{\cdot}m^{-3}$
Temperature–viscosity coefficient	$S0 = -1.1$
Temperature–density coefficient of lubricant	$D0 = -0.00\,065$.

If we need to change the above parameters, we need to recompile and relink the program before executing it.

Input parameters:

Input Character S. If S = "Y" or "y", calculate the example. If S = "N" or "n", then input the following data:

```
    Input: KT= If KT=2 consider temperature; if KT≠2 not consider
temperature
Input: W, US, FF=
If consider temperature, that is, KT=2, then input: NZ,CU=
    If change the iteration factors, input: Y, and then input: C1,C2=
```

Output parameters:

In the file FILEO.DAT, the following data will be output in columns:

```
Coordinate array Pressure array  Film thickness array
X(1)             P(1)            H(1)
X(2)             P(2)            H(2)
.....            .....           ......
X(N)             P(N)            H(N)
X(1)             T(1,1)          T(1,2)   T(1,3)   T(1,4)   T(1,5)
X(2)             T(2,1)          T(2,2)   T(2,3)   T(2,4)   T(2,5)
......           ......          ......   ......   ......   ......
X(N)             T(N,1)          T(N,2)   T(N,3)   T(N,4)   T(N,5)
```

```
PROGRAM GREASELINETHERMEHL
    CHARACTER*1 S,S1,S2
    CHARACTER*16 FILEO
    COMMON /COM1/ENDA,A1,A2,A3,Z,C1,C2,C3,CW
    COMMON /COM2/T0, EDA0/COM3/E1,PH,B,U1,U2,R,CT/COM4/X0,XE
```

```
COMMON /COM5/H2, P2,T2,ROM,HM,FM/COM6/FF
     DATA KT, S1, S2/0,1HY,1Hy/
     DATA FILEO/4HDATA/
     PAI=3.14159265
     Z=0.68
     P0=1.96E8
     N=129
     X0=-4.
     XE=1.4
     W=1.768E5
     E1=2.21E11
     R=0.02
     US=1.77
     C1=0.37
     C2=0.37
     NZ=5
     CU=0.25
     CT=0.35
     T2=0.0
     FF=0.85
     OPEN (8, FILE=FILEO, STATUS='UNKNOWN')
     WRITE (*,*)'Show the example or not (Y or N)?'
     READ (*,'(A)') S
     IF (S.EQ.S1.OR.S.EQ.S2)THEN
     KT=2
     GOTO 10
     ELSE
     WRITE (*,*)' Temperature is considered or not (Y or N) ?'
     READ (*,'(A)') S
     IF (S.EQ.S1.OR.S.EQ.S2) KT=2
     ENDIF
     WRITE (*,*)'W, US, FF='
     READ (*,*) W, US, FF
     IF (KT.EQ.2) THEN
     WRITE (*,*)'NZ, CU='
     READ (*,*) NZ,CU
     ENDIF
     WRITE (*,*)' Change iteration factors or not (Y or N) ?'
     READ (*,'(A)') S
     IF (S.EQ.S1.OR.S.EQ.S2) THEN
     WRITE (*,*)'C1, C2='
     READ (*,*) C1,C2
     ENDIF
```

```
10 CW=N+0.1
     LMAX=ALOG (CW) /ALOG(2.)
     N=2**LMAX+1
     LMIN= (ALOG(CW)-ALOG(SQRT(CW)))/ALOG(2.)
     LMAX=LMIN
     H00=0.0
     W1=W/ (E1*R)
     PH=E1*SQRT(0.5*W1/PAI)
     A1=(ALOG(EDA0)+9.67)
     A2=PH/P0
     A3=0.59/(PH*1.E-9)
     T2=0.0
     B=4.*R*PH/E1
     ALFA=Z*A1/P0
     G=ALFA*E1
     U=EDA0*US/(2.*E1*R)
     CC1=SQRT(2.*U)
     AM=2.*PAI*(PH/E1)**2/CC1
     AL=G*SQRT(CC1)
     CW= (PH/E1)*(B/R)
     C3=1.6*(R/B)**2*G**0.6*U**0.7*W1**(-0.13)
     ENDA=B**(2.+1/FF)*(PH/2/EDA0)**(1/FF)/R**(1+1/FF)/US/(2.
+1/FF)
     ENDA=1./ENDA
     U1=0.5*(2.+CU)*U
     U2=0.5*(2.-CU)*U
     CW=-1.13*C3
     WRITE(*,40)
40 FORMAT(2X,'Wait Please',//)
     CALL SUBAK(N)
     CALL MULTI(N,NZ,KT,LMIN,LMAX,H00)
     H2=H2*1.E-6
   P2=P2*1.E6
     Q=2.*ROM*HM*US
     FM=FM/W
     STOP
     END
     SUBROUTINE MULTI(N,NZ,KT,LMIN,LMAX,H00)
     DIMENSION X(1100),P(1100),H(1100),RO(1100),POLD(1100),
EPS(1100),EDA(1100),P0(2200),F(1100),F0(2200),R(1100),
R0(2200),G(10),T(22000)
     COMMON /COM1/ENDA,A1,A2,A3,Z,C1,C2,C3,CW/COM6/FF
     COMMON /COMK/K/COMT/LT,T1(1100)/COM3/E1,PH,B,U1,U2,RR,CT
     COMMON /COM5/H2,P2,T2,RM,HM,FM
     DATA MK,IT,KH,NMAX,PAI,G0/0,0,0,1100,3.14159265,1.570796325/
```

```
      LT=LMAX
      NX=N
      K=LMIN
      N0=(N-1)/2**(LMIN-1)
      CALL KNDX(K,N,N0,N1,NMAX,DX,X)
      DO 10 I=1,N
      T1(I)=1.0
      IF(ABS(X(I)).GE.1.0)P(I)=0.0
10 IF(ABS(X(I)).LT.1.0)P(I)=SQRT(1.-X(I)*X(I))
12 CALL HREE(N,DX,H00,G0,X,P,H,RO,EPS,EDA,F,0)
      IF(KH.NE.0)GOTO 14
      KH=1
      GOTO 12
14 CALL FZ(N,P,POLD)
      DO 100 L=LMIN,LMAX
      K=L
      G(K)=PAI/2.
      DO 18 I=1,N
      R(I)=0.0
      F(I)=0.0
      R0(N1+I)=0.0
18 F0(N1+I)=0.0
20 KK=2
      CALL ITER(N,KK,DX,H00,G0,X,P,H,RO,EPS,EDA,F,R,0)
      KK=1
      CALL ITER(N,KK,DX,H00,G0,X,P,H,RO,EPS,EDA,F,R,1)
      G(K-1)=G(K)
      DO 24 I=1,N
      IF(I.LT.N)G(K-1)=G(K-1)-0.5*DX*(P(I)+P(I+1))
24 P0(N1+I)=P(I)
      N2=N
      K=K-1
      CALL KNDX(K,N,N0,N1,NMAX,DX,X)
      CALL TRANS(N,N2,P,H,RO,EPS,EDA,R)
      CALL ITER(N,KK,DX,H00,G0,X,P,H,RO,EPS,EDA,F,R,2)
      DO 26 I=1,N
      IF(I.LT.N)G(K)=G(K)+0.5*DX*(P(I)+P(I+1))
26 F(I)=H(I)
      G0=G(K)
      CALL HREE(N,DX,H00,G0,X,P,H,RO,EPS,EDA,F,1)
      DO 28 I=1,N
      R0(N1+I)=R(I)
28 F0(N1+I)=F(I)
      IF(K.NE.1)GOTO 20
      KK=19
```

```
      CALL ITER(N,KK,DX,H00,G0,X,P,H,RO,EPS,EDA,F,R,0)
40 DO 42 I=1,N
42 P0(N1+I)=P(I)
      N2=N1
      K=K+1
      CALL KNDX(K,N,N0,N1,NMAX,DX,X)
      G0=G(K)
      DO 50 I=2,N,2
      I1=N1+I
      I2=N2+I/2
      P(I-1)=P0(I2)
      P(I)=P0(I1)+0.5*(P0(I2)+P0(I2+1)-P0(I1-1)-P0(I1+1))
50 IF(P(I).LT.0.0)P(I)=0.
      DO 52 I=1,N
      R(I)=R0(N1+I)
52 F(I)=F0(N1+I)
      CALL HREE(N,DX,H00,G0,X,P,H,RO,EPS,EDA,F,0)
      KK=1
      CALL ITER(N,KK,DX,H00,G0,X,P,H,RO,EPS,EDA,F,R,0)
      IF(K.LT.L)GOTO 40
100 CONTINUE
      MK=MK+1
      CALL ERROP(N,P,POLD,ERP)
      IF(ERP.GT.0.01*C2.AND.MK.LE.12)GOTO 14
      MK=8
      IF(KT.NE.2)GOTO 105
      CALL THERM(NX,NZ,DX,T,P,H)
      CALL ERROM(NX,NZ,T1,T,KT)
      IT=IT+1
    IF(KT.EQ.2.AND.IT.LT.25)GOTO 14
      KT=2
      IF(IT.GE.25)THEN
      WRITE(*,*)'Temperature is not convergent !!!'
      READ(*,*)
      ENDIF
105 IF(MK.GE.10)THEN
      WRITE(*,*)'Pressures are not convergent !!!'
      READ(*,*)
      ENDIF
      FM=FRICT(N,DX,H,P,EDA)
110 FORMAT(6(1X,E12.6))
      DO I=2,N-1
      IF(P(I).GE.P(I-1).AND.P(I).GE.P(I+1))THEN
      HM=H(I)*B*B/RR
      RM=RO(I)
```

```
      GOTO 120
      ENDIF
      ENDDO
120 H2=1.E5
      P2=1.E-10
      DO I=1,N
     H(I)=H(I)*B*B*1.E6/RR
      P(I)=P(I)*PH/1.E6
      IF(H(I).LT.H2)H2=H(I)
      IF(P(I).GT.P2)P2=P(I)
     ENDDO
     DO I=1,N
      WRITE(8,110)X(I),P(I),H(I)
      ENDDO
     IF(KT.EQ.2)THEN
      CALL OUPT(NX,NZ,X,T,T2)
      ENDIF
      RETURN
      END
      SUBROUTINE HREE(N,DX,H00,G0,X,P,H,RO,EPS,EDA,F0,KG)
      DIMENSION X(N),P(N),H(N),RO(N),EPS(N),EDA(N),F0(N)
      DIMENSION W(2200)
      COMMON /COM1/ENDA,A1,A2,A3,Z,C1,C2,C3,CW/COMK/K/COMT/LT,
T1(1100)
      COMMON /COM2/T0,EDA0,AK0,AK1,AK2,CV,CV1,CV2, RO0,RO1,RO2,
S0,D0/COMAK/AK(0:1100)/COM6/FF
      DATA KK,MK1,MK2,NW,PAI1/0,3,0,2200,0.318309886/
      IF(KK.NE.0)GOTO 3
      HM0=C3
3   W1=0.0
      DO 4 I=1,N
4   W1=W1+P(I)
      C3=(DX*W1)/G0
      DW=1.-C3
      IF(K.EQ.1)GOTO 6
      CALL DISP(N,NW,K,DX,P,W)
      GOTO 10
6   WX=W1*DX*ALOG(DX)
      DO 8 I=1,N
      W(I)=WX
      DO 8 J=1,N
      IJ=IABS(I-J)
8   W(I)=W(I)+AK(IJ)*P(J)*DX
10 HMIN=1.E3
      DO 30 I=1,N
```

```
      H0=0.5*X(I)*X(I)-PAI1*W(I)
      IF(KG.EQ.1)GOTO 20
      IF(H0+F0(I).LT.HMIN)HMIN=H0+F0(I)
      H(I)=H0
      GOTO 30
20 F0(I)=F0(I)-H00-H0
30 CONTINUE
      IF(KG.EQ.1)RETURN
      H0=H00+HMIN
      IF(KK.NE.0)GOTO 32
      KK=1
      H00=-H0+HM0
32 IF(H0.LE.0.0)GOTO 48
      IF(K.NE.1)GOTO 50
40 MK=MK+1
      IF(MK.LE.MK1)GOTO 50
      IF(MK.GE.MK2)MK=0
      IF(H0+CW*DW.GT.0.0)HM0=H0+CW*DW
      IF(H0+CW*DW.LE.0.0)HM0=HM0*C3
48 H00=HM0-HMIN
50 DO 60 I=1,N
60 H(I)=H00+H(I)+F0(I)
      IT=2**(LT-K)
      DO 100 I=1,N
      II=IT*(I-1)+1
      CT1=((T1(II)-0.455445545)/0.544554455)**S0
      CT2=D0*T0*(T1(II)-1.)
      EDA(I)=EXP(A1*(-1.+(1.+A2*P(I))**Z*CT1))
      RO(I)=(A3+1.34*P(I))/(A3+P(I))+CT2
      EPS(I)=RO(I)*H(I)**(2+1/FF)/ENDA/EDA(I)**(1/FF)
100 CONTINUE
      RETURN
      END
      SUBROUTINE ITER(N,KK,DX,H00,G0,X,P,H,RO,EPS,EDA,F0,R0,KG)
      DIMENSION X(N),P(N),H(N),RO(N),EPS(N),EDA(N),F0(N),R0(N)
      COMMON /COM1/ENDA,A1,A2,A3,Z,C1,C2,C3/COM6/FF/COMAK/AK
 (0:1100)
      DATA PAI/3.14159265/
      DX1=1./DX
      DX2=DX*DX
      DX3=1./DX2
      DX4=DX1/PAI
      DX5=DX1**(1.0+1/FF)
      DXL=DX*ALOG(DX)
      AK0=DX*AK(0)+DXL
```

```
      AK1=DX*AK(1)+DXL
      DO 100 K=1,KK
      RMAX=0.0
      D2=0.5*(EPS(1)+EPS(2))
      D3=0.5*(EPS(2)+EPS(3))
      D5=DX1*(RO(2)*H(2)-RO(1)*H(1))
      D7=DX4*(RO(2)*AK0-RO(1)*AK1)
      PP=0.
      DO 70 I=2,N-1
      D1=D2
      D2=D3
      D4=D5
      D6=D7
      IF(I+2.LE.N)D3=0.5*(EPS(I+1)+EPS(I+2))
      D5=DX1*(RO(I+1)*H(I+1)-RO(I)*H(I))
      D7=DX4*(RO(I+1)*AK0-RO(I)*AK1)
      AB1=(ABS(P(I+1)-P(I)))**(1/FF-1.0)
      AB2=(ABS(P(I)-P(I-1)))**(1/FF-1.0)
      IF(KG.NE.0)GOTO 30
      DD=(D1+D2)*DX3
      IF(DD.LT.0.1*ABS(D6))GOTO 10
      RI=-DX5*(D2*(P(I+1)-P(I))*AB1-D1*(P(I)-P(I-1))*AB2)+
D4+R0(I)
      DLDP=-DX3*(D1+D2)+D6
      RI=C1*RI/DLDP
      GOTO 20
10 RI=-DX5*(D2*(P(I+1)-P(I))*AB1-D1*(P(I)-PP)*AB2)+D4+R0(I)
      DLDP=-1/FF*DX5*(2*D1*AB1+D2*AB2)+2.*D6
      RI=C2*RI/DLDP
      IF(I.GT.2.AND.P(I-1)-RI.GT.0.0)P(I-1)=P(I-1)-RI
20 PP=P(I)
      P(I)=P(I)+RI
      IF(P(I).LT.0.0)P(I)=0.0
      IF(K.NE.KK)GOTO 70
      IF(RMAX.LT.ABS(RI).AND.P(I).GT.0.0)RMAX=ABS(RI)
      GOTO 70
30 IF(KG.EQ.2)GOTO 40
      R0(I)=-DX5*(D2*(P(I+1)-P(I))*AB1-D1*(P(I)-P(I-1))*AB2)+
D4+R0(I)
      GOTO 70
40 R0(I)=DX5*(D2*(P(I+1)-P(I))*AB1-D1*(P(I)-P(I-1))*AB2)-
D4+R0(I)
70 CONTINUE
      IF(KG.NE.0)GOTO 100
      CALL HREE(N,DX,H00,G0,X,P,H,RO,EPS,EDA,F0,0)
```

```
100 CONTINUE
      RETURN
      END
      SUBROUTINE DISP(N,NW,KMAX,DX,P1,W)
      DIMENSION P1(N),W(NW),P(2200),AK1(0:50),AK2(0:50)
      COMMON /COMAK/AK(0:1100)
      DATA NMAX,KMIN/2200,1/
      N2=N
      M=3+2*ALOG(FLOAT(N))
      K1=N+KMAX
      DO 10 I=1,N
 10 P(K1+I)=P1(I)
      DO 20 KK=KMIN,KMAX-1
      K=KMAX+KMIN-KK
      N1=(N2+1)/2
      CALL DOWNP(NMAX,N1,N2,K,P)
 20 N2=N1
      DX1=DX*2**(KMAX-KMIN)
      CALL WI(NMAX,N1,KMIN,KMAX,DX,DX1,P,W)
      DO 30 K=KMIN+1,KMAX
      N2=2*N1-1
      DX1=DX1/2.
      CALL AKCO(M+5,KMAX,K,AK1)
      CALL AKIN(M+6,AK1,AK2)
      CALL WCOS(NMAX,N1,N2,K,W)
      CALL CORR(NMAX,N2,K,M,1,DX1,P,W,AK1)
      CALL WINT(NMAX,N2,K,W)
      CALL CORR(NMAX,N2,K,M,2,DX1,P,W,AK2)
 30 N1=N2
      DO 40 I=1,N
 40 W(I)=W(K1+I)
      RETURN
      END
      SUBROUTINE DOWNP(NMAX,N1,N2,K,P)
      DIMENSION P(NMAX)
      K1=N1+K-1
      K2=N2+K-1
      DO 10 I=3,N1-2
      I2=2*I+K2
 10 P(K1+I)=(16.*P(I2)+9.*(P(I2-1)+P(I2+1))- (P(I2-3)+P(I2+3)))/32.
      P(K1+2)=0.25*(P(K2+3)+P(K2+5))+0.5*P(K2+4)
      P(K1+N1-1)=0.25*(P(K2+N2-2)+P(K2+N2))+ 0.5*P(K2+N2-1)
      RETURN
      END
      SUBROUTINE WCOS(NMAX,N1,N2,K,W)
```

```
      DIMENSION W(NMAX)
      K1=N1+K-1
      K2=N2+K
      DO 10 I=1,N1
      II=2*I-1
 10   W(K2+II)=W(K1+I)
      RETURN
      END
      SUBROUTINE WINT(NMAX,N,K,W)
      DIMENSION W(NMAX)
      K2=N+K
      DO 10 I=4,N-3,2
      II=K2+I
 10   W(II)=(9.*(W(II-1)+W(II+1))-(W(II-3)+W(II+3)))/16.
      I1=K2+2
      I2=K2+N-1
      W(I1)=0.5*(W(I1-1)+W(I1+1))
      W(I2)=0.5*(W(I2-1)+W(I2+1))
      RETURN
      END
      SUBROUTINE CORR(NMAX,N,K,M,I1,DX,P,W,AK)
      DIMENSION P(NMAX),W(NMAX),AK(0:M)
      K1=N+K
      IF(I1.EQ.2)GOTO 20
      DO 10 I=1,N,2
      II=K1+I
      J1=MAX0(1,I-M)
      J2=MIN0(N,I+M)
      DO 10 J=J1,J2
      IJ=IABS(I-J)
 10   W(II)=W(II)+AK(IJ)*DX*P(K1+J)
      RETURN
 20   DO 30 I=2,N,2
      II=K1+I
      J1=MAX0(1,I-M)
      J2=MIN0(N,I+M)
      DO 30 J=J1,J2
      IJ=IABS(I-J)
 30   W(II)=W(II)+AK(IJ)*DX*P(K1+J)
      RETURN
      END
      SUBROUTINE WI(NMAX,N,KMIN,KMAX,DX,DX1,P,W)
      DIMENSION P(NMAX),W(NMAX)
      COMMON /COMAK/AK(0:1100)
      K1=N+1
```

```
      K=2**(KMAX-KMIN)
      C=ALOG(DX)
      DO 10 I=1,N
      II=K1+I
      W(II)=0.0
      DO 10 J=1,N
      IJ=K*IABS(I-J)
   10 W(II)=W(II)+(AK(IJ)+C)*DX1*P(K1+J)
      RETURN
      END
      SUBROUTINE AKCO(KA,KMAX,K,AK1)
      DIMENSION AK1(0:KA)
      COMMON /COMAK/AK(0:1100)
      J=2**(KMAX-K)
      DO 10 I=0,KA
      II=J*I
   10 AK1(I)=AK(II)
      RETURN
      END
      SUBROUTINE AKIN(KA,AK1,AK2)
      DIMENSION AK1(KA),AK2(KA)
      DO 10 I=4,KA-3
   10 AK2(I)=(9.*(AK1(I-1)+AK1(I+1))-(AK1(I-3)+AK1(I+3)))/16.
      AK2(1)=(9.*AK1(2)-AK1(4))/8.
      AK2(2)=(9.*(AK1(1)+AK1(3))-(AK1(3)+AK1(5)))/16.
      AK2(3)=(9.*(AK1(2)+AK1(4))-(AK1(2)+AK1(6)))/16.
      DO 20 I=1,KA
   20 AK2(I)=AK1(I)-AK2(I)
      DO 30 I=1,KA,2
      I1=I+1
      AK1(I)=0.0
   30 IF(I1.LE.KA) AK1(I1)=AK2(I1)
      RETURN
      END
      SUBROUTINE SUBAK(MM)
      COMMON /COMAK/AK(0:1100)
      DO 10 I=0,MM
   10 AK(I)=(I+0.5)*(ALOG(ABS(I+0.5))-1.)-(I-0.5)*(ALOG(ABS(I-0.5))-1.)
      RETURN
      END
      FUNCTION FRICT(N,DX,H,P,EDA)
      DIMENSION H(N),P(N),EDA(N)
      COMMON /COM3/E1,PH,B,U1,U2,R,CT
      DATA TAU0/4.E7/
      TP=TAU0/PH
```

```
      TE=TAU0/E1
      BR=B/R
      FRICT=0.0
      DO I=1,N
      DP=0.0
      IF(I.LT.N)THEN
      DP=(P(I+1)-P(I))/DX
      TAU=0.5*H(I)*ABS(DP)*(BR/TP)+2.*ABS(U1-U2)*EDA(I)/
     (H(I)*BR**2*TE)
      FRICT=FRICT+TAU
      ENDIF
      ENDDO
      FRICT=FRICT*DX*B*TAU0
      RETURN
      END
      SUBROUTINE FZ(N,P,POLD)
      DIMENSION P(N),POLD(N)
      DO 10 I=1,N
   10 POLD(I)=P(I)
      RETURN
      END
      SUBROUTINE ERROP(N,P,POLD,ERP)
      DIMENSION P(N),POLD(N)
      SD=0.0
      SUM=0.0
      DO 10 I=1,N
      SD=SD+ABS(P(I)-POLD(I))
      POLD(I)=P(I)
   10 SUM=SUM+P(I)
      ERP=SD/SUM
      RETURN
      END
      SUBROUTINE KNDX(K,N,N0,N1,NMAX,DX,X)
      DIMENSION X(NMAX)
      COMMON /COM4/X0,XE
      N=2**(K-1)*N0
      DX=(XE-X0)/N
      N=N+1
      N1=N+K
      DO 10 I=1,N
   10 X(I)=X0+(I-1)*DX
      RETURN
      END
      SUBROUTINE TRANS(N1,N2,P,H,RO,EPS,EDA,R)
      DIMENSION P(N2),H(N2),RO(N2),EPS(N2),EDA(N2),R(N2)
```

```
      DO 10 I=1,N1
      II=2*I-1
      P(I)=P(II)
      H(I)=H(II)
      R(I)=R(II)
      RO(I)=RO(II)
      EPS(I)=EPS(II)
10    EDA(I)=EDA(II)
      RETURN
      END
      SUBROUTINE OUPT(NX,NZ,X,T,T2)
      DIMENSION X(NX),T(NX,NZ)
      DO I=1,NX
      DO K=1,NZ
      T(I,K)=303.*(T(I,K)-1.0)
      IF(T(I,K).GT.T2)T2=T(I,K)
      END DO
      END DO
      WRITE(8,20)NX,NX,NZ
20    FORMAT(15X,' T(1,1)-T(1,',I4,')-T(',I4,',',I2,') ')
      DO I=1,NX
      WRITE(8,30)X(I),(T(I,K),K=1,NZ)
      ENDDO
30    FORMAT(6(1X,E12.6))
      RETURN
      END
      SUBROUTINE ERROM(NX,NZ,T1,T,KT)
      DIMENSION T(NX,NZ),T1(NX)
      KT=2
      ERM=0.
      C1=1./FLOAT(NZ)
      DO 20 I=1,NX
      TT=0.
      DO 10 K=1,NZ
10    TT=TT+T(I,K)
      TT=C1*TT
      ER=ABS((TT-T1(I))/TT)
      IF(ER.GT.ERM)ERM=ER
20    T1(I)=TT
      IF(ERM.LT.0.003)KT=1
      RETURN
      END
      SUBROUTINE THERM(NX,NZ,DX,T,P,H)
      DIMENSION T(NX,NZ),P(NX),H(NX), T1(21),TI(21),TF(21),
  U(21),DU(21), W(21),EDA(21), RO(21),EDA1(21),EDA2(21),ROR(21),
  UU(21)
```

```
      DATA KK/0/
      IF(KK.NE.0)GOTO 4
      DO 2 K=1,NZ
2 T(1,K)=1.0
4 DO 30 I=2,NX
      KG=0
      DO 8 K=1,NZ
      TF(K)=T(I-1,K)
      IF(KK.NE.0)GOTO 6
      T1(K)=T(I-1,K)
      GOTO 8
6 T1(K)=T(I,K)
8 TI(K)=T1(K)
      P1=P(I)
      H1=H(I)
      DP=(P(I)-P(I-1))/DX
      CALL TBOUD(NX,NZ,I,CC1,CC2,T)
10 CALL EROEQ(NZ,T1,P1,EDA,RO,EDA1,EDA2,KG)
      CALL UCAL(NZ,DX,H1,EDA,RO,ROR,EDA1,EDA2,U,UU,DU,W,DP)
      CALL TCAL(NZ,DX,CC1,CC2,T1,TF,U,W,DU,H1,DP,EDA,RO)
   CALL ERRO(NZ,TI,T1,ETS)
   KG=KG+3
      IF(ETS.GT.1.E-4.AND.KG.LE.50)GOTO 10
      DO 20 K=1,NZ
      ROR(K)=RO(K)
      UU(K)=U(K)
20 T(I,K)=T1(K)
30 CONTINUE
      KK=1
      RETURN
      END
      SUBROUTINE TBOUD(NX,NZ,I,CC1,CC2,T)
      DIMENSION T(NX,NZ)
      CC1=0.
      CC2=0.
      DO 10 L=1,I-1
      DS=1./SQRT(FLOAT(I-L))
      IF(L.EQ.I-1)DS=1.1666667
      CC1=CC1+DS*(T(L,2)-T(L,1))
10 CC2=CC2+DS*(T(L,NZ)-T(L,NZ-1))
      RETURN
      END
      SUBROUTINE ERRO(NZ,T0,T,ETS)
      DIMENSION T0(NZ),T(NZ)
      ETS=0.0
```

```
      DO 10 K=1,NZ
      IF(T(K).LT.1.E-5)ETS0=1.
      IF(T(K).GE.1.E-5)ETS0=ABS((T(K)-T0(K))/T(K))
      IF(ETS0.GT.ETS)ETS=ETS0
10 T0(K)=T(K)
      RETURN
      END
      SUBROUTINE EROEQ(NZ,T,P,EDA,RO,EDA1,EDA2,KG)
      DIMENSION T(NZ),EDA(NZ),RO(NZ),EDA1(NZ),EDA2(NZ)
      COMMON /COM1/ENDA,A1,A2,A3,Z,O1,O2,O3
      COMMON /COM2/T0,EDA0,AK,AK1,AK2,CV,CV1,CV2,RO0,RO1,RO2,
S0,D0
    COMMON /COM3/E1,PH,B,U1,U2,R,CC/COM6/FF
      DATA A4,A5/0.455445545,0.544554455/
      IF(KG.NE.0)GOTO 20
      B1=(1.+A2*P)**Z
      B2=(A3+1.34*P)/(A3+P)
20 DO 30 K=1,NZ
      EDA(K)=EXP(A1*(-1.+B1*((T(K)-A4)/A5)**S0))
30 RO(K)=1+D0*T0*(T(K)-1.)
      CC1=0.5/(NZ-1.)
      CC2=1./(NZ-1.)
      C1=0.
      C2=0.
      DO 40 K=1,NZ
      IF(K.EQ.1)GOTO 32
      C1=C1+0.5/EDA(K)+0.5/EDA(K-1)
      C2=C2+CC1*((K-1.)/EDA(K)+(K-2.)/EDA(K-1))
32 EDA1(K)=C1*CC2
40 EDA2(K)=C2*CC2
      IF(KG.NE.2)RETURN
      C1=0.
      C2=0.
      C3=0.
      DO 50 K=1,NZ
      IF(K.EQ.1)GOTO 50
      C1=C1+0.5*(RO(K)+RO(K-1))
      C2=C2+0.5*(RO(K)*EDA1(K)+RO(K)*EDA1(K-1))
      C3=C3+0.5*(RO(K)*EDA2(K)+RO(K)*EDA2(K-1))
50 CONTINUE
      B1=12.*CC2*(C1*EDA2(NZ)/EDA1(NZ)-C2)
60 B2=2.*CC2/(U1+U2)*(C1*(U1-U2)/EDA1(NZ)+C3*U1)
      RETURN
      END
      SUBROUTINE UCAL(NZ,DX,H,EDA,RO,ROR,EDA1,EDA2,U,UU,DU,W,DP)
```

```
      DIMENSION U(NZ),DU(NZ),W(NZ),ROR(NZ),UU(NZ), EDA(NZ),
RO(NZ), EDA1(NZ),EDA2(NZ)
      COMMON /COM2/T0,EDA0,AK,AK1,AK2,CV,CV1,CV2, RO0,RO1,RO2,
S0,D0
      COMMON /COM3/E1,PH,B,U1,U2,R,CC
      DATA KK/0/
      IF(KK.NE.0)GOTO 20
      A1=U1
      A2=PH*(B/R)**3/E1
      A3=U2-U1
20 CC1=A2*DP*H
      CC2=CC1*H
      CC3=A3/H
      CC4=1./EDA1(NZ)
      DO 30 K=1,NZ
      U(K)=A1+CC2*(EDA2(K)-CC4*EDA2(NZ)*EDA1(K))+ A3*CC4*EDA1(K)
      IF(U(K).LT.0.0)U(K)=0.
30 DU(K)=CC1/EDA(K)*((K-1.)/(NZ-1.)-CC4*EDA2(NZ))+CC3*CC4/EDA(K)
      A4=B/((NZ-1)*R*DX)
      C1=A4*H
      IF(KK.EQ.0)GOTO 50
      DO 40 K=2,NZ-1
      W(K)=(RO(K-1)*W(K-1)+C1*(RO(K)*U(K)-ROR(K)*UU(K)))/RO(K)
40 CONTINUE
50 KK=1
      RETURN
      END
      SUBROUTINE TCAL(NZ,DX,CC1,CC2,T,TF,U,W,DU,H,DP,EDA,RO)
      DIMENSION T(NZ),TF(NZ),U(NZ),DU(NZ),W(NZ), EDA(NZ),RO(NZ),
A(4,21),D(21),AA(2,21)
      COMMON /COM2/T0,EDA0,AK,AK1,AK2,CV,CV1,CV2, RO0,RO1,RO2,S0,D0
      COMMON /COM3/E1,PH,B,U1,U2,R,CC
      DATA KK,CC5,PAI,TAU0/0,0.6666667,3.14159265,4.E7/
      IF(KK.NE.0)GOTO 5
      KK=1
      TAU=TAU0*B*B/(E1*R*R)
      A2=-CV*RO0*E1*B**3/(EDA0*AK*R)
      A3=-E1*PH*B**3*D0/(AK*EDA0*T0*R)
      A4=-(E1*R)**2/(AK*EDA0*T0)
      A5=0.5*R/B*A2
      A6=AK*SQRT(EDA0*R/(PAI*RO1*CV1*U1*E1*AK1*B**3))
      A7=AK*SQRT(EDA0*R/(PAI*RO2*CV2*U2*E1*AK2*B**3))
5 CC3=A6*SQRT(DX)
      CC4=A7*SQRT(DX)
      DZ=H/(NZ-1.)
```

```
      DZ1=1./DZ
      DZ2=DZ1*DZ1
      CC6=A3*DP
      DO 10 K=2,NZ-1
      A(1,K)=DZ2+DZ1*A5*RO(K)*W(K)
      A(2,K)=-2.*DZ2+A2*RO(K)*U(K)/DX+CC6*U(K)/RO(K)
      A(3,K)=DZ2-DZ1*A5*RO(K)*W(K)
      AE=ABS(EDA(K)*DU(K))
10 A(4,K)=A4*ABS(DU(K))*AE+A2*RO(K)*U(K)*TF(K)/DX
      A(1,1)=0.
      A(2,1)=1.+2.*DZ1*CC3*CC5
      A(3,1)=-2.*DZ1*CC3*CC5
      A(1,NZ)=-2.*DZ1*CC4*CC5
      A(2,NZ)=1.+2.*DZ1*CC4*CC5
      A(3,NZ)=0.
      A(4,1)=1.+CC1*CC3*DZ1
      A(4,NZ)=1.-CC2*CC4*DZ1
      CALL TRA3(NZ,D,A,AA)
      DO 20 K=1,NZ
20 T(K)=(1.-CC)*T(K)+CC*D(K)
30 CONTINUE
      RETURN
      END
      SUBROUTINE TRA3(N,D,A,B)
      DIMENSION D(N),A(4,N),B(2,N)
      C=1./A(2,N)
      B(1,N)=-A(1,N)*C
      B(2,N)=A(4,N)*C
      DO 10 I=1,N-1
      IN=N-I
      IN1=IN+1
      C=1./(A(2,IN)+A(3,IN)*B(1,IN1))
      B(1,IN)=-A(1,IN)*C
10 B(2,IN)=(A(4,IN)-A(3,IN)*B(2,IN1))*C
      D(1)=B(2,1)
      DO 20 I=2,N
20 D(I)=B(1,I)*D(I-1)+B(2,I)
      RETURN
      END
      BLOCK DATA
      COMMON /COM2/T0,EDA0,AK,AK1,AK2,CV,CV1,CV2, RO0,RO1,RO2,
S0,D0
      DATA T0, EDA0, AK, AK1, AK2, CV, CV1, CV2, RO0,RO1,RO2,S0,
D0/303.,0.08,0.14,46.,46.,2000.,470.,470.,890.,7850.,7850.,
-1.1,-0.00065/
      END
```

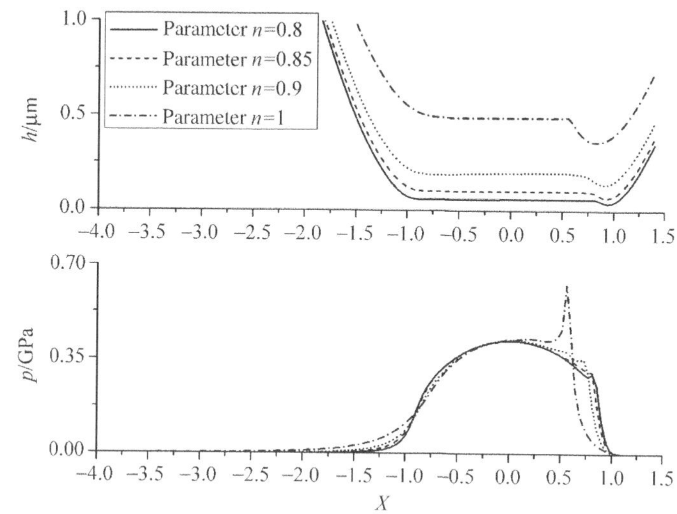

Figure 19.2 Influence of rheology parameter n on lubrication film thickness and pressure ($w = 100\,\text{kN}$, $u_0 = 0.87\,\text{m s}^{-1}$).

19.3.3 Calculation Results

The characteristics of thermal grease EHL in line contact can be analyzed according to the operation results. Figure 19.2 shows the influence of the rheology parameter n on film thickness and pressure distribution of thermal grease EHL in line contact. Although heat produces a certain influence on pressure and film thickness, the thermal and isothermal EHL give no significant difference in the basic characteristics.

Figure 19.3 shows that the temperature rises across the film under the given parameters. Curves A and E are the temperature rises in the two surfaces; the three

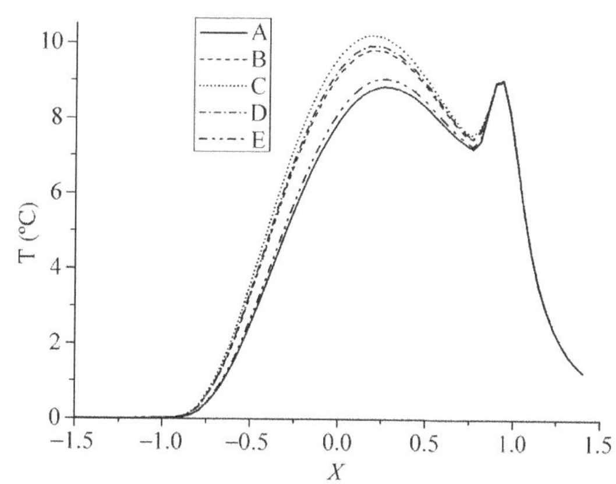

Figure 19.3 Temperature rise across the film ($w = 100\,\text{kN}$, $u_0 = 0.87\,\text{m s}^{-1}$, $n = 0.846$).

curves B, C, D are the temperature rises in the lubricant film, in which curve C is the temperature rise in the middle layer. From this figure, we can see that each temperature curve presents a peak at the secondary pressure peak, but the temperature peak is a little later than the pressure peak. We can also see that the temperature rises in the two surfaces are significantly lower than that in the middle of the film. The temperature rise of the down surface with a higher velocity is lower than that of the up surface with a lower velocity.

20

Numerical Method and Program for Thermal Grease EHL in Point Contact

20.1 Basic Equations

The basic equations for thermal grease EHL in a point contact can be obtained by adding the energy equation to the equations for isothermal grease EHL in point contact. The main equations for thermal grease EHL in point contact are as follows.

20.1.1 Reynolds Equation

The Reynolds equation for grease EHL based on the Ostwald rheological model in point contact can be written as

$$\frac{n}{2n+1}\left(\frac{1}{2}\right)^{\frac{n+1}{n}}\left\{\frac{\partial}{\partial x}\left[\rho h^{\frac{2n+1}{n}}\left(\frac{1}{\phi}\frac{\partial p}{\partial x}\right)^{\frac{1}{n}}\right]+\frac{\partial}{\partial y}\left[\rho h^{\frac{2n+1}{n}}\left(\frac{1}{\phi}\frac{\partial p}{\partial y}\right)^{\frac{1}{n}}\right]\right\}=U\frac{\partial(\rho h)}{\partial x} \quad (20.1)$$

where n is the rheological parameter ($n \leq 1$), ϕ is the plastic viscosity of grease, x is the coordinate in the motion direction, y is the horizontal coordinate perpendicular to the motion, ρ is the grease density, p is the pressure, h is the film thickness, U is the average velocity ($U=(u_1+u_2)/2$) and u_1 and u_2 are, respectively, the tangential velocities of the down and up surfaces.

Numerical Calculation of Lubrication: Methods and Programs, First Edition. Ping Huang.

20.1.2 Energy Equation

The three-dimensional energy equation, including the film thickness direction, can be written as

$$\rho c_p\left(u\frac{\partial T}{\partial x}+v\frac{\partial T}{\partial y}\right)=k\frac{\partial^2 T}{\partial z^2}-\frac{T}{\rho}\frac{\partial\rho}{\partial T}\left(u\frac{\partial p}{\partial x}+v\frac{\partial p}{\partial y}\right)+\phi\left[\left(\frac{\partial u}{\partial z}\right)^2+\left(\frac{\partial v}{\partial z}\right)^2\right] \tag{20.2}$$

where T is the grease temperature, u and v are the flow velocities of grease in the x and y directions, c_p is the specific heat at constant pressure, K is the thermal conductivity coefficient and z is the coordinate across the film.

Because the energy Equation 20.2 includes the term $\frac{\partial^2 T}{\partial z^2}$, the thermal boundary conditions of the down and up surfaces should be given. They are

$$\begin{aligned}
T(x,y,0) &= \frac{K}{\sqrt{\pi\rho_1 c_1 K_1 u_1}}\int_{-\infty}^{x}\left.\frac{\partial T}{\partial z}\right|_{x,y,0}\frac{ds}{\sqrt{x-s}}+T_0\\
T(x,y,h) &= \frac{K}{\sqrt{\pi\rho_2 c_2 K_2 u_2}}\int_{-\infty}^{x}\left.\frac{\partial T}{\partial z}\right|_{x,y,h}\frac{ds}{\sqrt{x-s}}+T_0
\end{aligned} \tag{20.3}$$

where T_0 is the initial temperature, the subscripts 1 and 2 correspond, respectively, to the down and up surfaces, ρ_1 and ρ_2 are the material densities, c_1 and c_2 are the material specific heats and K_1 and K_2 are the material heat transfer coefficients.

In order to calculate the temperature, not only must we know the average velocity U of the two surfaces, but also both surface velocities u_1 and u_2 should be calculated. If the slip–roll ratio s is known, we can use the following equations to calculate them.

$$\begin{cases} u_1 = 0.5\times(2+s)\times U \\ u_2 = 0.5\times(2-s)\times U \end{cases} \tag{20.4}$$

20.1.3 Film Thickness Equation

$$h(x,y) = h_0+\frac{x^2}{2R_x}+\frac{y^2}{2R_y}+v(x,y) \tag{20.5}$$

where h_0 is a film thickness parameter to be determined by calculation, R_x is the equivalent radius of the surfaces in the x direction $\left(\frac{1}{R_x}=\frac{1}{R_{1x}}+\frac{1}{R_{2x}}\right)$, R_y is the equivalent radius of the surfaces in the y direction $\left(\frac{1}{R_y}=\frac{1}{R_{1y}}+\frac{1}{R_{2y}}\right)$ and v is the elastic deformation.

20.1.4 Elastic Deformation Equation

$$v(x,y) = \frac{2}{\pi E}\iint_{\Omega} \frac{p(s,t)}{\sqrt{(x-s)^2+(y-t)^2}}dsdt \tag{20.6}$$

where E is the equivalent elastic modulus $\left[\frac{1}{E} = \frac{1}{2}\left(\frac{1-\nu_1^2}{E_1}+\frac{1-\nu_2^2}{E_2}\right)\right]$, ν_1 and ν_2 are the Poisson's ratios of the two surfaces and E_1 and E_2 are the elastic modulus of the two surfaces.

20.1.5 Roelands Viscosity–Pressure–Temperature Equation

$$\phi = \phi_0 \exp\left\{(\ln \phi_0 + 9.67)\left[(1+5.1\times 10^{-9}p)^{0.68}\times\left(\frac{T-138}{T_0-138}\right)^{-1.1} - 1\right]\right\} \tag{20.7}$$

where ϕ_0 is the viscosity of grease at $p=0$ and $T=T_0$, $p=0$ means the pressure is equal to the ambient pressure and T_0 is the ambient temperature.

20.1.6 Density–Pressure–Temperature Equation

$$\rho = \rho_0\left[1+\frac{0.6p}{1+1.7p}+D(T-T_0)\right] \tag{20.8}$$

where ρ_0 is the density of grease at $p=0$ and $T=T_0$, the unit of pressure p is GPa and $D=-0.00065\ \mathrm{K}^{-1}$.

20.2 Numerical Method

20.2.1 Dimensionless Equations

The dimensionless Reynolds equation for grease EHL in a point contact can be expressed as

$$\frac{\partial}{\partial X}\left(\varepsilon'\frac{\partial P}{\partial X}\right)+\alpha^2\frac{\partial}{\partial Y}\left(\varepsilon''\frac{\partial P}{\partial Y}\right)-\frac{\partial(\rho^* H)}{\partial X}=0 \tag{20.9}$$

where $\varepsilon' = \varepsilon\left|\frac{\partial P}{\partial X}\right|^{\frac{1-n}{n}}$, $\varepsilon'' = \varepsilon\left|\frac{\partial P}{\partial Y}\right|^{\frac{1-n}{n}}$, $\varepsilon = \frac{\lambda\rho^* H^{\frac{2n+1}{n}}}{\phi^{*\frac{1}{n}}}$, $\lambda = \frac{p_H^{\frac{1}{n}} a^{\frac{2n+1}{n}}}{2^{\frac{n+1}{n}}U\left(2+\frac{1}{n}\right)R_x^{\frac{n+1}{n}}\phi_0^{\frac{1}{n}}}$, R_x is the equivalent radius of the surface curvature in the x direction, P is the dimensionless pressure ($P=p/p_H$), p_H is the Hertzian maximum contact stress, H is the dimensionless film thickness $\left(H=\dfrac{hR_x}{a^2}\right)$, a is the half length of the ellipse contact area in the x direction, X is a dimensionless coordinate ($X=x/a$), Y is a dimensionless coordinate ($Y=y/b$), b is the half length of the ellipse contact area in the y direction, α is a ratio parameter ($\alpha=a/b$; and $\alpha=1$ in the following results), ρ^* is the dimensionless density ($\rho^*=\rho/\rho_0$), $\rho_0$is the density at room temperature and atmospheric pressure and ϕ^* is the dimensionless plastic viscosity of grease ($\phi^*=\phi/\phi_0$).

The boundary conditions are:

$$\text{At the inlet}\quad P(X_0,Y)=0$$
$$\text{At the outlet}\quad P(X_e,Y)=0 \text{ and } \frac{\partial P(X_e,Y)}{\partial X}=0$$
$$\text{At the two sides}\quad P|_{Y=\pm1}=0$$

The dimensionless energy equation is

$$A\rho^*u^*\left(\frac{\partial T}{\partial X}+\alpha\frac{\partial T}{\partial Y}\right)+B\frac{T^*}{\rho^*}\frac{\partial\rho^*}{\partial T^*}\left(u^*\frac{\partial P}{\partial X}+\alpha v^*\frac{\partial P}{\partial Y}\right)+C\frac{\partial^2 T}{\partial Z^2}+D\phi^*\left(\frac{\partial u^*}{\partial Z}\right)^2=0 \tag{20.10}$$

where T^* is the dimensionless temperature ($T^*=T/T_0$), $A=\rho_0c_pUT_0/b$, $B=Up_H/b$, $C=-KT_0R/b^2$, $D=-\phi_0\left(\dfrac{UR}{b^2}\right)^2$, u^* is the dimensionless flow velocity in the x direction ($u^*=u/U$), v^* is the dimensionless flow velocity in the y direction ($v^*=v/U$) and Z is the dimensionless coordinates cross the film thickness ($Z=z/h$).

The dimensionless film thickness equation is

$$H(X,Y)=H_0+\frac{X^2+Y^2}{2}+\frac{2}{\pi^2}\int_{X_0}^{X_e}\int_{Y_0}^{Y_e}\frac{P(S,T)dSdT}{\sqrt{(X-S)^2+(Y-T)^2}} \tag{20.11}$$

where H_0 is the dimensionless film thickness to be determined by calculation.

The dimensionless viscosity–pressure–temperature equation

$$\phi^*=\exp\left\{(\ln\phi_0+9.67)\left[(1+5.1\times10^{-9}p)^{0.68}\times\left(\frac{T-138}{T_0-138}\right)^{-1.1}-1\right]\right\} \tag{20.12}$$

where ϕ^* is the dimensionless viscosity of grease, $\phi^*=\phi/\phi_0$.

The dimensionless density–pressure–temperature equation

$$\rho^* = \left(1 + \frac{0.6p}{1 + 1.7p} + D(T - T_0)\right) \tag{20.13}$$

where ρ^* is the dimensionless viscosity of grease, $\rho^* = \rho/\rho_0$.

The dimensionless load equation

$$\int_{X_0}^{X_e}\int_{Y_0}^{Y_e} P(X, Y)dXdY = \frac{2}{3}\pi \tag{20.14}$$

20.2.2 Discrete Equations

Discrete dimensionless Reynolds equation

$$\frac{\left(\varepsilon'_{i-1/2,j}P_{i-1,j} + \varepsilon'_{i+1/2,j}P_{i+1,j} + \varepsilon''_{i,j-1/2}P_{i,j-1} + \varepsilon''_{i,j+1/2}P_{i,j+1} - \varepsilon_0 P_{i,j}\right)}{\Delta X^2}$$

$$-\frac{\left(\rho^*_{i,j}H_{i,j} - \rho^*_{i-1,j}H_{i-1,j}\right)}{\Delta X} = 0 \tag{20.15}$$

where $\varepsilon'_{i-1/2,j} = \frac{1}{2}\left(\varepsilon_{i,j} + \varepsilon_{i-1,j}\right)\left|\frac{P_{i,j} - P_{i-1,j}}{\Delta X}\right|^{\frac{1-n}{n}}$, $\varepsilon'_{i+1/2,j} = \frac{1}{2}\left(\varepsilon_{i,j} + \varepsilon_{i+1,j}\right)\left|\frac{P_{i+1,j} - P_{i,j}}{\Delta X}\right|^{\frac{1-n}{n}}$, $\varepsilon''_{i,j-1/2} = \frac{1}{2}\left(\varepsilon_{i,j} + \varepsilon_{i,j-1}\right)\left|\frac{P_{i,j} - P_{i,j-1}}{\Delta Y}\right|^{\frac{1-n}{n}}$, $\varepsilon''_{i,j+1/2} = \frac{1}{2}\left(\varepsilon_{i,j} + \varepsilon_{i,j+1}\right)\left|\frac{P_{i,j} - P_{i,j+1}}{\Delta Y}\right|^{\frac{1-n}{n}}$, $\varepsilon_0 = \varepsilon'_{i-1/2,j} + \varepsilon'_{i+1/2,j} + \varepsilon''_{i,j-1/2} + \varepsilon''_{i,j+1/2}$.

Since the grids are uniform, the dimensionless node spaces are the same, $\Delta Y = \Delta X$.

Discrete dimensionless energy equation

$$A\rho^* u^*\left(\frac{T_{i,j,k} - T_{i-1,j,k}}{\Delta X} + \alpha\frac{T_{i,j,k} - T_{i,j-1,k}}{\Delta Y}\right) + B\frac{T^*}{\rho^*}\frac{\partial\rho^*}{\partial T^*}\left(u^*\frac{P_{i,j} - P_{i-1,j}}{\Delta X} + \alpha v^*\frac{P_{i,j} - P_{i,j-1}}{\Delta Y}\right)$$

$$+ C\frac{T_{i,j,k+1} - 2T_{i,j,k} + T_{i,j,k-1}}{\Delta Z^2} + D\phi^*\left(\frac{u^*_{i,j,k+1} - u^*_{i,j,k}}{\Delta Z}\right)^2 = 0 \tag{20.16}$$

where k is the node number in the film thickness direction, ρ^*, ϕ^* and $\frac{\partial\rho^*}{\partial T^*}$ can be calculated analytically by expressions (so their differences are not needed), the film thickness is divided into k layers and ΔZ is the dimensionless distance.

Discrete dimensionless film thickness equation

$$H_{ij} = H_0 + \frac{X_i^2 + Y_j^2}{2} + \frac{2}{\pi^2}\sum_{k=1}^{n}\sum_{l=1}^{n} D_{ij}^{kl} P_{kl} \tag{20.17}$$

where D_{ij}^{kl} is the stiffness coefficient of elastic deformation.

Discrete dimensionless load equation

$$\Delta X \Delta Y \sum_{i=1}^{n}\sum_{j=1}^{n} P_{ij} = 2\pi/3 \tag{20.18}$$

The following convergent criterion of pressure is used to determine the convergency.

$$\frac{\sum_i \sum_j |P_{i,j}^{k+1} - P_{i,j}^{k}|}{\sum_i \sum_j P_{i,j}^{k+1}} \le \varepsilon_4 \tag{20.19}$$

where ε_4 is a given error.

Since the temperature in the energy equation, the viscosity in the viscosity–pressure–temperature equations and the elastic deformation in the film thickness equation are all dependent on the pressure distribution, the general numerical method is as follows.

We give an initial pressure (Hertz contact pressure) and temperature distribution (uniform temperature field), calculate the film thickness and viscosity, and then we substitute them into the Reynolds equation to solve the new pressure distribution. We revise repeatedly the pressure distribution obtained by previous iteration, then substitute the pressure into the energy equation to get the temperature distribution. To update the viscosity by the new temperature, we solve the pressure iteratively, repeat the process until the difference between the pressure distributions obtained from two successive iterations is very small and then stop the iteration. Thus, we get the final film thicknesss, pressure and temperature distribution.

20.3 Calculation Program

20.3.1 Calculation Diagram

The calculation diagram for thermal grease EHL in point contact is shown in Figure 20.1.

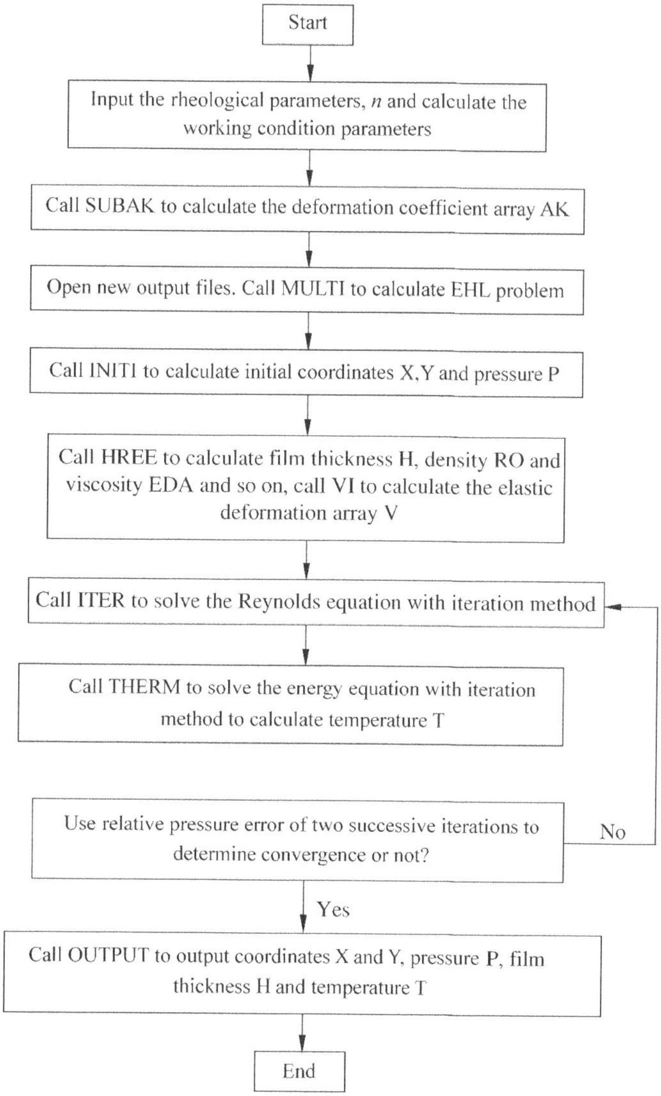

Figure 20.1 Calculation diagram for thermal grease EHL in point contact.

20.3.2 Calculation Program

In this program, the following parameters are preassigned.

Viscosity–pressure coefficient	Z = 0.68
Number of nodes	N = 65 × 65
Load	W0 = 39.24 N
Equivalent modulus of elasticity	E1 = 2.21E11 Pa
Initial plastic viscosity	EDA0 = 0.058 Pa·s

Equivalent curvature radius	RX = RY = 0.05 m
Average velocity	US = 1.5 m s^{-1}
Dimensionless starting coordinate in *x* direction	X0 = −2.5
Dimensionless ending coordinate in *x* direction	XE = 1.5
Dimensionless starting coordinate in *y* direction	Y0 = −2.0
Dimensionless ending coordinate in *y* direction	YE = 2.0
Number of layers across the film thickness direction	NZ = 5
Iteration coefficient of temperature	CT = 0.31
Sliding–rolling ratio	AKC = 1.0

In BLOCK DATA, the following parameters are given:

Initial temperature	T0 = 303 K
Thermal conductivity of grease	AK0 = 0.14 W·m^{-1}·K
Thermal conductivity of down surface	AK1 = 46 W·m^{-1}·K
Thermal conductivity of up surface	AK2 = 46 W·m^{-1}·K
Specific heat of grease	CV = 2000 J·kg^{-1}·K^{-1}
Specific heat of down surface	CV1 = 470 J·kg^{-1}·K^{-1}
Specific heat of up surface	CV2 = 470 J·kg^{-1}·K^{-1}
Initial density of grease	RO0 = 890 kg·m^{-3}
Density of down surface	RO1 = 7850 kg·m^{-3}
Density of up surface	RO2 = 7850 kg·m^{-3}
Viscosity–temperature coefficient	S0 = −1.1
Density–temperature coefficient	D0 = −0.00 065

The output parameters are as follows.

The pressure array P(I, J) will be saved in the file PRESS.DAT
The film thickness array H(I, J) will be saved in the file FILM.DAT
The temperature array T(I, J) will be saved in the file TEM.DAT.

Furthermore, after executing the program, one needs to input the rheological parameter n ($0.8 < n < 1$). If the user needs to change the above parameters' values, the program must be re-compiled and re-linked before execution.

```
      PROGRAM GREASEPOINTEHLT
      DIMENSION THETA(15),EALFA(15),EBETA(15)
      COMMON /COM1/Z,ENDA,AKC,HM0,HMC,EK,EAL,EBE,AD,AD1,KK1,KK2,
KK3,KK4,FN,FN1,FF
      COMMON /COM2/W0,E1,RX,B,PH,US,U1,U2,T1,T2,CT
      COMMON /COM3/T0,EDA0,AK,AK1,AK2,CV,CV1,CV2,RO0,RO1,RO2,S0,D0
      COMMON /COM4/A1,A2,A3,LMIN
      DATA PAI,Z,AKC,AD,AD1/3.14159265,0.68,1.0,0.0,0.0/
      DATA T0,EDA0,AK,AK1,AK2,CV,CV1,CV2,RO0,RO1,RO2,S0,D0/303.,
 0.058,0.14,46.,46.,2000.,470.,470.,890.,7850.,7850.,-1.1,-0.00065/
```

```
      DATA N,NZ,RX,RY,X0,XE,W0,E1,US,CT/65,5,0.05,0.05,-2.5,1.5,
39.24,2.21E11,1.5,0.31/
      DATA THETA/10.,20.,30.,35.,40.,45.,50.,55.,60.,65.,70.,75.,80.,
85.,90./
      DATA EALFA/6.612,3.778,2.731,2.397,2.136,1.926,1.754,1.611,
1.486,1.378,1.284,1.202,1.128,1.061,1.0/
      DATA EBETA/0.319,0.408,0.493,0.53,0.567,0.604,0.641,0.678,
0.717,0.759,0.802,0.846,0.893,0.944,1.0/
      DATA KK1,KK2,KK3,KK4,EAL,EBE/0,0,0,0,1.0,1.0/
      WRITE(*,*)'n<=1 INPUT n=?'
      READ(*,*)FN
      FN1=1.0/FN
      FF=1.0/FN-1.0
      WRITE(*,*)"FF=",FF
      EK=RX/RY
      AA=0.5*(1./RX+1./RY)
      BB=0.5*ABS(1./RX-1./RY)
      CC=ACOS(BB/AA)*180.0/PAI
      DO I=1,15
      IF(CC.LT.THETA(I))THEN
      WRITE(*,*)I
      EAL=EALFA(I-1)+(CC-THETA(I-1))*(EALFA(I)-EALFA(I-1))/(THETA(I)-
THETA(I-1))
      EBE=EBETA(I-1)+(CC-THETA(I-1))*(EBETA(I)-EBETA(I-1))/(THETA(I)-
THETA(I-1))
      GOTO 10
      ENDIF
      ENDDO
10    EA=EAL*(1.5*W0/AA/E1)**(1./3.0)
      EB=EBE*(1.5*W0/AA/E1)**(1./3.0)
      PH=1.5*W0/(EA*EB*PAI)
      OPEN(8,FILE='FILM.DAT',STATUS='UNKNOWN')
      OPEN(9,FILE='PRESS.DAT',STATUS='UNKNOWN')
      OPEN(10,FILE='TEM.DAT',STATUS='UNKNOWN')
      WRITE(*,*)"N,X0,XE,PH,E1,EDA0,RX,US"
      WRITE(*,*)N,X0,XE,PH,E1,EDA0,RX,US
      H00=0.0
      MM=N-1
      LMIN=ALOG(N-1.)/ALOG(2.)-1.99
      U=EDA0*(US/2.)**FN/(E1*RX**FN)
      U1=0.5*(2.+AKC)*U
      U2=0.5*(2.-AKC)*U
      A1=ALOG(EDA0)+9.67
      A2=5.1E-9*PH
      A3=0.59/(PH*1.E-9)
```

```
      B=PAI*PH*RX/E1
      W=2.*PAI*PH/(3.*E1)*(B/RX)**2
      ALFA=Z*5.1E-9*A1
      G=ALFA*E1
      AHM=1.0-EXP(-0.68*1.03)
      AHC=1.0-0.61*EXP(-0.73*1.03)
      HM0=3.63*(RX/B)**2*G**0.49*U**0.68*W**(-0.073)*AHM
      HMC=2.69*(RX/B)**2*G**0.53*U**0.67*W**(-0.067)*AHC
      ENDA=2.*U*(3.+FF)*2.0**(1.0+FF)*(E1/PH)**(1.0+FF)*(RX/B)
**(3.0+FF)
      WRITE(*,*)"ENDA=",ENDA
      UTL=EDA0*US*RX/(B*B*2.E7)
      W0=2.0*PAI*EA*EB*PH/3.0
      T1=PH*B/RX
      T2=EDA0*US*RX/(B*B)
      WRITE(*,*)' Wait please'
      CALL SUBAK(MM)
      CALL MULTI(N,NZ,X0,XE,H00)
      STOP
      END
      SUBROUTINE MULTI(N,NZ,X0,XE,H00)
      DIMENSION X(65),Y(65),H(4500),RO(4500),EPS(4500),EDA(4500),
P(4500),POLD(4500),T(65,65,5)
      COMMON /COM1/Z,ENDA,AKC,HM0,HMC,EK,EAL,EBE,AD,AD1,KK1,KK2,
KK3,KK4,FN,FN1,FF
      COMMON /COMT/TE(65,65)
      DATA MK,KTK,G00/200,1,2.0943951/
      G0=G00*EAL*EBE
      NX=N
      NY=N
      NN=(N+1)/2
      DO I=1,N
      DO J=1,N
      TE(I,J)=1.0
      DO K=1,5
      T(I,J,K)=1.0
      ENDDO
      ENDDO
      ENDDO
      CALL INITI(N,DX,X0,XE,X,Y,P,POLD)
      CALL HREE(N,DX,H00,G0,X,Y,H,RO,EPS,EDA,P)
      M=0
 14   KK=15
      CALL ITER(N,KK,DX,H00,G0,X,Y,H,RO,EPS,EDA,P)
      CALL ERP(N,ER,P,POLD)
```

```
      ER=ER/KK
      WRITE(*,*)'ER=',ER
      CALL THERM(NX,NY,NZ,DX,P,H,T)
      CALL ERROM(NX,NY,NZ,T,ERM)
      M=M+1
      IF(M.LT.MK.AND.ER.GT.1.E-5)GOTO 14
      CALL OUPT(N,DX,X,Y,H,P,EDA,TMAX)
      RETURN
      END
      SUBROUTINE INITI(N,DX,X0,XE,X,Y,P,POLD)
      DIMENSION X(N),Y(N),P(N,N),POLD(N,N)
      NN=(N+1)/2
      DX=(XE-X0)/(N-1.)
      Y0=-0.5*(XE-X0)
      DO 5 I=1,N
      X(I)=X0+(I-1)*DX
      Y(I)=Y0+(I-1)*DX
 5    CONTINUE
      DO 10 I=1,N
      D=1.-X(I)*X(I)
      DO 10 J=1,NN
      C=D-Y(J)*Y(J)
      IF(C.LE.0.0)P(I,J)=0.0
10    IF(C.GT.0.0)P(I,J)=SQRT(C)
      DO 20 I=1,N
      DO 20 J=NN+1,N
      JJ=N-J+1
20    P(I,J)=P(I,JJ)
      DO I=1,N
      DO J=1,N
      POLD(I,J)=P(I,J)
      ENDDO
      ENDDO
      RETURN
      END
      SUBROUTINE HREE(N,DX,H00,G0,X,Y,H,RO,EPS,EDA,P)
      DIMENSION X(N),Y(N),P(N,N),H(N,N),RO(N,N),EPS(N,N),EDA(N,N)
      DIMENSION W(150,150),P0(150,150),ROU(65,65)
      COMMON /COM1/Z,ENDA,AKC,HM0,HMC,EK,EAL,EBE,AD,AD1,KK1,KK2,
KK3,KK4,FN,FN1,FF
      COMMON /COM2/W0,E1,RX,B,PH,US,U1,U2,T1,T2,CT
      COMMON /COM3/T0,EDA0,AK,AK1,AK2,CV,CV1,CV2,RO0,RO1,RO2,S0,D0
      COMMON /COM4/A1,A2,A3,LMIN
      COMMON /COMT/TE(65,65)
      DATA KR,NW,PAI,PAI1,DELTA/0,150,3.14159265,0.2026423,0.0/
```

```
      NN=(N+1)/2
      CALL VI(NW,N,DX,P,W)
      HMIN=1.E3
      DO 30 I=1,N
      DO 30 J=1,NN
      RAD=X(I)*X(I)+EK*Y(J)*Y(J)
      W1=0.5*RAD+DELTA
      ZZ=0.5*AD1*AD1+X(I)*ATAN(AD*PAI/180.0)
      IF(W1.LE.ZZ)W1=ZZ
      H0=W1+W(I,J)
      IF(H0.LT.HMIN)HMIN=H0
 30   H(I,J)=H0
      IF(KK.EQ.0)THEN
      KG1=0
      H01=-HMIN+HM0
      DH=0.005*HM0
      H02=-HMIN
      H00=0.5*(H01+H02)
      ENDIF
      W1=0.0
      DO 32 I=1,N
      DO 32 J=1,N
 32   W1=W1+P(I,J)
      W1=DX*DX*W1/G0
      DW=1.-W1
      IF(KK.EQ.0)THEN
      KK=1
      GOTO 50
      ENDIF
      IF(DW.LT.0.0)THEN
      KG1=1
      H00=AMIN1(H01,H00+DH)
      ENDIF
      IF(DW.GT.0.0)THEN
      KG2=2
      H00=AMAX1(H02,H00-DH)
      ENDIF
 50   DO 60 I=1,N
      DO 60 J=1,NN
      H(I,J)=H00+H(I,J)
      CT1=((TE(I,J)-0.455445545)/0.544554455)**S0
      CT2=D0*T0*(TE(I,J)-1.)
      IF(P(I,J).LT.0.0)P(I,J)=0.0
      EDA1=EXP(A1*(-1.+(1.+A2*P(I,J))**Z*CT1))
      EDA(I,J)=EDA1
```

```
      RO(I,J)=1.+CT2
      EPS(I,J)=ENDA*RO(I,J)*H(I,J)**(2.+FN1)/(EDA(I,J)**FN1)
 60   CONTINUE
      DO 70 J=NN+1,N
      JJ=N-J+1
      DO 70 I=1,N
      H(I,J)=H(I,JJ)
      RO(I,J)=RO(I,JJ)
      EDA(I,J)=EDA(I,JJ)
 70   EPS(I,J)=EPS(I,JJ)
      RETURN
      END
      SUBROUTINE ITER(N,KK,DX,H00,G0,X,Y,H,RO,EPS,EDA,P)
      DIMENSION X(N),Y(N),P(N,N),H(N,N),RO(N,N),EPS(N,N),EDA(N,N)
      DIMENSION D(70),A(350),B(210),ID(70)
      COMMON /COM1/Z,ENDA,AKC,HM0,HMC,EK,EAL,EBE,AD,AD1,KK1,KK2,
     KK3,KK4,FN,FN1,FF
      COMMON /COMAK/AK(0:65,0:65)
      DATA KG1,PAI1,C1,C2/0,0.2026423,0.27,0.27/
      IF(KG1.NE.0)GOTO 2
      KG1=1
      AK00=AK(0,0)
      AK10=AK(1,0)
      AK20=AK(2,0)
      BK00=AK00-AK10
      BK10=AK10-0.25*(AK00+2.*AK(1,1)+AK(2,0))
      BK20=AK20-0.25*(AK10+2.*AK(2,1)+AK(3,0))
 2    NN=(N+1)/2
      MM=N-1
      DX1=1./DX
      DX2=DX*DX
      DX3=1./DX2
      DO 100 K=1,KK
      PMAX=0.0
      DO 70 J=2,NN
      J0=J-1
      J1=J+1
      IA=1
 8    MM=N-IA
      IF(P(MM,J0).GT.1.E-6)GOTO 20
      IF(P(MM,J).GT.1.E-6)GOTO 20
      IF(P(MM,J1).GT.1.E-6)GOTO 20
      IA=IA+1
      IF(IA.LT.N)GOTO 8
      GOTO 70
```

```
20   IF(MM.LT.N-1)MM=MM+1
     DPDX1=ABS((P(2,J)-P(1,J))*DX1)**(FF)
     D2=0.5*(EPS(1,J)+EPS(2,J))*DPDX1
     DO 50 I=2,MM
     I0=I-1
     I1=I+1
     II=5*I0
     DPDX2=ABS((P(I1,J)-P(I,J))*DX1)**(FF)
     DPDY1=ABS((P(I,J)-P(I,J0))*DX1)**(FF)
     DPDY2=ABS((P(I,J1)-P(I,J))*DX1)**(FF)
     D1=D2
     D2=0.5*(EPS(I1,J)+EPS(I,J))*DPDX2
     D4=0.5*(EPS(I,J0)+EPS(I,J))*DPDY1
     D5=0.5*(EPS(I,J1)+EPS(I,J))*DPDY2
     P1=P(I0,J)
     P2=P(I1,J)
     P3=P(I,J)
     P4=P(I,J0)
     P5=P(I,J1)
     D3=D1+D2+D4+D5
     IF(H(I,J).LE.0.0)THEN
     ID(I)=0
     A(II+1)=0.0
     A(II+2)=0.0
     A(II+3)=1.0
     A(II+4)=0.0
     A(II+5)=1.0
     A(II-4)=0.0
     GOTO 50
     ENDIF
     ID(I)=1
     IF(J.EQ.NN)P5=P4
     A(II+1)=PAI1*(RO(I0,J)*AK10-RO(I,J)*AK20)
     A(II+2)=DX3*D1+PAI1*(RO(I0,J)*AK00-RO(I,J)*AK10)
     A(II+3)=-DX3*D3+PAI1*(RO(I0,J)*AK10-RO(I,J)*AK00)
     A(II+4)=DX3*D2+PAI1*(RO(I0,J)*AK20-RO(I,J)*AK10)
     A(II+5)=-DX3*(D1*P1+D2*P2+D4*P4+D5*P5-D3*P3)+DX1*(RO(I,J)*H(I,J)-RO
 (I0,J)*H(I0,J))
50   CONTINUE
     CALL TRA4(MM,D,A,B)
     DO 60 I=2,MM
     IF(ID(I).EQ.1)P(I,J)=P(I,J)+C1*D(I)
     IF(P(I,J).LT.0.0)P(I,J)=0.0
     IF(PMAX.LT.P(I,J))PMAX=P(I,J)
60   CONTINUE
```

```
70   CONTINUE
     DO 80 J=1,NN
     JJ=N+1-J
     DO 80 I=1,N
80   P(I,JJ)=P(I,J)
     CALL HREE(N,DX,H00,G0,X,Y,H,RO,EPS,EDA,P)
100  CONTINUE
     RETURN
     END
     SUBROUTINE TRA4(N,D,A,B)
     DIMENSION D(N),A(5,N),B(3,N)
     C=1./A(3,N)
     B(1,N)=-A(1,N)*C
     B(2,N)=-A(2,N)*C
     B(3,N)=A(5,N)*C
     DO 10 I=1,N-2
     IN=N-I
     IN1=IN+1
     C=1./(A(3,IN)+A(4,IN)*B(2,IN1))
     B(1,IN)=-A(1,IN)*C
     B(2,IN)=-(A(2,IN)+A(4,IN)*B(1,IN1))*C
10   B(3,IN)=(A(5,IN)-A(4,IN)*B(3,IN1))*C
     D(1)=0.0
     D(2)=B(3,2)
     DO 20 I=3,N
20   D(I)=B(1,I)*D(I-2)+B(2,I)*D(I-1)+B(3,I)
     RETURN
     END
     SUBROUTINE VI(NW,N,DX,P,V)
     DIMENSION P(N,N),V(NW,NW)
     COMMON /COMAK/AK(0:65,0:65)
     PAI1=0.2026423
     DO 40 I=1,N
     DO 40 J=1,N
     H0=0.0
     DO 30 K=1,N
     IK=IABS(I-K)
     DO 30 L=1,N
     JL=IABS(J-L)
30   H0=H0+AK(IK,JL)*P(K,L)
40   V(I,J)=H0*DX*PAI1
     RETURN
     END
     SUBROUTINE SUBAK(MM)
     COMMON /COMAK/AK(0:65,0:65)
```

```
      S(X,Y)=X+SQRT(X**2+Y**2)
      DO 10 I=0,MM
      XP=I+0.5
      XM=I-0.5
      DO 10 J=0,I
      YP=J+0.5
      YM=J-0.5
      A1=S(YP,XP)/S(YM,XP)
      A2=S(XM,YM)/S(XP,YM)
      A3=S(YM,XM)/S(YP,XM)
      A4=S(XP,YP)/S(XM,YP)
      AK(I,J)=XP*ALOG(A1)+YM*ALOG(A2)+XM*ALOG(A3)+YP*ALOG(A4)
10    AK(J,I)=AK(I,J)
      RETURN
      END
      SUBROUTINE ERP(N,ER,P,POLD)
      DIMENSION P(N,N),POLD(N,N)
      ER=0.0
      SUM=0.0
      NN=(N+1)/2
      DO 10 I=1,N
      DO 10 J=1,NN
      ER=ER+ABS(P(I,J)-POLD(I,J))
      SUM=SUM+P(I,J)
10    CONTINUE
      ER=ER/SUM
      DO I=1,N
      DO J=1,N
      POLD(I,J)=P(I,J)
      ENDDO
      ENDDO
      RETURN
      END
      SUBROUTINE ERROM(NX,NY,NZ,T,ERM)
      DIMENSION T(NX,NY,NZ)
      COMMON /COMT/TE(65,65)
      ERM=0.
      C1=1./FLOAT(NZ)
      DO 20 I=2,NX
      DO 20 J=2,NY
      TT=0.
      DO 10 K=1,NZ
10    TT=TT+T(I,J,K)
      TT=C1*TT
      ER=ABS((TT-TE(I,J))/TT)
```

```
      IF(ER.GT.ERM)ERM=ER
20    TE(I,J)=TT
      RETURN
      END
      SUBROUTINE OUPT(N,DX,X,Y,H,P,EDA,TMAX)
      DIMENSION X(N),Y(N),H(N,N),P(N,N),EDA(N,N)
      COMMON /COM1/Z,ENDA,AKC,HM0,HMC,EK,EAL,EBE,AD,AD1,KK1,KK2,
KK3,KK4,FN,FN1,FF
      COMMON /COM2/W0,E1,RX,B,PH,US,U1,U2,T1,T2,CT
      COMMON /COMT/TE(65,65)
      A=0.0
      WRITE(8,40)A,(Y(I),I=1,N)
      DO I=1,N
      WRITE(8,40)X(I),(H(I,J),J=1,N)
      ENDDO
      WRITE(9,40)A,(Y(I),I=1,N)
      DO I=1,N
      WRITE(9,40)X(I),(P(I,J),J=1,N)
      ENDDO
40    FORMAT(66(E12.6,1X))
      WRITE(10,60)A,(Y(I),I=1,N)
      TMAX=0.0
      DO I=1,N
      WRITE(10,60)X(I),(273.0*(TE(I,JJ)-1.),JJ=1,N)
      DO J=1,N
      IF(TMAX.LT.273.0*(TE(I,J)-1.))TMAX=273.*(TE(I,J)-1.)
      ENDDO
      ENDDO
60    FORMAT(66(E12.6,1X))
      HMIN=1.E3
      PMAX=0.0
   DO J=1,N
   DO I=2,N
      IF(H(I,J).LT.HMIN)HMIN=H(I,J)
      IF(P(I,J).GT.PMAX)PMAX=P(I,J)
      ENDDO
      ENDDO
      HMIN=HMIN*B*B/RX
      PMAX=PMAX*PH
      RETURN
      END
      SUBROUTINE THERM(NX,NY,NZ,DX,P,H,T)
      DIMENSION T(NX,NY,NZ),T1(21),TI(21),U(21),DU(21),UU(21),V
(21),DV(21),VV(21),W(21),EDA(21),RO(21),EDA1(21),EDA2(21),ROR
(21),P(NX,NX),H(NX,NX),TFX(21),TFY(21)
```

```
      COMMON /COM1/Z,ENDA,AKC,HM0,HMC,EK,EAL,EBE,AD,AD1,KK1,KK2,
     KK3,KK4,FN,FN1,FF
      IF(KK.NE.0)GOTO 4
      DO 2 K=1,NZ
      DO 1 J=1,NY
1     T(1,J,K)=1.0
      DO 2 I=1,NX
2     T(I,1,K)=1.0
4     DO 30 I=2,NX
      DO 30 J=2,NY
      KG=0
      DO 6 K=1,NZ
      TFX(K)=T(I-1,J,K)
      TFY(K)=T(I,J-1,K)
      IF(KK.NE.0)GOTO 5
      T1(K)=T(I-1,J,K)
      GOTO 6
5     T1(K)=T(I,J,K)
6     TI(K)=T1(K)
      P1=P(I,J)
      H1=H(I,J)
      DPX=(P(I,J)-P(I-1,J))/DX
      DPY=(P(I,J)-P(I,J-1))/DX
      CALL TBOUD(NX,NY,NZ,I,J,CC1,CC2,T)
10    CALL EROEQ(NZ,T1,P1,H1,DPX,DPY,EDA,RO,EDA1,EDA2,KG)
      CALL UCAL(NZ,DX,H1,EDA,RO,ROR,EDA1,EDA2,U,UU,DU,V,VV,DV,W,
     DPX,DPY)
      CALL TCAL(NZ,DX,CC1,CC2,T1,TFX,TFY,U,V,W,DU,DV,H1,DPX,DPY,EDA,RO)
      CALL ERRO(NZ,TI,T1,ETS)
      KG=KG+3
      IF(ETS.GT.1.E-4.AND.KG.LE.50)GOTO 10
      DO 20 K=1,NZ
      ROR(K)=RO(K)
      UU(K)=U(K)
      VV(K)=V(K)
20    T(I,J,K)=T1(K)
30    CONTINUE
      KK=1
      RETURN
      END
      SUBROUTINE TBOUD(NX,NY,NZ,I,J,CC1,CC2,T)
      DIMENSION T(NX,NY,NZ)
      CC1=0.
      CC2=0.
      DO 10 L=1,I-1
```

```
      DS=1./SQRT(FLOAT(I-L))
      IF(L.EQ.I-1)DS=1.1666667
      CC1=CC1+DS*(T(L,J,2)-T(L,J,1))
 10   CC2=CC2+DS*(T(L,J,NZ)-T(L,J,NZ-1))
      RETURN
      END
      SUBROUTINE ERRO(NZ,T0,T,ETS)
      DIMENSION T0(NZ),T(NZ)
      ETS=0.0
      DO 10 K=1,NZ
      IF(T(K).LT.1.E-5)ETS0=1.
      IF(T(K).GE.1.E-5)ETS0=ABS((T(K)-T0(K))/T(K))
      IF(ETS0.GT.ETS)ETS=ETS0
 10   T0(K)=T(K)
      RETURN
      END
      SUBROUTINE EROEQ(NZ,T,P,H,DPX,DPY,EDA,RO,EDA1,EDA2,KG)
      DIMENSION T(NZ),EDA(NZ),RO(NZ),EDA1(NZ),EDA2(NZ)
      COMMON /COM1/Z,ENDA,AKC,HM0,HMC,EK,EAL,EBE,AD,AD1,KK1,KK2,
KK3,KK4,FN,FN1,FF
      COMMON /COM2/W0,E1,RX,B,PH,US,U1,U2,T1,T2,CT
      COMMON /COM3/T0,EDA0,AK,AK1,AK2,CV,CV1,CV2,RO0,RO1,RO2,S0,D0
      COMMON /COM4/A1,A2,A3,LMIN
      DATA A4,A5/0.455445545,0.544554455/
      IF(KG.NE.0)GOTO 20
      B1=(1.+A2*P)**Z
      B2=(A3+1.34*P)/(A3+P)
 20   DO 30 K=1,NZ
      EDA3=EXP(A1*(-1.+B1*((T(K)-A4)/A5)**S0))
      EDA(K)=EDA3
 30   RO(K)=B2+D0*T0*(T(K)-1.)
      CC1=0.5/(NZ-1.)
      CC2=1./(NZ-1.)
      C1=0.
      C2=0.
      DO 40 K=1,NZ
      IF(K.EQ.1)GOTO 32
      C1=C1+0.5/EDA(K)+0.5/EDA(K-1)
      C2=C2+CC1*((K-1.)/EDA(K)+(K-2.)/EDA(K-1))
 32   EDA1(K)=C1*CC2
 40   EDA2(K)=C2*CC2
      RETURN
      END
      SUBROUTINE UCAL(NZ,DX,H,EDA,RO,ROR,EDA1,EDA2,U,UU,DU,V,VV,
DV,W,DPX,DPY)
```

```
      DIMENSION U(NZ),UU(NZ),DU(NZ),V(NZ),VV(NZ),DV(NZ),W(NZ),ROR(NZ),
EDA(NZ),RO(NZ),EDA1(NZ),EDA2(NZ)
      COMMON /COM1/Z,ENDA,AKC,HM0,HMC,EK,EAL,EBE,AD,AD1,KK1,KK2,
KK3,KK4,FN,FN1,FF
      COMMON /COM2/W0,E1,R,B,PH,US,U1,U2,T1,T2,CC
      COMMON /COM3/T0,EDA0,AK,AK1,AK2,CV,CV1,CV2,RO0,RO1,RO2,S0,D0
      IF(KK.NE.0)GOTO 20
      A1=U1
      A2=PH*(B/R)**3/E1
      A3=U2-U1
20    CUA=A2*DPX*H
      CUB=CUA*H
      CVA=A2*DPY*H
      CVB=CVA*H
      CC3=A3/H
      CC4=1./EDA1(NZ)
      DO 30 K=1,NZ
      U(K)=A1+CUB*(EDA2(K)-CC4*EDA2(NZ)*EDA1(K))+A3*CC4*EDA1(K)
      V(K)=CVB*(EDA2(K)-CC4*EDA2(NZ)*EDA1(K))
      DU(K)=CUA/EDA(K)*((K-1.)/(NZ-1.)-CC4*EDA2(NZ))+CC3*CC4/EDA(K)
30    DV(K)=CVA/EDA(K)*((K-1.)/(NZ-1.)-CC4*EDA2(NZ))
      A4=B/((NZ-1)*R*DX)
      C1=A4*H
      IF(KK.EQ.0)GOTO 50
      DO 40 K=2,NZ-1
      W(K)=(RO(K-1)*W(K-1)+C1*(RO(K)*(U(K)+V(K))-ROR(K)*(UU(K)
+VV(K))))/RO(K)
40    CONTINUE
50    KK=1
      RETURN
      END
      SUBROUTINE TCAL(NZ,DX,CC1,CC2,T,TFX,TFY,U,V,W,DU,DV,H,DPX,
DPY,EDA,RO)
      DIMENSION T(NZ),U(NZ),DU(NZ),V(NZ),DV(NZ),W(NZ),EDA(NZ),RO
(NZ),A(4,21),D(21),AA(2,21),TFX(NZ),TFY(NZ)
      COMMON /COM1/Z,ENDA,AKC,HM0,HMC,EK,EAL,EBE,AD,AD1,KK1,KK2,
KK3,KK4,FN,FN1,FF
      COMMON /COM2/W0,E1,R,B,PH,US,U1,U2,T1,T2,CC
      COMMON /COM3/T0,EDA0,AK,AK1,AK2,CV,CV1,CV2,RO0,RO1,RO2,S0,D0
      DATA CC5,PAI/0.6666667,3.14159265/
      IF(KK.NE.0)GOTO 5
      KK=1
      A2=-CV*RO0*E1*B**3/(EDA0*AK*R)
      A3=-E1*PH*B**3*D0/(AK*EDA0*T0*R)
      A4=-(E1*R)**2/(AK*EDA0*T0)
```

```
      A5=0.5*R/B*A2
      A6=AK*SQRT(EDA0*R/(PAI*RO1*CV1*U1*E1*AK1*B**3))
      A7=AK*SQRT(EDA0*R/(PAI*RO2*CV2*U2*E1*AK2*B**3))
 5    CC3=A6*SQRT(DX)
      CC4=A7*SQRT(DX)
      DZ=H/(NZ-1.)
      DZ1=1./DZ
      DZ2=DZ1*DZ1
      CC6=A3*DPX
      CC7=A3*DPY
      DO 10 K=2,NZ-1
      A(1,K)=DZ2+DZ1*A5*RO(K)*W(K)
      A(2,K)=-2.*DZ2+A2*RO(K)*(U(K)+V(K))/DX+(CC6*U(K)+CC7*V
(K))/RO(K)
      A(3,K)=DZ2-DZ1*A5*RO(K)*W(K)
10    A(4,K)=A4*EDA(K)*(DU(K)**2+DV(K)**2)+A2*RO(K)*(U(K)*TFX
(K)+V(K)*TFY(K))/DX
      A(1,1)=0.
      A(2,1)=1.+2.*DZ1*CC3*CC5
      A(3,1)=-2.*DZ1*CC3*CC5
      A(1,NZ)=-2.*DZ1*CC4*CC5
      A(2,NZ)=1.+2.*DZ1*CC4*CC5
      A(3,NZ)=0.
      A(4,1)=1.+CC1*CC3*DZ1
      A(4,NZ)=1.-CC2*CC4*DZ1
      CALL TRA3(NZ,D,A,AA)
      DO 20 K=1,NZ
      T(K)=(1.-CC)*T(K)+CC*D(K)
20    IF(T(K).LT.1.)T(K)=1.
30    CONTINUE
      RETURN
      END
      SUBROUTINE TRA3(N,D,A,B)
      DIMENSION D(N),A(4,N),B(2,N)
      C=1./A(2,N)
      B(1,N)=-A(1,N)*C
      B(2,N)=A(4,N)*C
      DO 10 I=1,N-1
      IN=N-I
      IN1=IN+1
      C=1./(A(2,IN)+A(3,IN)*B(1,IN1))
      B(1,IN)=-A(1,IN)*C
10    B(2,IN)=(A(4,IN)-A(3,IN)*B(2,IN1))*C
      D(1)=B(2,1)
      DO 20 I=2,N
```

```
20   D(I)=B(1,I)*D(I-1)+B(2,I)
     RETURN
     END
```

20.3.3 Calculation Results

Under the given working conditions in the program, the film thickness, the pressure distribution and temperature distribution are given in Figure 20.2.

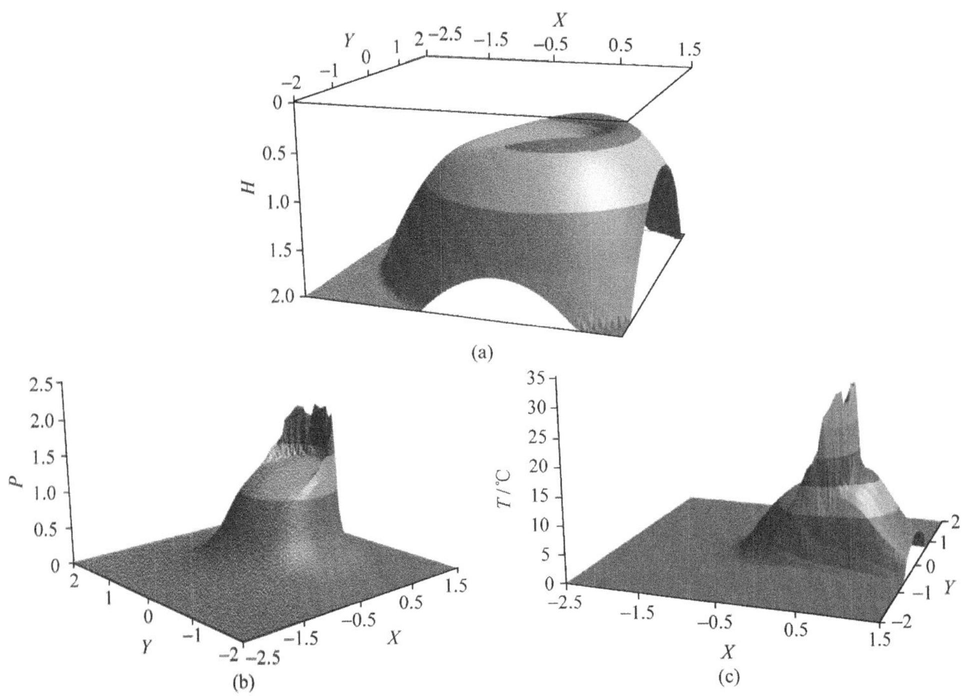

Figure 20.2 Film thickness, pressure and temperature distribution for thermal grease EHL in point contact ($n = 0.9$). (a) Film thickness. (b) Pressure distribution. (c) Temperature distribution.

Part Four

Calculation Programs for Lubrication Analysis in Engineering

21

Lubrication Calculation Program for Herringbone Grooved Journal Bearing of Micro Motor

21.1 Basic Theory of Lubrication Calculation of Herringbone Groove Bearing

21.1.1 Journal Center Position and Film Thickness

A rotating journal brings lubricating oil into the convergence clearance and produces hydrodynamic pressure. The resultant force of the pressure balances the load on the journal. The position is as shown in Figure 21.1.

The position of journal center O_2 can be determined through two parameters: the deviation angle ψ and the eccentricity e. The deviation angle ψ is the angle between the center line connecting O_1O_2 and the action line of the load W.

Figure 21.1 shows that the film thickness h is a function of the angle θ and can be approximately expressed as

$$h = e\cos\theta + c = c(1 + \varepsilon\cos\theta) \tag{21.1}$$

where e is the eccentricity, c is the clearance between the two radii ($c = R_1 - R_2$) and ε is the eccentricity ratio ($\varepsilon = e/c$).

In Equation 21.1, the film thickness of the bearing is approximately the cosine function with an error of only 0.1%. Here, a journal bearing is calculated with hydrodynamic lubrication, and it is assumed that the viscosity of lubricant is a constant.

Numerical Calculation of Lubrication: Methods and Programs, First Edition. Ping Huang.

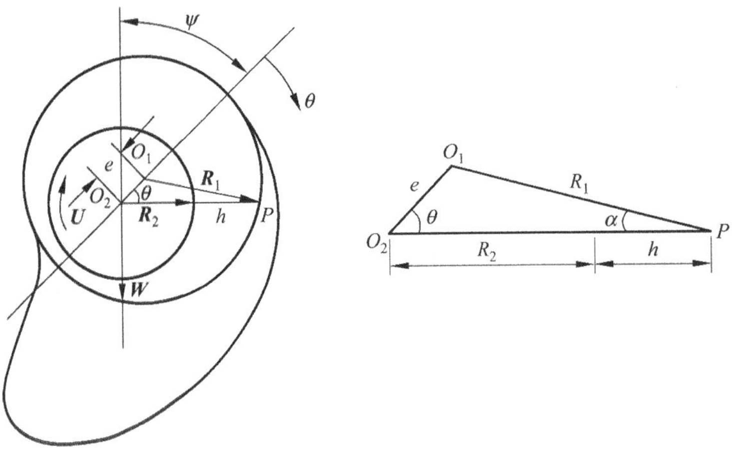

Figure 21.1 Position of journal center.

21.1.2 Reynolds Equation

The film thickness h can be expressed as a function of x. The bearing surface along the circumferential direction is unfolded in Figure 21.2.

By substituting $x = R\theta$ and $dx = Rd\theta$ into the Reynolds Equation 1.1, it becomes

$$\frac{\partial}{\partial\theta}\left(h^3\frac{\partial p}{\partial\theta}\right)+R^2\frac{\partial}{\partial y}\left(h^3\frac{\partial p}{\partial y}\right)=6\eta R^2\left[\left(\omega-2\omega_L-2\frac{d\psi}{dt}\right)\frac{dh}{d\theta}+2c\frac{d\varepsilon}{dt}\cos\theta\right] \tag{21.2}$$

where R is the bearing radius, ω is the angular velocity of the journal, ω_L is the angular velocity of the journal center to the bearing center, η is the viscosity of lubricant and y is the coordinate in the axial direction of the bearing.

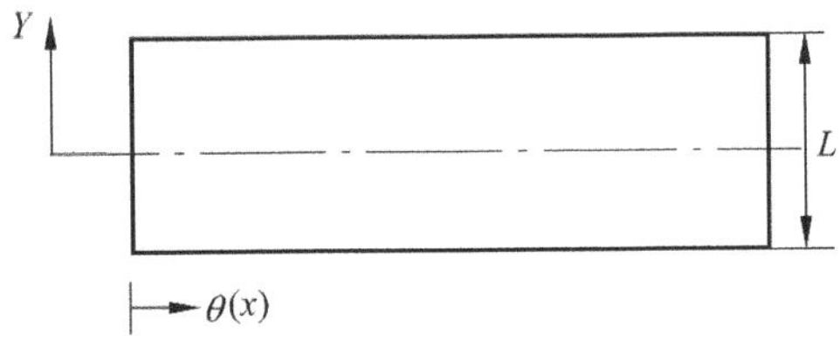

Figure 21.2 Bearing unfolded along the circumferential direction.

If gravity is far greater than the centrifugal force, the static load Reynolds equation becomes

$$\frac{\partial}{\partial \theta}\left(h^3 \frac{\partial p}{\partial \theta}\right) + R^2 \frac{\partial}{\partial y}\left(h^3 \frac{\partial p}{\partial y}\right) = 6\eta R^2 \omega \frac{dh}{d\theta} \tag{21.3}$$

If the centrifugal force is far greater than the gravity of the motor, $\frac{d\psi}{dt} = 0$, $\frac{d\varepsilon}{dt} = 0$ and $\omega_L = \omega$, the dynamic load Reynolds equation will be

$$\frac{\partial}{\partial \theta}\left(h^3 \frac{\partial p}{\partial \theta}\right) + R^2 \frac{\partial}{\partial y}\left(h^3 \frac{\partial p}{\partial y}\right) = -6\eta R^2 \omega \frac{dh}{d\theta} \tag{21.4}$$

If we let $y = YL, \alpha = (R/L)^2, h = c(1 + \varepsilon \cos\theta) = Hc, p = P\frac{6U\eta R}{c^2}$ and $U = R\omega$, where L is the bearing length, the dimensionless static load Reynolds equation is

$$\frac{\partial}{\partial \theta}\left(H^3 \frac{\partial P}{\partial \theta}\right) + \alpha \frac{\partial}{\partial Y}\left(H^3 \frac{\partial P}{\partial Y}\right) = \frac{dH}{d\theta} \tag{21.5}$$

The dimensionless dynamic load Reynolds equation is

$$\frac{\partial}{\partial \theta}\left(H^3 \frac{\partial P}{\partial \theta}\right) + \alpha \frac{\partial}{\partial Y}\left(H^3 \frac{\partial P}{\partial Y}\right) = -\frac{dH}{d\theta} \tag{21.6}$$

21.1.3 Boundary Conditions

21.1.3.1 Axial Direction

In the edges, $P|_{Y=\pm 1/2} = 0$.

In the middle section, $\frac{\partial P}{\partial Y}|_{Y=0} = 0$.

21.1.3.2 Circumferential Direction

The circumferential boundary conditions: the pressure at the starting point is equal to the pressure at the ending point, that is, $P|_{\theta=0} = P|_{\theta=2\pi}$.

In addition, because the oil film cannot bear a negative pressure, in the iterative process, if $P < 0$, we set $P = 0$.

21.1.4 Flux Calculation

The maximum flux can be approximately estimated with the maximum clearance.

$$Q_0 \approx \frac{1}{2} ULc(1+\varepsilon) \tag{21.7}$$

The leak flux is

$$Q_L = \frac{1}{12\eta} \int_0^1 \left(h^3 \frac{\partial p}{\partial y} \right) |_{y=\pm L/2} dx \tag{21.8}$$

21.1.5 Temperature Calculation

When the journal is rotating, the friction power it brings will be converted to heat, which causes the lubricating oil temperature to rise. If the average temperature of the oil exceeds the assumed value, the load carrying capacity drops. Therefore, the oil temperature rise Δt must be limited within the allowed range.

The temperature can be determined by the thermal equilibrium condition that the bearing friction heat H in unit time is equal to the sum of the heat H_1 taken away by the oil flow and the heat H_2 conducted and radiated by the bearing surface, that is

$$H = H_1 + H_2 \tag{21.9}$$

The heat in the bearing is transformed by the power of friction loss, such that the heat H generated in the bearing per second can be calculated by

$$H = fpU \tag{21.10}$$

The heat H_1 taken away by the oil is equal to

$$H_1 = Q_L \rho c_p (t_o - t_i) \tag{21.11}$$

where Q_L is the flux consumption (according to the fuel consumption coefficient), c_p is the specific heat capacity of the oil, t_o is the outlet temperature of the oil and t_i is the inlet temperature of the oil.

The heat H_2 conducted and radiated to the surrounding media through the surfaces can be approximately calculated with the following expression.

$$H_2 = \alpha_s \pi d B (t_o - t_i) \tag{21.12}$$

where α_s is bearing surface heat transfer coefficient, d is the diameter of the bearing and B is the width of the bearing.

By substituting Equations 21.10–21.12 into Equation 21.9, we have

$$fpU = Q_L \rho c(t_o - t_i) + \alpha_s \pi dB(t_o - t_i) \tag{21.13}$$

Therefore, the lubricant temperature rise Δt can be obtained as follows.

$$\Delta t = t_o - t_i = \frac{\left(\frac{f}{\phi} p_m\right)}{c_p \rho \left(\frac{Q_L}{\phi UBd}\right) + \frac{\pi \alpha_s}{\phi U}} \tag{21.14}$$

where $Q_L/\phi UBd$ is the consumption coefficient (a dimensionless number), ϕ is the clearance ratio of the diameter (i.e., $\phi = 2c/d$), U is the tangential velocity ($U = R\omega$), f is the friction coefficient ($f = \frac{\pi \eta \omega}{\psi p} + 0.55\psi\xi$; ξ is a coefficient varying with the wide-diameter ratio; if $D/L < 1$, ξ is $(D/L)^{3/2}$ and if $D/L > 1$, ξ is 1) and p_m is the average pressure of the bearing ($p_m = w/LD$).

We need to point out that Equation 21.14 can only be used to calculate the average temperature. In fact, the temperature at different points is not the same. In oil flowing from the bearing into the outflow, the temperature is increasing gradually, so in different parts the oil viscosity is also different. Study results show that the viscosity of the average temperature can be used to calculate the load carrying capacity. The average temperature of the lubricating oil is $t_m = (t_i + t_o)/2$, and the temperature rise is $\Delta t = t_o - t_i$. Therefore, the calculation of the average temperature of the lubricating oil t_m is equal to

$$t_m = t_i + \frac{\Delta t}{2} \tag{21.15}$$

In order to guarantee the bearing capacity, the average temperature is not more than 75 °C. In the design process, the average temperature t_m is usually given first, and then we calculate Δt according to Equation 21.14 to check the inlet temperature t_i, namely

$$t_i = t_m - \frac{\Delta t}{2} \tag{21.16}$$

21.2 Program for Performance Calculation

21.2.1 Lubrication Performances

According to the given parameters, such as rotational speed, radial load, bearing inner diameter, bearing clearance, bearing width and lubricating oil viscosity, we can calculate the following performances.

1. If the eccentricity ratio is given, the relationship of load with speed and the relationship of load with oil film thickness can be obtained.
2. With a given rotational speed and given eccentricity ratio, the flux and its leakage flux can be obtained.
3. With the given maximum eccentricity ratio, the maximum load and deviation angle can be calculated.
4. Under given conditions, such as rotational speed, eccentricity ratio and so on, the temperature rise caused by lubricating oil shear stress can be calculated.

Figure 21.3 is the diagram of the performance calculation program.

21.2.2 Calculation Program

```
PROGRAM HBFA
USE DFLOGM
INCLUDE 'RESOURCE.FD'
CALL DoDialog()
STOP
END PROGRAM
SUBROUTINE DoDialog()
USE DFLOGM
INCLUDE 'RESOURCE.FD'
INTEGER retint,LL
LOGICAL retlog
TYPE (dialog) dlg
EXTERNAL DISPLAY
EXTERNAL CLEAR
EXTERNAL CHOOSE
EXTERNAL CHOOSE1,CHOOSE2
IF ( .not. DlgInit( IDD_DIALOG1, dlg ) ) THEN
WRITE (*,*) 'Error: dialog not found'
ELSE
retlog=dlgset(dlg,idc_EDIT_DD,"2.992")
retlog=dlgsetsub(dlg,idc_EDIT_DD,CLEAR,dlg_change)
retlog=dlgset(dlg,idc_EDIT_D,"3.0")
retlog=dlgsetsub(dlg,idc_EDIT_D,CLEAR,dlg_change)
retlog=dlgset(dlg,idc_EDIT_AL,"2.4")
retlog=dlgsetsub(dlg,idc_EDIT_AL,CLEAR,dlg_change)
retlog=dlgset(dlg,idc_EDIT_AN,"5000")
retlog=dlgsetsub(dlg,idc_EDIT_AN,CLEAR,dlg_change)
retlog=dlgset(dlg,idc_EDIT_M,"10")
```

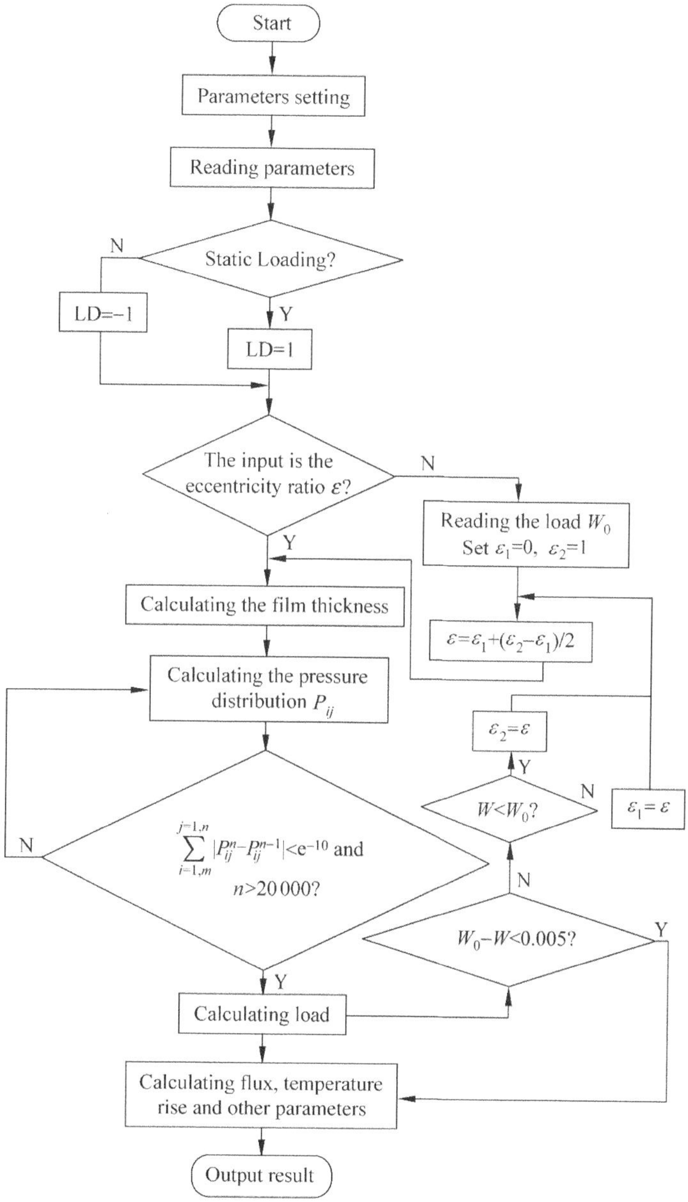

Figure 21.3 Diagram of HBFA.

```
retlog=dlgsetsub(dlg,idc_EDIT_M,CLEAR,dlg_change)
retlog=dlgset(dlg,idc_EDIT_ALF,"30")
retlog=dlgsetsub(dlg,idc_EDIT_ALF,CLEAR,dlg_change)
retlog=dlgset(dlg,idc_EDIT_RLGL,"1.0")
```

```
retlog=dlgsetsub(dlg,idc_EDIT_RLGL,CLEAR,dlg_change)
retlog=dlgset(dlg,idc_EDIT_CG,"0.004")
retlog=dlgsetsub(dlg,idc_EDIT_CG,CLEAR,dlg_change)
retlog=dlgset(dlg,idc_EDIT_DEDA,"24.5")
retlog=dlgsetsub(dlg,idc_EDIT_DEDA,CLEAR,dlg_change)
retlog=dlgset(dlg,idc_EDIT_RO,"0.945")
retlog=dlgsetsub(dlg,idc_EDIT_RO,CLEAR,dlg_change)
retlog=dlgset(dlg,idc_EDIT_AS,"80")
retlog=dlgsetsub(dlg,idc_EDIT_AS,CLEAR,dlg_change)
retlog=dlgset(dlg,idc_EDIT_CO,"1800")
retlog=dlgsetsub(dlg,idc_EDIT_CO,CLEAR,dlg_change)
retlog=dlgsetsub(dlg,idc_EDIT_SUM,CLEAR,dlg_change)
retlog=dlgsetsub(dlg,idc_EDIT_EPSON,CLEAR,dlg_change)
retlog=dlgsetsub(dlg,idc_CALCULATE,DISPLAY)
retlog=dlgsetsub(dlg,idc_CLEAR,CLEAR)
retlog=dlgset(dlg,idc_EDIT_EPSON,.TRUE.)
retlog=dlgset(dlg,idc_EDIT_SUM,.FALSE.)
retlog=dlgsetsub(dlg,IDC_RADIO_EPSON,CHOOSE)
retlog=dlgsetsub(dlg,IDC_RADIO_SUM,CHOOSE)
retlog=dlgset(dlg,idc_CHECK_STATIC,.TRUE.,DLG_STATE)
retlog=dlgset(dlg,idc_CHECK_DYNAMIC,.FALSE.,DLG_STATE)
retlog=dlgsetsub(dlg,IDC_CHECK_STATIC,CHOOSE1)
retlog=dlgsetsub(dlg,IDC_CHECK_DYNAMIC,CHOOSE2)
retint=DLGMODAL(dlg)
CALL DLGUNINIT(dlg)
END IF
LL=0
RETURN
END SUBROUTINE DoDialog
SUBROUTINE CHOOSE(dlg,control_name,calltype)
USE DFLOGM
INCLUDE 'resource.fd'
TYPE (dialog) dlg
LOGICAL retlog,PUSHED_state
retlog = DLGGET (dlg, IDC_RADIO_EPSON, pushed_state)
IF (PUSHED_STATE) THEN
retlog=dlgset(dlg,idc_EDIT_SUM,.false.,dlg_enable)
retlog=dlgset(dlg,idc_EDIT_EPSON,.TRUE.,dlg_enable)
ELSE
retlog=dlgset(dlg,idc_EDIT_EPSON,.false.,dlg_enable)
retlog=dlgset(dlg,idc_EDIT_SUM,.TRUE.,dlg_enable)
ENDIF
END SUBROUTINE CHOOSE
SUBROUTINE CHOOSE1(dlg,control_name,calltype)
USE DFLOGM
INCLUDE 'resource.fd'
TYPE (dialog) dlg
LOGICAL retlog,PUSHED_state
retlog=dlgset(dlg,idc_CHECK_STATIC,.TRUE.,DLG_STATE)
retlog=dlgset(dlg,idc_CHECK_DYNAMIC,.FALSE.,DLG_STATE)
retlog=dlgset(dlg,idc_EDIT_SUM2," ")
retlog=dlgset(dlg,idc_EDIT_EPSON2," ")
retlog=dlgset(dlg,idc_EDIT_E," ")
```

```
retlog=dlgset(dlg,idc_EDIT_Q1," ")
retlog=dlgset(dlg,idc_EDIT_Q," ")
retlog=dlgset(dlg,idc_EDIT_DT," ")
retlog=dlgset(dlg,idc_EDIT_AI," ")
END SUBROUTINE CHOOSE1
SUBROUTINE CHOOSE2(dlg,control_name,calltype)
USE DFLOGM
INCLUDE 'resource.fd'
TYPE (dialog) dlg
LOGICAL retlog,PUSHED_state
retlog=dlgset(dlg,idc_CHECK_DYNAMIC,.TRUE.,DLG_STATE)
retlog=dlgset(dlg,idc_CHECK_STATIC,.FALSE.,DLG_STATE)
retlog=dlgset(dlg,idc_EDIT_SUM2," ")
retlog=dlgset(dlg,idc_EDIT_EPSON2," ")
retlog=dlgset(dlg,idc_EDIT_E," ")
retlog=dlgset(dlg,idc_EDIT_Q1," ")
retlog=dlgset(dlg,idc_EDIT_Q," ")
retlog=dlgset(dlg,idc_EDIT_DT," ")
retlog=dlgset(dlg,idc_EDIT_AI," ")
END SUBROUTINE CHOOSE2
SUBROUTINE CLEAR(dlg,control_name,calltype)
USE DFLOGM
INCLUDE 'resource.fd'
TYPE (dialog) dlg
LOGICAL retlog
INTEGER retint,control_name,calltype
retlog=dlgset(dlg,idc_EDIT_SUM2," ")
retlog=dlgset(dlg,idc_EDIT_EPSON2," ")
retlog=dlgset(dlg,idc_EDIT_E," ")
retlog=dlgset(dlg,idc_EDIT_Q1," ")
retlog=dlgset(dlg,idc_EDIT_Q," ")
retlog=dlgset(dlg,idc_EDIT_DT," ")
retlog=dlgset(dlg,idc_EDIT_AI," ")
RETURN
END SUBROUTINE CLEAR
SUBROUTINE DISPLAY(dlg,control_name,calltype)
USE dflogm
INCLUDE 'resource.fd'
TYPE (dialog) dlg
LOGICAL retlog
INTEGER retint,control_name,calltype
INTEGER N,N1,N2,M,K1,K2,I,J,K,I1,IK,L
REAL*4 X(121),Y(121),H(121,121),P(121,121),PD(121,121),HH(121)
REAL*4 HXY(121,121),HXF(121,121),HXB(121,121),HYF(121,121),HYB
(121,121)
REAL*4 PI,EPSON,U,EDA,ALD,AL,AN,DD,D,R,ALF,RLGL,CG,CCG,PI2,C,E,DX
REAL*4 ALX,ALY,RATIO,ALFA,ALENDA,PESAI
REAL*4 CO,RO,AS,DEDA
REAL*4 OMIGA,TEMP,TEMP1,HMIN,SUM1,Q,Q1
REAL*4 A,B,SUM
REAL*4 PX,PY,AI,KEXI,PA,F,DT,C1
LOGICAL PUSHED_STATE
DATA PI/3.14159265/
```

```
      CHARACTER(256) text,DG
      OPEN(8,FILE='pressure.DAT',STATUS='UNKNOWN')
      OPEN(9,FILE='height.DAT',STATUS='UNKNOWN')
      OPEN(13,FILE='result.DAT',STATUS='UNKNOWN')
2     FORMAT(20X,A12,I2.2,':',I2.2,':',I2.2,'.',I3.3)
      retlog=dlgget(dlg,idc_edit_DD,text)
      READ(text,*) DD
      retlog=dlgget(dlg,idc_edit_D,text)
      read(text,*) D
      retlog=dlgget(dlg,idc_edit_AL,text)
      read(text,*) AL
      retlog=dlgget(dlg,idc_edit_AN,text)
      read(text,*) AN
      retlog=dlgget(dlg,idc_edit_M,text)
      read(text,*) M
      retlog=dlgget(dlg,idc_edit_ALF,text)
      read(text,*) ALF
      retlog=dlgget(dlg,idc_edit_RLGL,text)
      read(text,*) RLGL
       retlog=dlgget(dlg,idc_edit_CG,text)
      read(text,*) CG
      retlog=dlgget(dlg,idc_EDIT_DEDA,text)
      read(text,*) DEDA
      retlog=dlgget(dlg,idc_EDIT_RO,text)
      read(text,*) RO
      retlog=dlgget(dlg,idc_EDIT_AS,text)
      read(text,*) AS
      retlog=dlgget(dlg,idc_EDIT_CO,text)
      read(text,*) CO
      retlog = DLGGET (dlg, IDC_CHECK_STATIC, pushed_state)
      IF (PUSHED_STATE) THEN
      LD=1
      ELSE
      LD=-1
      ENDIF
!     WRITE(*,*)'LD=',LD
!     WRITE(*,*)'DD,D,AL,AN='
!     WRITE(*,*) DD,D,AL,AN
!     WRITE(*,*)'EPSON,RLGL,ALF,CG='
!     WRITE(*,*) EPSON,RLGL,ALF,CG
!     WRITE(*,*)'DEDA,RO,CO,AS='
!     WRITE(*,*) DEDA,RO,CO,AS
      EDA=DEDA*RO*1.0E-3
      RO=RO*1.0E3
      DD=DD*1.0E-3
      D=D*1.0E-3
      AL=AL*1.0E-3
      CG=CG*1.0E-3
      N=121
      N1=N-1
      N2=N1/2+1
      PI2=2.0*PI
      R=D/2
      ALD=AL/D
```

```
      C=(D-DD)/2
      PESAI=(D-DD)/D
      ALF=ALF*PI/180.0
      K2=120/M
      K1=(RLGL/(1+RLGL))*K2
      IF(ABS(K1-(RLGL/(1+RLGL))*K2).GT.0.5) K1=K1+1
      CCG=CG/C
      DX=1.0/N1
      ALX=PI2*R
      ALY=AL
      RATIO=ALX/ALY
      ALFA=RATIO**2
      OMIGA=AN*PI2/60.0
      U=OMIGA*R
      TEMP=1./N1
      DO I=1,N
      X(I)=(I-1)*TEMP
      Y(I)=-0.5+X(I)
      ENDDO
      Y(N2)=0.0
      DO I=1,N
      DO J=1,N
      P(I,J)=0.0
      ENDDO
      ENDDO
      retlog = DLGGET (dlg, IDC_RADIO_EPSON, pushed_state)
      IF (PUSHED_STATE) THEN
      retlog=dlgget(dlg,idc_edit_EPSON,text)
      read(text,*) EPSON
      WRITE(*,*) 'EPSON=',EPSON
      CALL SUBH
      CALL SUBP
      CALL SUBM
      GO TO 70
      ELSE
      GO TO 50
      ENDIF
50    retlog=dlgget(dlg,idc_EDIT_SUM,text)
      read(text,*) SUM
      SUM0=SUM*(N*N)*C**2/(6.0*U*EDA*ALX*ALX*ALY)
      EPSON=0.5
      EPSON1=0.0
      EPSON2=1.0
      WRITE(*,*)'LOAD=',SUM
60    CALL SUBH
      CALL SUBP
      CALL SUBM
      IF(ABS((SUM1-SUM0)/SUM0).LE.0.005)GOTO 70
      IF(SUM1.GT.SUM0)THEN
      EPSON2=EPSON
      EPSON=0.5*(EPSON1+EPSON)
      ELSE
      EPSON1=EPSON
      EPSON=0.5*(EPSON+EPSON2)
```

```
      ENDIF
      GOTO 60
70    E=EPSON*C*1.0E3
      HMIN=C*(1.0-EPSON)
      AI=180.0*ATAN(PY/PX)/PI
      Q1=0.0
      Q=0.0
      DO I=1,N1
      Q=H(I,1)**3*(4.0*P(I,2)-P(I,3))+H(I,N)**3*(4.0*P(I,N-1)-P(I,N-2))
      IF(Q.LT.0.0)Q=0.0
      Q1=Q1+Q
      ENDDO
      Q1=Q1*0.25*U*ALY*C
      Q=0.5*U*ALY*(C*(1+EPSON)+CG*K1/K2)
      DDT=0.0
      ALENDA=6.0*U*EDA*ALX/C**2
      SUM1=ALENDA*ALX*ALY*SUM1/(N*N)
      KEXI=1.0
      IF(ALD.GT.1.0)KEXI=(D/ALY)**1.5
      PA=SUM1/(D*ALY)
      F=PI*EDA*OMIGA/PESAI/PA+0.55*PESAI*KEXI
      DT=(F/PESAI)*PA/(CO*RO*Q/(PESAI*U*ALY*D)+PI*AS/PESAI/U)
!     WRITE(*,*)DT,F,PA,CO,RO,AS
      WRITE(8,40)Y(1),(Y(I),I=1,N)
      DO I=1,N
      WRITE(8,40)X(I)*360,(P(I,J)*ALENDA,J=1,N)
      ENDDO
      WRITE(9,40)Y(1),(Y(I),I=1,N)
      DO I=1,N
      WRITE(9,40)X(I)*360,(H(I,J)*C,J=1,N)
      ENDDO
40    FORMAT(122(E12.6,1X))
4     FORMAT(10(G12.4))
!     WRITE(*,*)'EPSON,E=',EPSON,E
!     WRITE(*,*)'M,CG,AL=',M,CG,AL
!     WRITE(*,*)'SUM1,AI,Q,Q1=',SUM1,AI,Q,Q1
      IF(LL.EQ.0) WRITE(13,*)'ε e W ψ Flux Leakage ΔT Loading'
      LL=LL+1
      IF(LD.EQ.1) THEN
      TEXT='STATIC'
      ELSE
      TEXT='DYNAMIC'
      ENDIF
      WRITE(13,4)EPSON,E,SUM1,AI,Q,Q1,DT,TEXT
!     WRITE(*,*)'DD,D,C,PESAI=',DD,D,C,PESAI
      WRITE(TEXT,'(G12.4)') EPSON
      retlog=dlgset(dlg,idc_edit_EPSON2,trim(adjustl(TEXT)))
      WRITE(TEXT,'(G12.4)') SUM1
      retlog=dlgset(dlg,idc_edit_SUM2,trim(adjustl(TEXT)))
      write(TEXT,'(G12.4)') AI
      retlog=dlgset(dlg,idc_edit_AI,trim(adjustl(TEXT)))
      write(TEXT,'(G12.4)') Q
      retlog=dlgset(dlg,idc_EDIT_Q,trim(adjustl(text)))
```

```
      WRITE(TEXT,'(G12.4)') DT
      retlog=dlgset(dlg,idc_EDIT_DT,trim(adjustl(text)))
      WRITE(TEXT,'(G12.4)') Q1
      retlog=dlgset(dlg,idc_EDIT_Q1,trim(adjustl(text)))
      WRITE(TEXT,'(G12.4)') E
      retlog=dlgset(dlg,idc_EDIT_E,trim(adjustl(text)))
      CONTAINS
      SUBROUTINE SUBH
      DO I=1,N
      DO J=1,N
      H(I,J)=1.0+EPSON*COS(PI2*X(I))
      ENDDO
      ENDDO
      IF(K1.EQ.0)GOTO 20
      DO I=1,N
      HH(I)=0.0
      ENDDO
      DO I=1,N-K2,K2
      DO L=1,K1
      HH(I+L-1)=CCG
      ENDDO
      ENDDO
      HH(N)=HH(1)
      DO I=1,N
      ENDDO
      TEMP=1./RATIO
      DO J=1,N
      DO I=1,N1
      I1=I+ABS(J-N2)*(1./TAN(ALF/2))*TEMP
      IF(I1.GT.N)THEN
      I1=I1-N1
      ENDIF
      H(I,J)=H(I,J)+HH(I1)
      ENDDO
      H(N,J)=H(1,J)
      ENDDO
20    DO I=1,N
      I1=I-1
      I2=I+1
      IF(I.EQ.1)I1=N1
      IF(I.EQ.N)I2=2
      DO J=2,N1
      HXF(I,J)=(0.5*(H(I2,J)+H(I,J)))**3
      HXB(I,J)=(0.5*(H(I1,J)+H(I,J)))**3
      HYF(I,J)=ALFA*(0.5*(H(I,J+1)+H(I,J)))**3
      HYB(I,J)=ALFA*(0.5*(H(I,J-1)+H(I,J)))**3
      HXY(I,J)=1.0/(HXF(I,J)+HXB(I,J)+HYF(I,J)+HYB(I,J))
      ENDDO
      ENDDO
      RETURN
      END SUBROUTINE SUBH
      SUBROUTINE SUBP
      DO I=1,N
```

```
          DO J=1,N
          PD(I,J)=P(I,J)
          ENDDO
          ENDDO
          IK=0
          TEMP=0.5*DX
10        C1=0.0
          DO I=1,N1
          I1=I-1
          I2=I+1
          IF(I1.EQ.0)I1=N1
          IF(I2.EQ.N)I2=1
          DO J=2,N1
          P(I,J)=(HXF(I,J)*P(I2,J)+HXB(I,J)*P(I1,J)+HYF(I,J)*P(I,J+1)+HYB
(I,J)*P(I,J-1)-LD*TEMP*(H(I2,J)-H(I1,J)))*HXY(I,J)
          IF(P(I,J).LE.0.0)P(I,J)=0.0
          C1=C1+ABS(P(I,J)-PD(I,J))
          PD(I,J)=P(I,J)
          ENDDO
          ENDDO
          DO J=2,N1
          P(N,J)=P(1,J)
          PD(N,J)=PD(1,J)
          ENDDO
          IK=IK+1
          IF(C1.GT.1.E-20.AND.IK.LE.20000)GOTO 10
!         WRITE(*,*)'ER=',IK,C1
          RETURN
          END SUBROUTINE SUBP
          SUBROUTINE SUBM
          PX=0.0
          PY=0.0
          TEMP=PI/60.0
          DO I=1,N1
          AI=(I-1)*TEMP
          DO J=1,N
          PX=PX-P(I,J)*COS(AI)
          PY=PY+P(I,J)*SIN(AI)
          ENDDO
          ENDDO
          SUM1=SQRT(PX*PX+PY*PY)
          RETURN
          END SUBROUTINE SUBM
          END SUBROUTINE DISPLAY
```

21.3 Calculation Results

The parameters given in the program are shown in Table 21.1.

In Figures 21.4–21.6, the static and dynamic film thicknesses and pressure distributions with and without grooves are given, respectively.

Table 21.1 Parameters of calculation

Parameter	Value and unit
Motor speed n	5000 rpm
Load W	1 N
Motor shaft diameter d	2.992 mm
Bearing bore diameter D	3.000 mm
Bearing width L	2.4 mm
Lubricant viscosity η	0.02 315 Pa·s

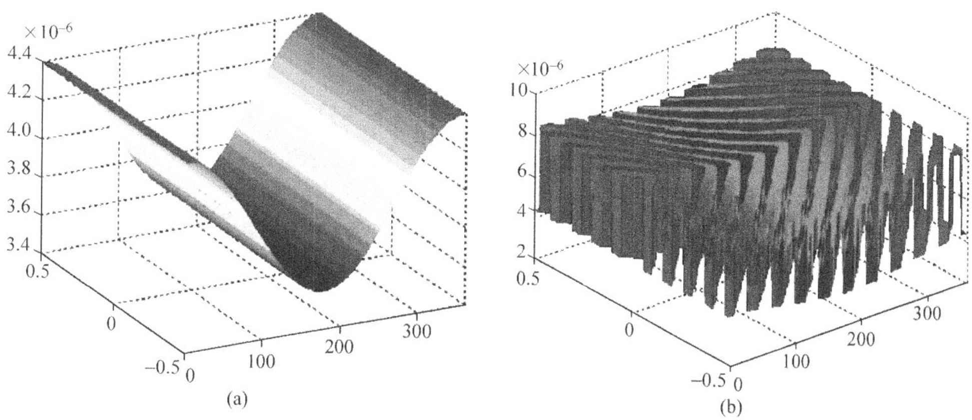

Figure 21.4 Film thickness (a) with and (b) without groove ($\varepsilon = 0.1$).

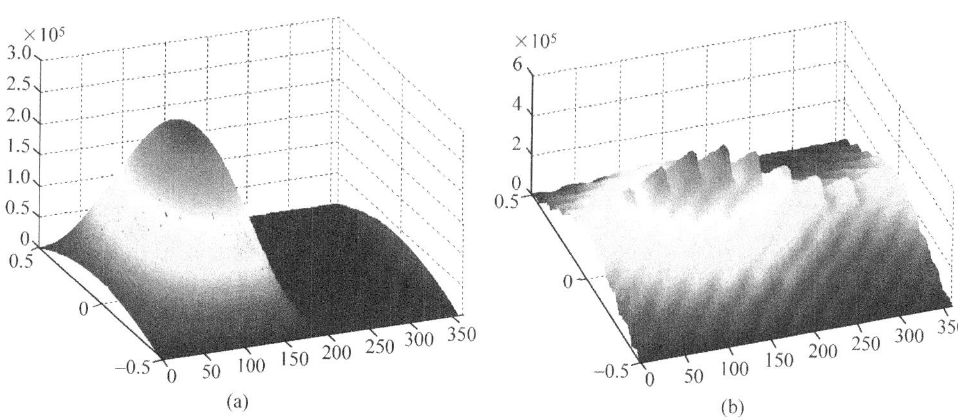

Figure 21.5 Pressure distribution (a) with and (b) without groove ($\varepsilon = 0.1$).

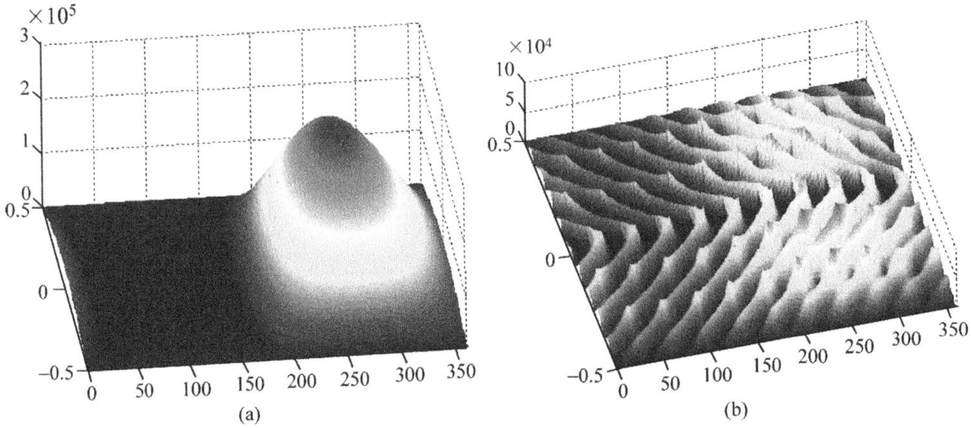

Figure 21.6 Dynamic pressure distribution (a) with and (b) without groove ($\varepsilon = 0.1$).

The following conclusions can be obtained from Figures 21.4–21.6.

1. The figures show that the static and dynamic pressure distributions without grooves are same as the theoretical. The static pressure distribution is at $0 \sim 180°$ and the maximum is at about 80°. The dynamic pressure is at $180 \sim 360°$ and the maximum is at about 270°.
2. If we consider the grooves, the pressure fluctuates with the grooves.

If the groove angle α is 30°, the ditch ridge ratio $\gamma = L_r/L_g = 1$, and the groove depth C_g is 0.004 mm, the variation of the load carrying capacity with eccentricity ratio $\varepsilon = e/C$ is as shown in Figure 21.7. The static load is slightly larger than the dynamic

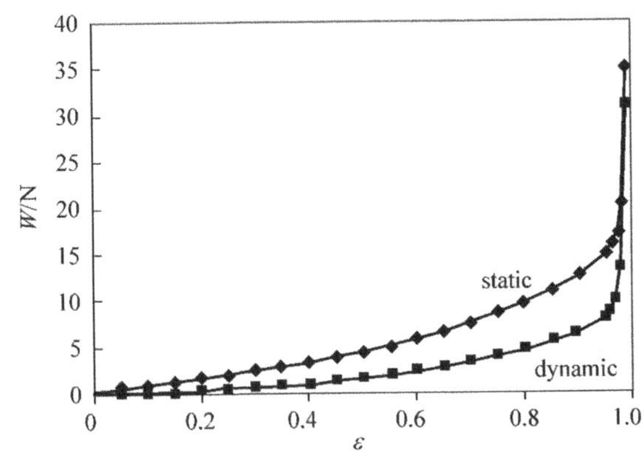

Figure 21.7 Load carrying capacity W varying with eccentricity ratio ε.

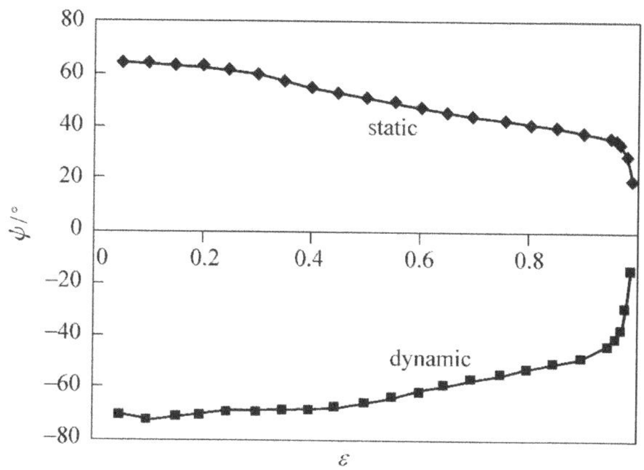

Figure 21.8 Deviation angle ψ varying with eccentricity ratio ε.

load. However, the tendencies of the curves are the same. When ε is equal to 0.1, the load carrying capacity is about 0.8628 N. When ε is close to 1, the load increases greatly. For the static load, when ε is equal to 0.99, the load carrying capacity is up to 34.7 N. When ε tends to 1, the load carrying capacity tends to infinity.

Figure 21.8 shows that the deviation angle ψ varies with the eccentricity ratio ε. When ε tends to 1, the deviation angle tends to 0°. The static deviation angle is greater than 0, but the dynamic deviation angle is less than 0. The curves are nearly reversely symmetric to the horizontal axis.

Figure 21.9 shows that the total flux Q_0 and the leakage flux Q_L vary with the eccentricity ratio ε. The static and dynamic fluxes have the same tendency. The two flows increase linearly with increasing eccentricity ratio ε. When the eccentricity ratio

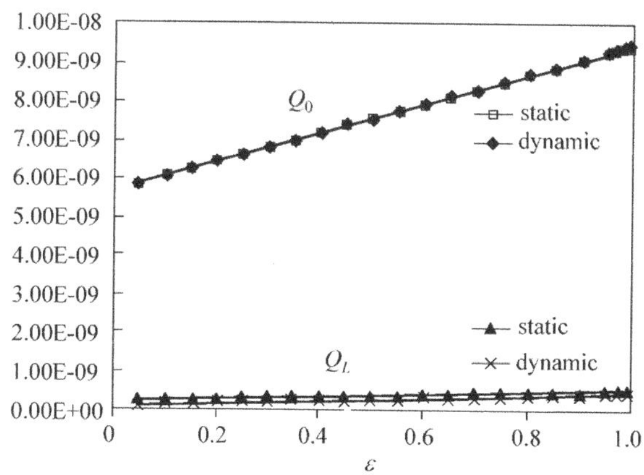

Figure 21.9 Total flux Q_0 and leakage flux Q_L varying with eccentricity ratio ε.

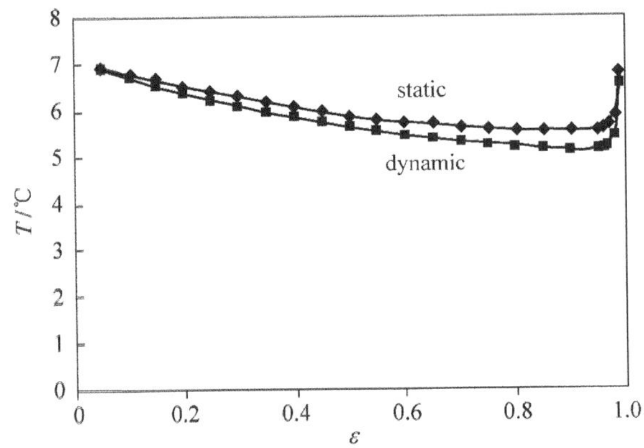

Figure 21.10 Temperature rise ΔT varying with eccentricity ratio ε.

ε is equal to zero, the fluxes are at their minimum. When the eccentricity ratio ε is close to 1, the fluxes are at their maximum. The leakage flux is far less than the total flux.

Figure 21.10 shows that the temperature rise ΔT varies with the eccentricity ratio ε. The trends of the static and dynamic temperature rises are similar and their values are only slightly different. The dynamic temperature rise is slightly smaller than the static temperature rise. When ε is close to zero, the temperature rise is larger, about 7 °C. With increasing ε, the temperature rises reduce gradually at first, but when ε is close to 1, they sharply increase because the film thickness becomes quite thin and the shear rate significantly increases. However, the overall temperature rise is less than 10 °C.

In Figures 21.11–21.14, the curves of W, ψ, Q_O, Q_L and ΔT varying with the rotational speed n are given, respectively. Here, the groove angle α is 30°, the ditch ridge ratio $\gamma = L_r/L_g = 1$, the eccentricity ratio ε is equal to 0.1. The changes of these

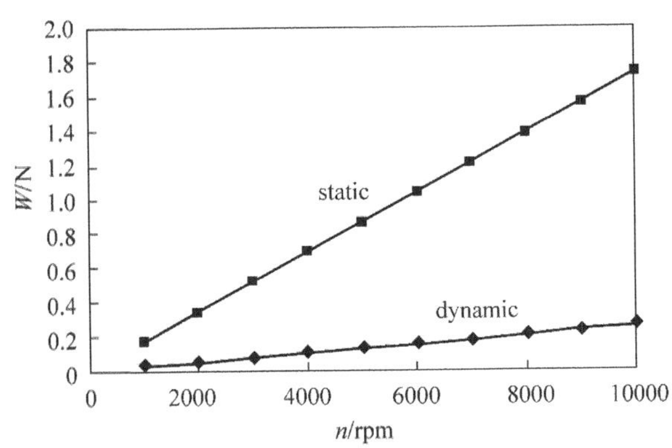

Figure 21.11 Curves of load with speed.

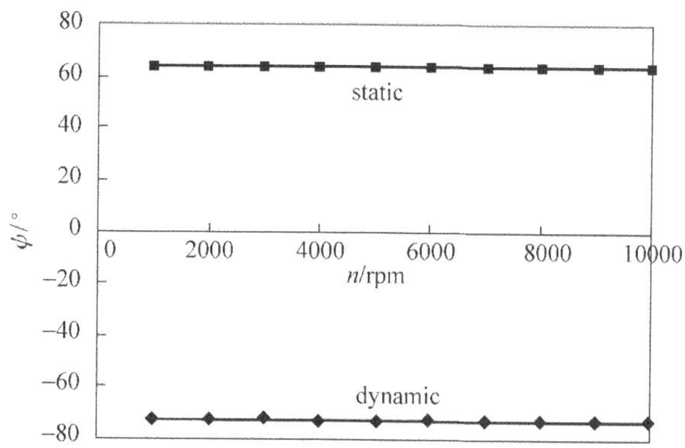

Figure 21.12 Curves of deviation angle with speed.

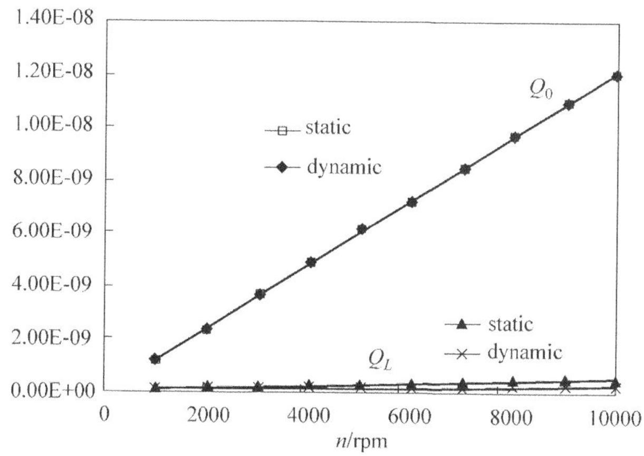

Figure 21.13 Curves of flux and leakage flux with speed.

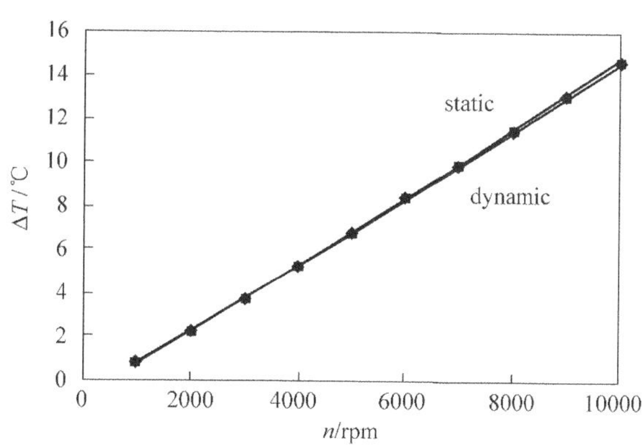

Figure 21.14 Curves of temperature rise with speed.

performance parameters (except the unchanged deviation angle) are linear with the rotational speed in the range of 1000–10 000 rpm.

It can be seen that the load changes linearly with the speed. The greater the speed, the greater the load. The slope of the static load is greater than that of the dynamic load. However, the speed has almost no effect on the deviation angle. The effect of the speed on the flow and the temperature rise is also linear. The effect of the speed is the same on both the static and the dynamic flow and temperature.

21.4 Instruction for HBFA Software Package

21.4.1 Package Contents

The files included in the package are shown in Table 21.2.

After execution, some new files will be generated, as shown in Table 21.3.

21.4.2 Program Installation

In order to run the program package, one must first copy the HBFA file package shown in Table 21.2 to the current disk directory.

Table 21.2 Files including in package

File	File type
HBFA.exe	Execution file
Resource.h	H file
Resource.fd	Source file
Resource.hm	HM file
Script1.rc	Resource template
HBFA.F90	Source program
HBFA.dsp	Project file
HBFA.dsw	Project workspace
HBFA.PLG	PLG file
Pressure.m and Height.m	Matlab file
Bitmap3 and bitmap5	Figure files

Table 21.3 New files generated after execution

File	File Type
Height.dat	Data on film thickness distribution
Pressure.dat	Data on pressure distribution
Result.dat	Calculation results
HBFA.opt	OPT files
Debug	Debug file package

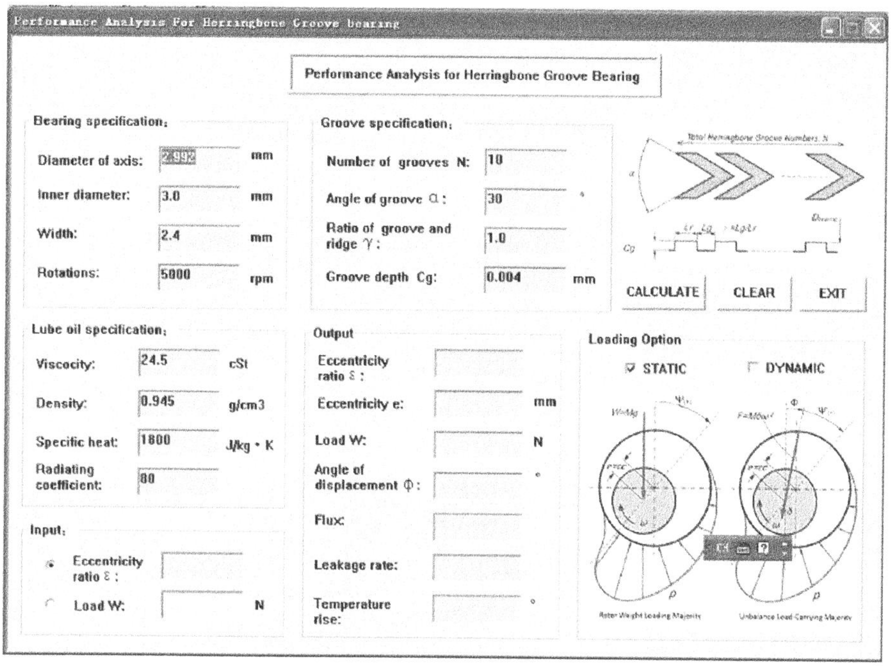

Figure 21.15 Main interface.

21.4.3 *Program Operation*

21.4.3.1 Executing the Program

After clicking the executable file HBFA.exe, the interface shown in Figure 21.15 will appear. Or we can double-click HBFA.dsw, and the Compaq Visual Fortran operation interface will appear. Then, we click Fileview to select the source file and double-click HBFA.F90, and the results will be as shown in Figure 21.16. Then, we compile, link and execute the program, and the interface shown in Figure 21.15 will re-appear.

21.4.3.2 Parameters Setting

As shown in Figure 21.15, the main interface has seven regions. They are: Bearing specification, Lube oil specification, Groove specification, Input, Output, Loading the option and the ditch ridge schematic.

In the first column, the bearing size parameters are mainly required, including motor shaft diameter, bearing bore diameter, width and motor speed.

The second column is used to set the physical properties of the lubricant, including viscosity, density, specific heat and temperature coefficient.

The third column is used to set the internal groove size and shape parameters of the herringbone bearing, including number of grooves, angle, ratio of groove to ridge and groove depth.

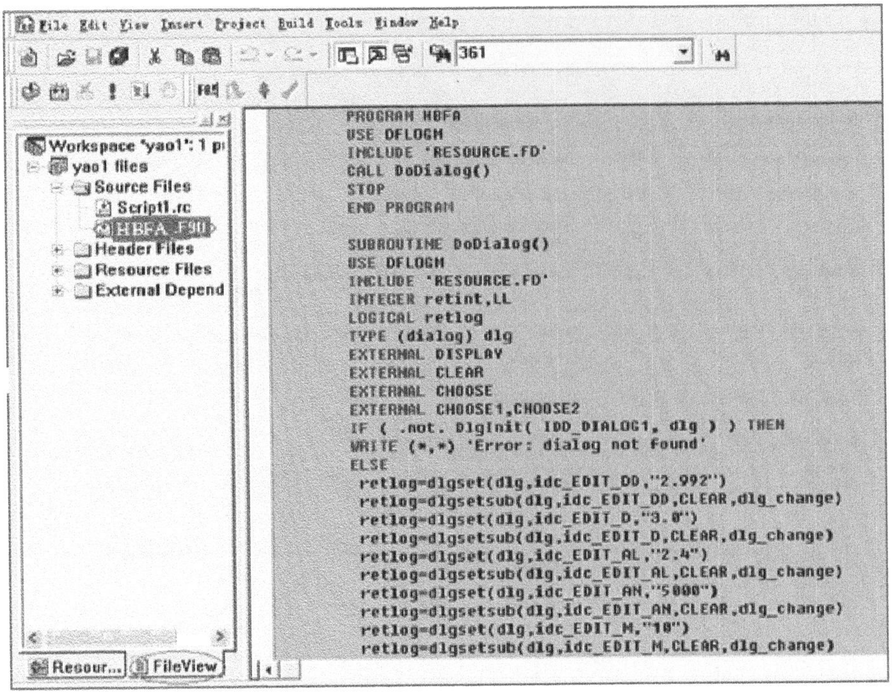

Figure 21.16 Fortran program running interface.

In order to input data conveniently, the units that people often used are chosen. The unit of length is mm; the force unit is N; the motor speed unit is rpm; the viscosity is the dynamic viscosity whose unit is mm^2/s; and the unit of density is g/cm^3. The parameters in the first three columns can be modified according to the user's need.

The fourth column of the parameter selection is mainly used in both cases of herringbone groove journal bearing performance analysis. One is to obtain the load with a given eccentricity ratio, and the other is to obtain the eccentricity ratio with a given load. ε is the ratio of the eccentricity e and the radius clearance C, $\varepsilon = e/C$ ($C = R_1 - R_2$, where R_1 is the radius of the bearing and R_2 is the radius of the motor shaft). The eccentricity ratio range is $0 \sim 1$. If the eccentricity ratio is given, the user can click the eccentricity ratio and input its value. This is the default state, as shown in Figure 21.17. If the load is given, the user can click the load and input its value, as shown in Figure 21.18.

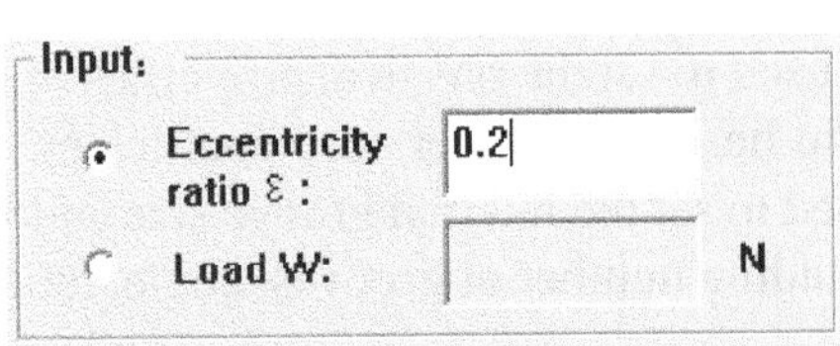

Figure 21.17 Selecting eccentricity ratio or load.

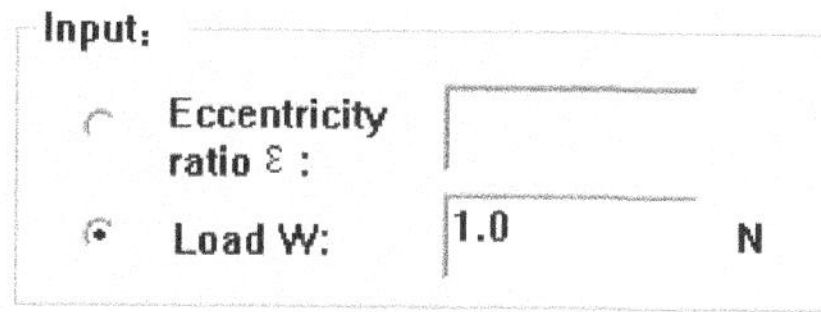

Figure 21.18 Selecting load state.

In the "Loading Option" column, two loading conditions are given: one is the dynamic load, that is, the centrifugal force produced with rotation, as shown on the right figure; and the other is the static load, namely, under the action of gravity, as shown on the left figure. When the bearing load is mainly gravity, select "STATIC", otherwise select "DYNAMIC" when the bearing is under a dynamic load.

After completion of the four-column parameters setting, click the "CALCULATE" button, as shown in Figure 21.15 and then enter the numerical simulation process. In the calculation program, the finite difference method and the iterative method are used to simulate the pressure.

In the "Output" column, several important characteristic parameters will be output, including the eccentricity ratio, the load, the deviation angle, the flux and the temperature rise after the calculation results reach the setting accuracy of convergence or the number of iterations. A group of output data is shown in Figure 21.19.

If the user needs to modify parameters to carry out their calculation again, press the "CLEAR" button to remove the original output data or modify the parameters directly so that the output column will be automatically emptied.

In addition to the several characteristics above, three text files will be produced in the current directory, including height.txt, pressure.txt and result.txt. Height.txt and

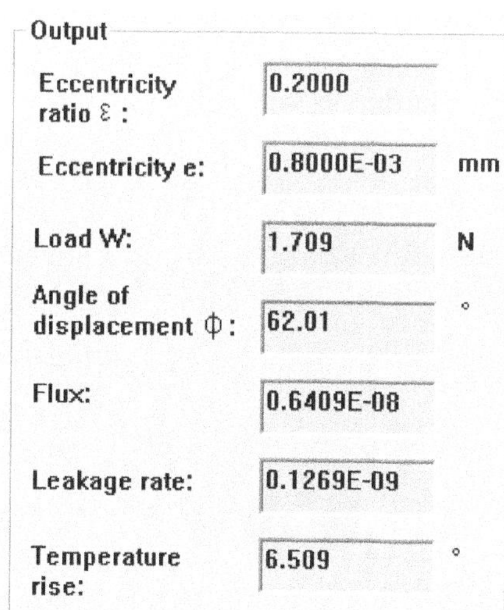

Figure 21.19 Results output.

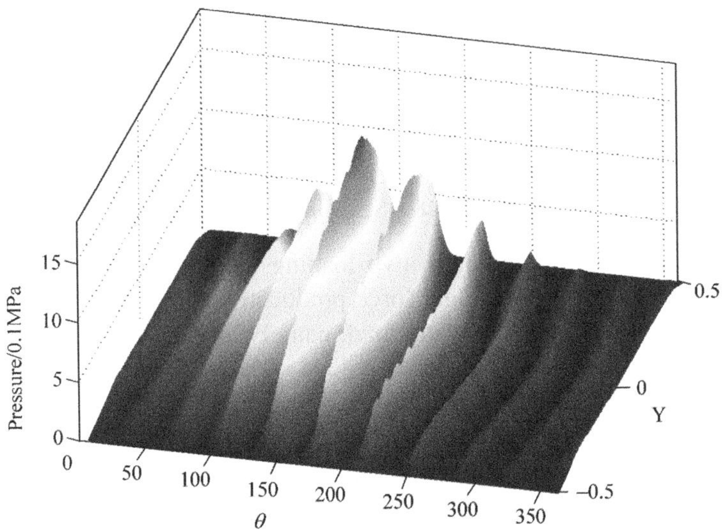

Figure 21.20 Pressure distribution.

pressure.txt are the data for film thickness and pressure distribution, respectively, while result.txt gives detailed calculation results for the given parameters.

Selecting the pressure.m file, the expansion figure of the pressure can be obtained, as shown in Figure 21.20.

Selecting the height.m file, the expansion figure of the film thickness can be obtained, as shown in Figure 21.21.

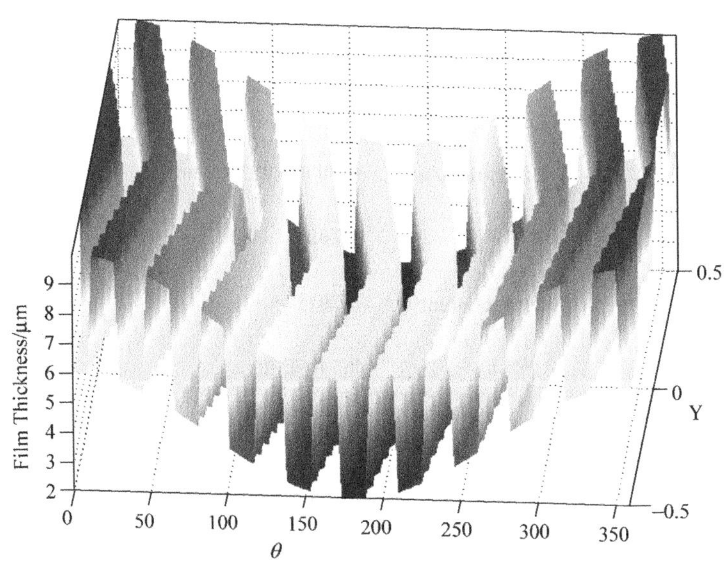

Figure 21.21 Film thickness.

22

Lubrication Optimization Program of Herringbone Grooved Journal Bearing of Micro Motor

22.1 Method of Optimization Calculation

22.1.1 Requirements of Parameter Optimization

According to the given conditions, the following parameters for a herringbone grooved journal bearing must be given to optimize the maximum load capacity or minimum flux as a target. They are

Groove angle	α
Ratio of groove and ridge	L_r/L_g
Groove depth	C_g

The computation program is developed according to the specific conditions of the parameters, the target function and the constraint conditions. The above three parameters are optimized and their optimal values are found.

22.1.2 Optimization Model

The target function is

$$\min f(x) \tag{22.1}$$

Numerical Calculation of Lubrication: Methods and Programs, First Edition. Ping Huang.

To meet the constraint conditions

$$
\begin{aligned}
&\alpha_{\min} \le \alpha \le \alpha_{\max} \\
&(L_r/L_g)_{\min} \le L_r/L_g \le (L_r/L_g)_{\max} \\
&(C_g)_{\min} \le C_g \le (C_g)_{\max}
\end{aligned}
\tag{22.2}
$$

where $f(x)$ may be the load W or the flux Q_L.

1. If the target function is load W, the eccentricity ratio ε is a constant value.
2. If the target function is flux Q_L, the eccentricity ratio ε will be a constant, else when $w = w_0$, the eccentricity ratio ε is a variable.

22.1.3 Optimization Methods and Steps

A simple and practical method is adopted here to save time and reduce computational tasks. It is:

Coordinates rotation method + interval decreasing method

The specific steps are:

1. Give a set of random initial values of α, L_r/L_g and C_g.
2. Search for optimization along one coordinate which is divided into 11 nodes.
3. Change the coordinate to the optimum until all coordinates have been finished.
4. Give the optimal parameters and the target function after finishing the search.
5. Check whether this needs to be calculated again.
6. If yes, build up a new cubical space with the optimization point as the center and one-fifth of the original interval as the next seeking space. Return to step 1 to continue the optimization.
7. If no, finish the calculation.

Note: if needed, choose a different point as the start point.
The search process is shown in Figure 22.1.

22.2 Program Layout of Optimization Calculation

22.2.1 Optimization Program Diagram

The optimization program diagram of HBOA is shown in Figure 22.2.

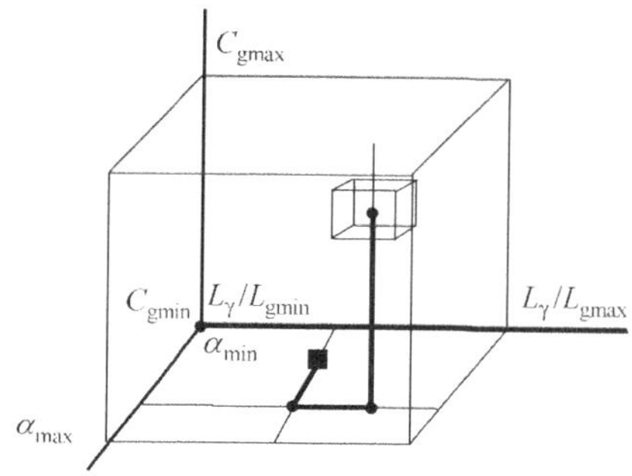

Figure 22.1 Sketch map of optimization process.

22.2.2 *Calculation Program*

```
PROGRAM HBOA
USE DFLOGM
INCLUDE 'RESOURCE.FD'
CALL DoDialog()
STOP
END PROGRAM
SUBROUTINE DoDialog()
USE DFLOGM
INCLUDE 'RESOURCE.FD'
INTEGER retint
LOGICAL retlog
TYPE (dialog) dlg
EXTERNAL DISPLAY
EXTERNAL CLEAR
EXTERNAL CHOOSE
EXTERNAL DISPLAY1
IF ( .not. DlgInit( IDD_DIALOG1, dlg)) THEN
WRITE (*,*) 'Error: dialog not found'
ELSE
  retlog=dlgset(dlg,idc_EDIT_DD,"2.992")
  retlog=dlgsetsub(dlg,idc_EDIT_DD,CLEAR,dlg_change)
  retlog=dlgset(dlg,idc_EDIT_D,"3.0")
  retlog=dlgsetsub(dlg,idc_EDIT_D,CLEAR,dlg_change)
  retlog=dlgset(dlg,idc_EDIT_AL,"2.4")
  retlog=dlgsetsub(dlg,idc_EDIT_AL,CLEAR,dlg_change)
  retlog=dlgset(dlg,idc_EDIT_AN,"5000")
  retlog=dlgsetsub(dlg,idc_EDIT_AN,CLEAR,dlg_change)
  retlog=dlgset(dlg,idc_EDIT_M,"10")
  retlog=dlgsetsub(dlg,idc_EDIT_M,CLEAR,dlg_change)
```

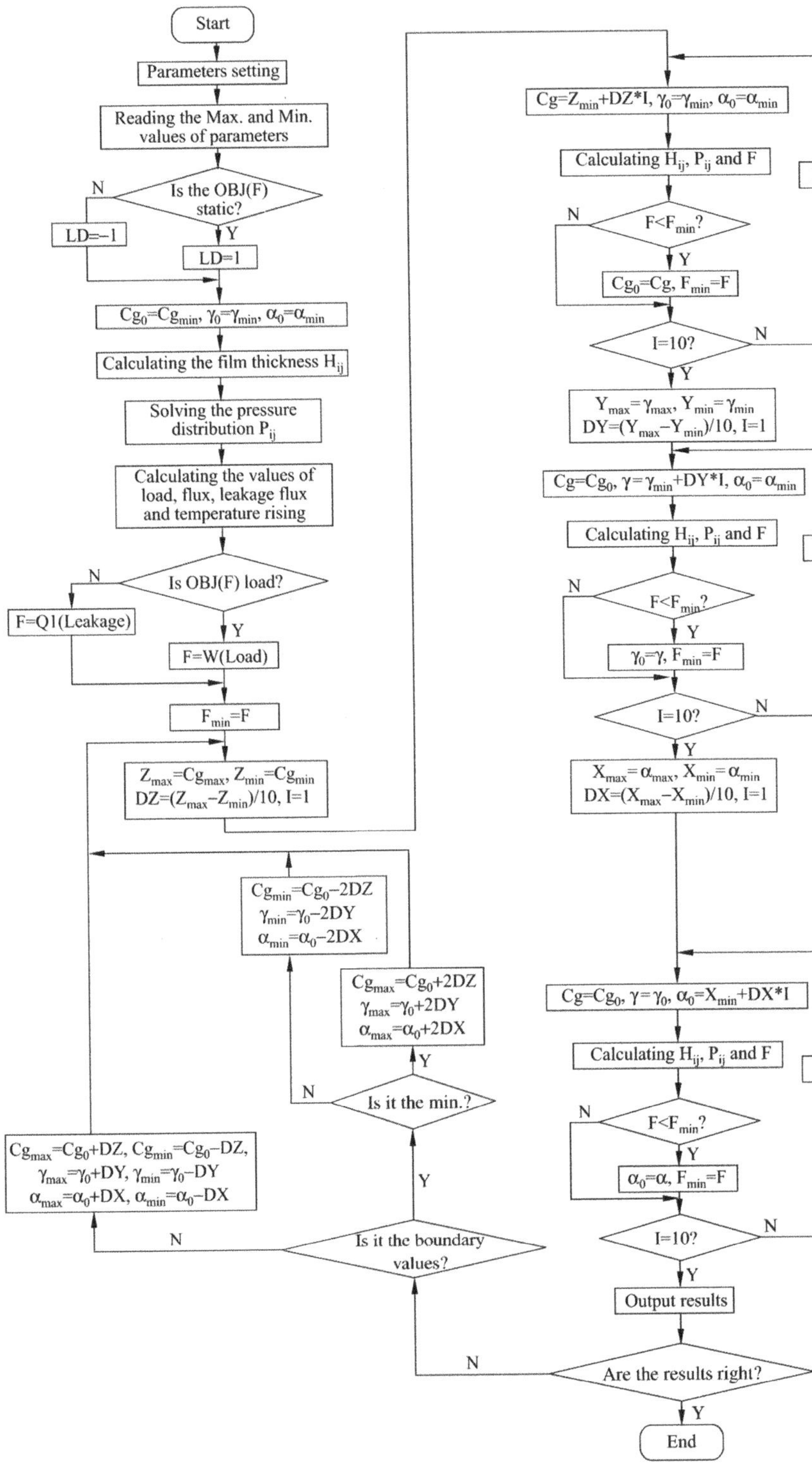

Figure 22.2 Optimization program diagram of HBOA.

```
    retlog=dlgset(dlg,idc_EDIT_MAXALF,"60")
    retlog=dlgsetsub(dlg,idc_EDIT_MAXALF,CLEAR,dlg_change)
    retlog=dlgset(dlg,idc_EDIT_MINALF,"20")
    retlog=dlgsetsub(dlg,idc_EDIT_MINALF,CLEAR,dlg_change)
    retlog=dlgset(dlg,idc_EDIT_MAXRLGL,"1.3")
    retlog=dlgsetsub(dlg,idc_EDIT_MAXRLGL,CLEAR,dlg_change)
    retlog=dlgset(dlg,idc_EDIT_MINRLGL,"0.7")
    retlog=dlgsetsub(dlg,idc_EDIT_MINRLGL,CLEAR,dlg_change)
    retlog=dlgset(dlg,idc_EDIT_MAXCG,"0.01")
    retlog=dlgsetsub(dlg,idc_EDIT_MAXCG,CLEAR,dlg_change)
    retlog=dlgset(dlg,idc_EDIT_MINCG,"0.003")
    retlog=dlgsetsub(dlg,idc_EDIT_MINCG,CLEAR,dlg_change)
    retlog=dlgset(dlg,idc_EDIT_DEDA,"24.5")
    retlog=dlgsetsub(dlg,idc_EDIT_DEDA,CLEAR,dlg_change)
    retlog=dlgset(dlg,idc_EDIT_RO,"0.945")
    retlog=dlgsetsub(dlg,idc_EDIT_RO,CLEAR,dlg_change)
    retlog=dlgset(dlg,idc_EDIT_AS,"80")
    retlog=dlgsetsub(dlg,idc_EDIT_AS,CLEAR,dlg_change)
    retlog=dlgset(dlg,idc_EDIT_CO,"1800")
    retlog=dlgsetsub(dlg,idc_EDIT_CO,CLEAR,dlg_change)
    retlog=dlgsetsub(dlg,idc_CALCULATE,DISPLAY)
    retlog=dlgsetsub(dlg,idc_continue,DISPLAY1)
    retlog=dlgset(dlg,idc_CONTINUE,.false.,dlg_enable)
    retlog=dlgsetsub(dlg,idc_CLEAR,CLEAR)
    retlog=dlgset(dlg,idc_EDIT_EPSON,.TRUE.)
    retlog=dlgset(dlg,idc_EDIT_SUM,.FALSE.)
    retlog=dlgset(dlg,idc_RADIO_LOAD,.TRUE.)
    retlog=dlgset(dlg,idc_RADIO_FLUX,.FALSE.)
    retlog1=dlgset(dlg,idc_RADIO_SUM,.FALSE.,dlg_enable)
    retlog=dlgsetsub(dlg,IDC_RADIO_LOAD,CHOOSE)
    retlog=dlgsetsub(dlg,IDC_RADIO_FLUX,CHOOSE)
    retlog=dlgsetsub(dlg,IDC_RADIO_SUM,CHOOSE)
    retlog=dlgsetsub(dlg,IDC_RADIO_EPSON,CHOOSE)
    retlog=dlgset(dlg,IDC_RADIO_DYNAMIC,.FALSE.,DLG_STATE)
    retlog=dlgset(dlg,IDC_RADIO_STATIC,.TRUE.,DLG_STATE)
    retint=DLGMODAL(dlg)
    CALL DLGUNINIT(dlg)
  END IF
  RETURN
  END SUBROUTINE DoDialog

  SUBROUTINE CHOOSE(dlg,control_name,calltype)
  USE DFLOGM
  INCLUDE 'resource.fd'
  TYPE (dialog) dlg
```

```
INTEGER retint,LF
LOGICAL retlog,retlog1,PUSHED_state
retlog=dlgget(dlg,idc_RADIO_LOAD,PUSHED_state)
IF(PUSHED_state) THEN
LF=1
retlog=dlgset(dlg,idc_RADIO_SUM,.false.,dlg_enable)
retlog=dlgset(dlg,idc_EDIT_SUM,.false.,dlg_enable)
retlog=dlgset(dlg,idc_EDIT_EPSON,.TRUE.,dlg_enable)
ELSE
LF=0
retlog=dlgset(dlg,idc_RADIO_SUM,.TRUE.,dlg_enable)
ENDIF
retlog=dlgget(dlg,idc_RADIO_EPSON,PUSHED_state)
IF(PUSHED_STATE.AND.LF.EQ.0) THEN
retlog1=dlgset(dlg,idc_EDIT_SUM,.FALSE.,dlg_enable)
retlog1=dlgset(dlg,idc_EDIT_EPSON,.TRUE.,dlg_enable)
ELSE
IF(LF.EQ.0) THEN
retlog=dlgset(dlg,idc_EDIT_EPSON,.FALSE.,dlg_enable)
retlog=dlgset(dlg,idc_EDIT_SUM,.TRUE.,dlg_enable)
ELSE
retlog=dlgset(dlg,idc_RADIO_SUM,.false.,dlg_enable)
retlog=dlgset(dlg,idc_EDIT_SUM,.false.,dlg_enable)
retlog=dlgset(dlg,idc_EDIT_EPSON,.TRUE.,dlg_enable)
ENDIF
ENDIF
END SUBROUTINE CHOOSE

SUBROUTINE CLEAR(dlg,control_name,calltype)
USE DFLOGM
INCLUDE 'resource.fd'
TYPE (dialog) dlg
LOGICAL retlog
INTEGER retint,control_name,calltype
retlog=dlgset(dlg,idc_EDIT_SUM2," ")
retlog=dlgset(dlg,idc_EDIT_EPSON2," ")
retlog=dlgset(dlg,idc_EDIT_E," ")
retlog=dlgset(dlg,idc_EDIT_ALF2," ")
retlog=dlgset(dlg,idc_EDIT_Q," ")
retlog=dlgset(dlg,idc_EDIT_RLGL2," ")
retlog=dlgset(dlg,idc_EDIT_CG2," ")
RETURN
END SUBROUTINE CLEAR

SUBROUTINE DISPLAY1(dlg,control_name,calltype)
USE dflogm
```

```
INCLUDE 'resource.fd'
TYPE (dialog) dlg
LOGICAL retlog, pushed_state
INTEGER retint,control_name,calltype
INTEGER LCON
retlog=dlgset(dlg,idc_EDIT_SUM2," ")
retlog=dlgset(dlg,idc_EDIT_EPSON2," ")
retlog=dlgset(dlg,idc_EDIT_E," ")
retlog=dlgset(dlg,idc_EDIT_ALF2," ")
retlog=dlgset(dlg,idc_EDIT_Q," ")
retlog=dlgset(dlg,idc_EDIT_RLGL2," ")
retlog=dlgset(dlg,idc_EDIT_CG2," ")
retlog=dlgset(dlg,idc_RADIO_continue,.true.)
CALL DISPLAY(dlg,control_name,calltype)
END SUBROUTINE DISPLAY1

SUBROUTINE DISPLAY(dlg,control_name,calltype)
USE dflogm
INCLUDE 'resource.fd'
TYPE (dialog) dlg
LOGICAL retlog, pushed_state
INTEGER retint,control_name,calltype,LLL
INTEGER LCON
CHARACTER(256) text
DATA PI/3.14159265/
OPEN(8,FILE='pressure.DAT',STATUS='UNKNOWN')
OPEN(9,FILE='height.DAT',STATUS='UNKNOWN')
OPEN(14,FILE='result.DAT',STATUS='UNKNOWN')
OPEN(13,FILE='ORESULT.DAT',STATUS='UNKNOWN')
LLL=0
retlog=dlgget(dlg,idc_radio_continue,pushed_state)
IF(pushed_state) GO TO 100
retlog=dlgget(dlg,idc_edit_MAXALF,text)
read(text,*) ALFMAX
retlog=dlgget(dlg,idc_edit_MINALF,text)
read(text,*) ALFMIN
retlog=dlgget(dlg,idc_edit_MAXRLGL,text)
read(text,*) RLGLMAX
retlog=dlgget(dlg,idc_edit_MINRLGL,text)
read(text,*) RLGLMIN
retlog=dlgget(dlg,idc_edit_MAXCG,text)
read(text,*) CGMAX
retlog=dlgget(dlg,idc_edit_MINCG,text)
read(text,*) CGMIN
retlog=dlgget(dlg,idc_edit_DD,text)
```

```
read(text,*) DD
retlog=dlgget(dlg,idc_edit_D,text)
read(text,*) D
retlog=dlgget(dlg,idc_edit_AL,text)
read(text,*) AL
retlog=dlgget(dlg,idc_edit_AN,text)
read(text,*) AN
retlog=dlgget(dlg,idc_edit_M,text)
read(text,*) M
retlog=dlgget(dlg,idc_EDIT_DEDA,text)
read(text,*) DEDA
retlog=dlgget(dlg,idc_EDIT_RO,text)
read(text,*) RO
retlog=dlgget(dlg,idc_EDIT_AS,text)
read(text,*) AS
retlog=dlgget(dlg,idc_EDIT_CO,text)
read(text,*) CO
retlog = DLGGET (dlg, IDC_RADIO_FLUX, pushed_state)
IF (PUSHED_STATE) THEN
LF=0
ELSE
LF=1
ENDIF
retlog = DLGGET (dlg, IDC_RADIO_EPSON, pushed_state)
IF (PUSHED_STATE) THEN
retlog=dlgget(dlg,idc_edit_EPSON,text)
read(text,*) EPSON
W0=0.0
ME=1
ELSE
retlog=dlgget(dlg,idc_edit_SUM,text)
read(text,*) W0
EPSON=0.2
ME=0
ENDIF
retlog = DLGGET (dlg, IDC_RADIO_STATIC, pushed_state)
IF(PUSHED_STATE) THEN
LD=1
ELSE
LD=-1
END IF
!   WRITE(*,*)'LF,ME,LD=',LF,ME,LD

Q1=0.0
SUM=0.0
DD=DD*1.0E-3
```

```
      D=D*1.0E-3
      AL=AL*1.0E-3
      RO=RO*1.0E3
      DEDA=DEDA*1.0E-6

      X1=ALFMAX*PI/180
      X0=ALFMIN*PI/180
      Y1=RLGLMAX
      Y0=RLGLMIN
      Z1=CGMAX*1.0E-3
      Z0=CGMIN*1.0E-3
 10   DX=(X1-X0)/10.0
      DY=(Y1-Y0)/10.0
      DZ=(Z1-Z0)/10.0
      CALL RANDOM(RVAL)
      I=IFIX(RVAL*10)+1
      XMIN=X0+I*(X1-X0)/10.0
      CALL RANDOM(RVAL)
      J=IFIX(RVAL*10)+1
      YMIN=Y0+J*(Y1-Y0)/10.0
      CALL RANDOM(RVAL)
      K=IFIX(RVAL*10)+1
      ZMIN=Z0+K*(Z1-Z0)/10.0
      WRITE(*,*)'XMIN=',XMIN,'YMIN=',YMIN,'ZMIN=',ZMIN

      FF=F(XMIN,YMIN,ZMIN,DD,D,AL,AN,RO,W0,DEDA,AS,CO,LF,ME,LD,
 M,EPSON,Q1,SUM)
      FMIN=FF
      WRITE(*,*) 'FMIN=', FMIN
      DO MM=1,11
      Z=Z0+DZ*(MM-1)
      FF=F(XMIN,YMIN,Z,DD,D,AL,AN,RO,W0,DEDA,AS,CO,LF,ME,LD,M,
 EPSON,Q1,SUM)
      WRITE(*,*)'XMIN,YMIN,Z,FF=',XMIN,YMIN,Z,FF
      IF(FF.LT.FMIN)THEN
      K=MM
      ZMIN=Z
      FMIN=FF
      ENDIF
      ENDDO
      WRITE(*,*)'XMIN=',XMIN,'YMIN=',YMIN,'ZMIN=',ZMIN,'FMIN=',FMIN
      DO MM=1,11
      Y=Y0+DY*(MM-1)
      FF=F(XMIN,Y,ZMIN,DD,D,AL,AN,RO,W0,DEDA,AS,CO,LF,ME,LD,M,
 EPSON,Q1,SUM)
      WRITE(*,*)'XMIN,Y,ZMIN,FF=',XMIN,Y,ZMIN,FF
```

```
    IF(FF.LT.FMIN)THEN
    J=MM
    YMIN=Y
    FMIN=FF
    ENDIF
    ENDDO
    WRITE(*,*)'XMIN=',XMIN,'YMIN=',YMIN,'ZMIN=',ZMIN,'
FMIN=',FMIN
    DO MM=1,11
    X=X0+DX*(MM-1)
    FF=F(X,YMIN,ZMIN,DD,D,AL,AN,RO,W0,DEDA,AS,CO,LF,ME,LD,M,
EPSON,Q1,SUM)
    WRITE(*,*)'X,YMIN,ZMIN,FF=',X,YMIN,ZMIN,FF
    IF(FF.LT.FMIN)THEN
    I=MM
    XMIN=X
    FMIN=FF
    ENDIF
    ENDDO

    FMIN=F(XMIN,YMIN,ZMIN,DD,D,AL,AN,RO,W0,DEDA,AS,CO,LF,ME,
LD,M,EPSON,Q1,SUM)

    WRITE(*,*)'XMIN=',XMIN,'YMIN=',YMIN,'ZMIN=',ZMIN,'

FMIN=',FMIN

    retlog=dlgset(dlg,idc_continue,.true.,dlg_enable)
    GO TO 200

100 retlog=dlgset(dlg,idc_RADIO_continue,.false.)
    IF(I.EQ.1)X1=X0+2.*DX
    IF(I.EQ.11)X0=X1-2.*DX
    IF(I.NE.1.AND.I.NE.11)THEN
    X0=XMIN-DX
    X1=XMIN+DX
    ENDIF
    IF(J.EQ.1)Y1=Y0+2.*DY
    IF(J.EQ.11)Y0=Y1-2.*DY
    IF(J.NE.1.AND.J.NE.11)THEN
    Y0=YMIN-DY
    Y1=YMIN+DY
    ENDIF
    IF(K.EQ.1)Z1=Z0+2.*DZ
    IF(K.EQ.11)Z0=Z1-2.*DZ
    IF(K.NE.1.AND.K.NE.11)THEN
    Z0=ZMIN-DZ
    Z1=ZMIN+DZ
```

```
      ENDIF
      WRITE(*,*)'X,Y,Z=',X0,X1,Y0,Y1,Z0,Z1
      GOTO 10
200   WRITE(TEXT,'(G12.4)') XMIN*180/PI
      retlog=dlgset(dlg,idc_edit_ALF2,trim(adjustl(TEXT)))
      WRITE(TEXT,'(G12.4)') YMIN
      retlog=dlgset(dlg,idc_edit_RLGL2,trim(adjustl(TEXT)))
      WRITE(TEXT,'(G12.4)') ZMIN*1.0E3
      retlog=dlgset(dlg,idc_edit_CG2,trim(adjustl(TEXT)))
      WRITE(TEXT,'(G12.4)') EPSON
      retlog=dlgset(dlg,idc_edit_EPSON2,trim(adjustl(TEXT)))
      IF (LF.EQ.1) THEN
      SUM=-FMIN
      WRITE(TEXT,'(G12.4)') -FMIN
      retlog=dlgset(dlg,idc_edit_SUM2,trim(adjustl(TEXT)))
      WRITE(TEXT,'(G12.4)') Q1
      retlog=dlgset(dlg,idc_edit_Q,trim(adjustl(TEXT)))
      ELSE
      WRITE(TEXT,'(G12.4)') FMIN
      retlog=dlgset(dlg,idc_edit_Q,trim(adjustl(TEXT)))
      WRITE(TEXT,'(G12.4)') SUM
      retlog=dlgset(dlg,idc_edit_SUM2,trim(adjustl(TEXT)))
      ENDIF
      IF(LD.EQ.1) THEN
      TEXT='STATIC'
      ELSE
      TEXT='DYNAMIC'
      ENDIF
      WRITE(13,*)'a γ Cg ε W0 LOAD LEAKAGE LOADING'
      WRITE(13,4) XMIN*180/PI,YMIN,ZMIN,EPSON,W0,SUM,Q1,TEXT
4     FORMAT(20(G12.4))
      END SUBROUTINE DISPLAY

      FUNCTION F(ALF,RLGL,CG,DD,D,AL,AN,RO,W0,DEDA,AS,CO,LF,
                 ME,LD,M,EPSON,Q1,SUM) RESULT(FF)
      REAL*4 X(121),Y(121),H(121,121),P(121,121),PD(121,121),HH(121)
      REAL*4 HXY(121,121),HXF(121,121),HXB(121,121),HYF
(121,121),HYB(121,121)
      REAL*4,INTENT(OUT):: Q1,SUM,EPSON
      CHARACTER(10) TEXT
      DATA N,N1,N2/121,120,61/
      DATA PI/3.14159265/
!     WRITE(*,*) 'ALF,RLGL,CG=',ALF*180/PI,RLGL,CG
!     WRITE(*,*) 'DD,D,AL,EPSON=',DD,D,AL,EPSON
!     WRITE(*,*) 'LDF=',LD
      PI2=2.0*PI
```

```
      EDA=DEDA*RO
      R=D/2.0
      ALD=AL/D
      C=(D-DD)/2.0
      PESAI=(D-DD)/D
      K2=120/M
      DX=1.0/N1
      ALX=PI2*R
      ALY=AL
      RATIO=ALX/ALY
      ALFA=RATIO**2
      OMIGA=AN*PI2/60.0
      U=OMIGA*R
      AKEXI=1.0
      IF(ALD.GT.1.0)AKEXI=(D/ALY)**1.5
      ALENDA=6.0*U*EDA*ALX/C**2
      HMIN=C*(1.0-EPSON)
      TEMP=1./N1
      DO I=1,N
      X(I)=(I-1)*TEMP
      Y(I)=-0.5+X(I)
      ENDDO
      Y(N2)=0.0
      K1=(RLGL/(1+RLGL))*K2
      IF(ABS(K1-(RLGL/(1+RLGL))*K2).GT.0.5) K1=K1+1
      CCG=CG/C
      DO I=1,N
      DO J=1,N
      P(I,J)=0.0
      ENDDO
      ENDDO

      IF (ME.EQ.1) THEN
      CALL SUBH(N,N1,EPSON,X,Y,H,HH,PI2,K1,K2,CCG,RATIO,ALF,
   ALFA,N2, HXF,HXB,HYF,HYB,HXY)
      CALL SUBP(N,N1,PD,P,H,DX,HXF,HXB,HYF,HYB,HXY,LD)
      CALL SUBM(N,N1,PI,P,PX,PY,SUM1)
      SUM0=ALENDA*ALX*ALY*SUM1/(N*N)
      GO TO 30
      ELSE
      EPSON=0.5
      EPSON1=0.0
      EPSON2=1.0
      GO TO 6
      ENDIF
```

```
6    CALL SUBH(N,N1,EPSON,X,Y,H,HH,PI2,K1,K2,CCG,RATIO,ALF,
ALFA,N2,HXF,HXB,HYF,HYB,HXY)
     CALL SUBP(N,N1,PD,P,H,DX,HXF,HXB,HYF,HYB,HXY,LD)
     CALL SUBM(N,N1,PI,P,PX,PY,SUM1)
     SUM0=ALENDA*ALX*ALY*SUM1/(N*N)
!    WRITE(*,*) ME,W0,SUM0,EPSON
     IF(ABS((SUM0-W0)/W0).LE.0.005) GOTO 30
     IF(W0.LT.SUM0)THEN
     EPSON2=EPSON
     EPSON=0.5*(EPSON1+EPSON)
     ELSE
     EPSON1=EPSON
     EPSON=0.5*(EPSON+EPSON2)
     ENDIF

     GOTO 6

30   E=EPSON*C
     HMIN=C*(1.0-EPSON)
     Q1=0.0
     DO I=1,N1
     Q=H(I,1)**3*(4.0*P(I,2)-P(I,3))+H(I,N)**3*(4.0*P(I,N-1)-P
(I,N-2))
     IF(Q.LT.0.0)Q=0.0
     Q1=Q1+Q
     ENDDO
     Q1=Q1*0.25*U*ALY*C
     Q=0.5*U*ALY*(C*(1+EPSON)+CG*K1/K2)
     DDT=0.0
     AI=180.0*ATAN(PY/PX)/PI
     PA=SUM0/(D*ALY)
     FS=PI*EDA*OMIGA/PESAI/PA+0.55*PESAI*AKEXI
     DT=(FS/PESAI)*PA/(CO*RO*Q/(PESAI*U*ALY*D)+PI*AS/PESAI/U)
     IF(LD.EQ.1) THEN
     TEXT='STATIC'
     ELSE
     TEXT='DYNAMIC'
     ENDIF

     WRITE(14,*)'α γ Cg ε e W0 LOAD FLUX LEAKAGE ΔT LOADING'

     WRITE(14,4) ALF*180/PI,RLGL,CG,EPSON,E,W0,SUM0,Q,Q1,DT,TEXT
40   FORMAT(122(E12.6,1X))
4    FORMAT(20(G12.4))
     SUM=SUM0
!    WRITE(*,*) 'Q1,SUM=',Q1,SUM
     IF(LF.EQ.1) THEN
```

```
      FF=-SUM0
      ELSE
      FF=Q1
      ENDIF
!     WRITE(*,*)'FF=',FF
      RETURN
      END

      SUBROUTINE SUBH(N,N1,EPSON,X,Y,H,HH,PI2,K1,K2,CCG,RATIO,
ALF,ALFA,N2, HXF,HXB,HYF,HYB,HXY)
      REAL*4 X(N),Y(N),H(N,N),HH(N),HXF(N,N),HXB(N,N),HYF(N,N),
HYB(N,N),HXY(N,N)
      DO I=1,N
      DO J=1,N
      H(I,J)=1.0+EPSON*COS(PI2*X(I))
      ENDDO
      ENDDO
      IF(K1.EQ.0)GOTO 20
      DO I=1,N
      HH(I)=0.0
      ENDDO
      DO I=1,N-K2,K2
      DO L=1,K1
      HH(I+L-1)=CCG
      ENDDO
      ENDDO
      HH(N)=HH(1)
      TEMP=1./RATIO
      DO J=1,N
      DO I=1,N1
      I1=I+ABS(J-N2)*(1./TAN(ALF/2))*TEMP
      IF(I1.GT.N)THEN
      I1=I1-N1
      ENDIF
      H(I,J)=H(I,J)+HH(I1)
      ENDDO
      H(N,J)=H(1,J)
      ENDDO
 20   DO I=1,N
      I1=I-1
      I2=I+1
      IF(I.EQ.1)I1=N1
      IF(I.EQ.N)I2=2
      DO J=2,N1
      HXF(I,J)=(0.5*(H(I2,J)+H(I,J)))**3
      HXB(I,J)=(0.5*(H(I1,J)+H(I,J)))**3
```

```
    HYF(I,J)=ALFA*(0.5*(H(I,J+1)+H(I,J)))**3
    HYB(I,J)=ALFA*(0.5*(H(I,J-1)+H(I,J)))**3
    HXY(I,J)=1.0/(HXF(I,J)+HXB(I,J)+HYF(I,J)+HYB(I,J))
    ENDDO
    ENDDO
    RETURN
    END SUBROUTINE SUBH

    SUBROUTINE SUBP(N,N1,PD,P,H,DX,HXF,HXB,HYF,HYB,HXY,LD)
    REAL*4 X(N),Y(N),PD(N,N),P(N,N),H(N,N),HXF(N,N),HXB(N,N),
HYF(N,N),HYB(N,N),HXY(N,N)
    DO I=1,N
    DO J=1,N
    PD(I,J)=P(I,J)
    ENDDO
    ENDDO
    IK=0
    TEMP=0.5*DX
10  C1=0.0
    DO I=1,N1
    I1=I-1
    I2=I+1
    IF(I1.EQ.0)I1=N1
    IF(I2.EQ.N)I2=1
    DO J=2,N1
    P(I,J)=(HXF(I,J)*P(I2,J)+HXB(I,J)*P(I1,J)+HYF(I,J)*P
(I,J+1)+HYB(I,J)*P(I,J-1)-LD*TEMP*(H(I2,J)-H(I1,J)))*HXY(I,J)
    IF(P(I,J).LE.0.0)P(I,J)=0.0
    C1=C1+ABS(P(I,J)-PD(I,J))
    PD(I,J)=P(I,J)
    ENDDO
    ENDDO
    DO J=2,N1
    P(N,J)=P(1,J)
    PD(N,J)=PD(1,J)
    ENDDO
    IK=IK+1
    IF(C1.GT.1.E-20.AND.IK.LE.20000)GOTO 10
!   WRITE(*,*)'IK,ER=',IK,C1
    RETURN
    END SUBROUTINE SUBP
    SUBROUTINE SUBM(N,N1,PI,P,PX,PY,SUM1)
    DIMENSION P(N,N)
    PX=0.0
    PY=0.0
    TEMP=PI/60.0
```

```
DO I=1,N1
AI=(I-1)*TEMP
DO J=1,N
PX=PX-P(I,J)*COS(AI)
PY=PY+P(I,J)*SIN(AI)
ENDDO
ENDDO
SUM1=SQRT(PX*PX+PY*PY)
RETURN
END SUBROUTINE SUBM
```

22.2.3 *Parameters in Program*

The main parameters in the program are shown in Table 22.1.

22.3 Optimization Calculation Examples

The calculation parameters used in the following examples are as follows.

Rotation n	5000 rpm
Diameter of axis d	2.992 mm
Inner diameter D	3.000 mm
Bearing width L	2.4 mm
Viscosity of oil η	0.02315 Pa s
α_{min}	20°
α_{max}	60°
$(L_r/L_g)_{\text{min}}$	0.7
$(L_r/L_g)_{\text{max}}$	1.3
$(C_g)_{\text{min}}$	0.003 mm
$(C_g)_{\text{max}}$	0.01 mm

22.3.1 *Example 1: Optimization Calculation for Static Load*

The given eccentricity ratio ε is 0.1; the initial conditions are $\alpha=24°$, $L_r/L_g=0.76$ and $C_g=0.0079$; and the initial load $W=0.6630$ N.

The first loop optimization results are:

When $\alpha=44°$, $L_r/L_g=0.76$ and $C_g=0.003$, the maximum load $W=1.054$ N and the flux $Q_L=0.1457\times10^{-9}\,\text{m}^3\,\text{s}^{-1}$.

The second loop load optimization results are:

When $\alpha=44°$, $L_r/L_g=0.82$ and $C_g=0.003$, the maximum load $W=1.054$ N and the flux $Q_L=0.1457\times10^{-9}\,\text{m}^3\,\text{s}^{-1}$.

This shows that the results of L_r/L_g are different, but the optimization results are same. That means that L_r/L_g has no or little effect on the load.

Table 22.1 Parameters in program

Parameter	Parameter specification	Parameter	Parameter specification
AI	Angle of displacement	P(121,121)	Pressure at each node
AL	Length of bearing	PI	π
ALD	AL/D	PI2	2π
ALENDA	Dimensionless Coefficient of pressure	PA	Average pressure
ALF	Angle of groove α	PX	Force in x direction
ALFA	$RATIO^2 = (ALX/ALY)^2$	PY	Force in y direction
ALX	Length of x interval	PESAI	φ
ALY	Length of y interval	Q	Flux
AN	Rotational speed	Q1	Leakage flux
AS	Coefficient of heat transfer	R	Diameter of bearing
C	Radius clearance	RLGL	Ratio of groove and ridge
C1	Error	RO	Density of oil
CG	Groove depth	RATIO	ALX/ALY
CCG	Coefficient of groove depth CG/C	SUM	Load
CO	Specific heat	SUM0	Initial load
D	Inner diameter	U	Linear velocity
DD	Diameter of axis	X(121)	X coordinate value
DEDA	Viscosity	Y(121)	Y coordinate value
DX	Clearance in x direction	ALFMAX	Max. groove angle
DT	Temperature rise	ALFMIN	Min. groove angle
E	Eccentricity	CGMAX	Max. groove depth
EDA	Dynamic viscosity	CGMIN	Min. groove depth
EPSON	Eccentricity ratio ε	RLGLMAX	Max. ratio of groove and ridge
F	Shear force	RLGLMIN	Min. ratio of groove and ridge
H(121,121)	Film thickness	XMIN	Optimization of groove angle
HMIN	Minimum of film thickness	YMIN	Optimization of ratio of groove and ridge
K1	Nodes occupied by grooves	ZMIN	Optimization of groove depth
K2	Nodes occupied by one groove	FF	Value of target function
KEXI	ζ	WO	Given load
M	Sum of ridges	LOADING	State of loading
N	Node number		
N1	N-1		
N2	$N1/2+1$		
OMIGA	Angular velocity		

22.3.2 Example 2: Optimization Calculation for Static Flux (Eccentricity Ratio ε is Constant)

Under the given conditions, the eccentricity ratio $\varepsilon = 0.2$, the initial conditions are $\alpha = 24°$, $L_r/L_g = 0.76$, $C_g = 0.0079$ and $Q_L = 1.5824 \times 10^{-10}\,\mathrm{m^3\,s^{-1}}$.

The first loop flux optimization results are:

When $\alpha = 20°$, $L_r/L_g = 1.24$ and $C_g = 0.003$, the minimum flux $Q_L = 0.9403 \times 10^{-10}\,\mathrm{m^3\,s^{-1}}$, the load $W = 1.334$ N.

The second loop flux optimization results are:

When $\alpha = 20.80°$, $L_r/L_g = 1.3$ and $C_g = 0.003$, the minimum flux $Q_L = 0.9230 \times 10^{-10}\,\mathrm{m^3\,s^{-1}}$, the load $W = 1.359$ N and $\varepsilon = 0.1114$.

22.3.3 Example 3: Optimization Calculation for Static Flux (Load W is Constant)

Under the given conditions, the static load $W_0 = 1$ N and the initial conditions are $\alpha = 24°$, $L_r/L_g = 0.76$, $C_g = 0.0079$ and $Q_L = 1.6344 \times 10^{-10}\,\mathrm{m^3\,s^{-1}}$.

The first loop flux optimization results are:

When $\alpha = 20°$, $L_r/L_g = 0.88$ and $C_g = 0.003$, the minimum flux $Q_L = 0.7411 \times 10^{-10}\,\mathrm{m^3\,s^{-1}}$, the load $W = 0.9988$ N and $\varepsilon = 0.1230$.

The second loop flux optimization results are:

When $\alpha = 20°$, $L_r/L_g = 0.82$ and $C_g = 0.003$, the minimum flux $Q_L = 0.7332 \times 10^{-10}\,\mathrm{m^3\,s^{-1}}$, the load $W = 1.003$ N and $\varepsilon = 0.1113$.

22.3.4 Example 4: Optimization Calculation for Dynamic Load

Under the given conditions, the eccentricity ratio $\varepsilon = 0.1$ and the initial conditions are $\alpha = 24°$, $L_r/L_g = 0.76$ and $C_g = 0.0079$ and $W = 0.1071$ N.

The first load optimization results are:

When $\alpha = 20°$, $L_r/L_g = 0.76$ and $C_g = 0.003$, the maximum load $W = 0.2281$ N and the flux $Q_L = 0.1308 \times 10^{-9}\,\mathrm{m^3\,s^{-1}}$.

The second loop load optimization results are:

When $\alpha = 23.20°$, $L_r/L_g = 0.8200$ and $C_g = 0.003$, the maximum load $W = 0.2303$ N and the flux $Q_L = 0.1410 \times 10^{-9}\,\mathrm{m^3\,s^{-1}}$.

22.3.5 Example 5: Optimization Calculation for Dynamic Flux (Eccentricity ε is Constant)

Under the given conditions, the eccentricity ratio $\varepsilon = 0.3$ and the initial conditions are $\alpha = 24°$, $L_r/L_g = 0.76$, $C_g = 0.0079$ and $Q_L = 4.6646 \times 10^{-10}\,\mathrm{m^3\,s^{-1}}$.

The first loop flux optimization results are:

When $\alpha = 20°$, $L_r/L_g = 0.76$ and $C_g = 0.003$, the minimum flux $Q_L = 0.1907 \times 10^{-9}\,\mathrm{m^3\,s^{-1}}$, the load $W = 1.311$ N.

The second loop flux optimization results are:

When $\alpha = 20.8°$, $L_r/L_g = 0.82$ and $C_g = 0.003$, the minimum flux $Q_L = 0.1853 \times 10^{-9}\,m^3\,s^{-1}$, the load $W = 1.286\,N$.

22.3.6 *Example 6: Optimization Calculation for Dynamic Flux (Load W is Constant)*

Under the given conditions, the dynamic load $W_0 = 3\,N$ and the initial conditions are $\alpha = 24°$, $L_r/L_g = 0.76$, $C_g = 0.0079$ and $Q_L = 5.4022 \times 10^{-10}\,m^3\,s^{-1}$.

The first loop flux optimization results are:

When $\alpha = 20°$, $L_r/L_g = 0.76$ and $C_g = 0.003$, the minimum flux $Q_L = 0.2608 \times 10^{-9}\,m^3\,s^{-1}$, the load $W = 3.002\,N$ and $\varepsilon = 0.5039$.

The second loop flux optimization results are:

When $\alpha = 20.80°$, $L_r/L_g = 0.82$ and $C_g = 0.003$, the min flux $Q_L = 0.2497 \times 10^{-9}\,m^3\,s^{-1}$, the load $W = 3.011\,N$ and $\varepsilon = 0.5078$.

22.4 Instructions for HBOA Software Package

22.4.1 *Program Package*

The files included in the package are shown in Table 22.2.

The files produced after execution are shown in Table 22.3.

Table 22.2 Files in package

File	File type
HBOA.exe	Execute file
Resource.h	H file
Resource.fd	Source file
Resource.hm	HM file
Script1.rc	Resource template
HBOA.F90	Program source file
HBOA.dsp	Project file
HBOA.dsw	Project workspace

Table 22.3 Files generated after execution

File	File type
Result.dat	The result
Oresult.dat	Optimization result
HBOA.plg	PLG file
Script1.asp	ASP file
Debug	Debug document

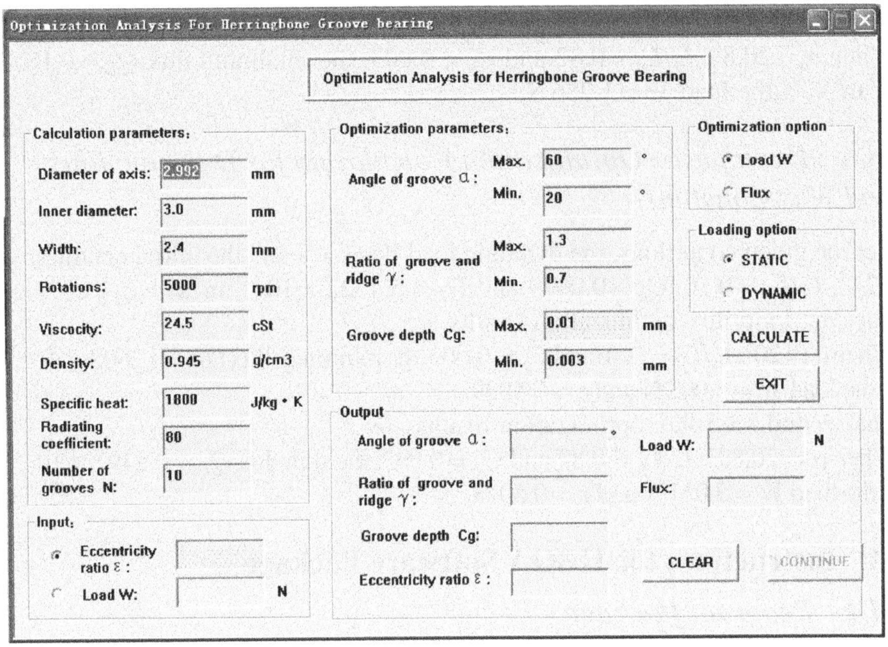

Figure 22.3 Main interface of the program.

22.4.2 Program Execution

22.4.2.1 Running the Program

The interface shown in Figure 22.3 will appear after clicking the executable file HBOA.exe. Or, double-click HBOA.dsw and the Compaq Visual Fortran operation interface will appear. Then click Fileview, select the source file and double-click HBOA.F90. The results are shown in Figure 22.4. After we compile, link and execute the program, the interface in Figure 22.3 will re-appear.

22.4.2.2 Setting the Parameters

As shown in Figure 22.3, the main interface has six regions: Calculation parameters, Optimization parameters, Optimization option, Input, Loading option and Output.

The first column "Calculation parameters" is used to set the parameters of the bearing, the lubricant and the part of the grooves.

The second column "Optimization parameters" is used to set the maximum and minimum values for parameters, including groove angle, ratio of groove and ridge and groove depth.

In order to input the data conveniently, common units are chosen. The unit of length is mm, the force unit is N, the unit of motor speed is rpm, the unit of viscosity is the

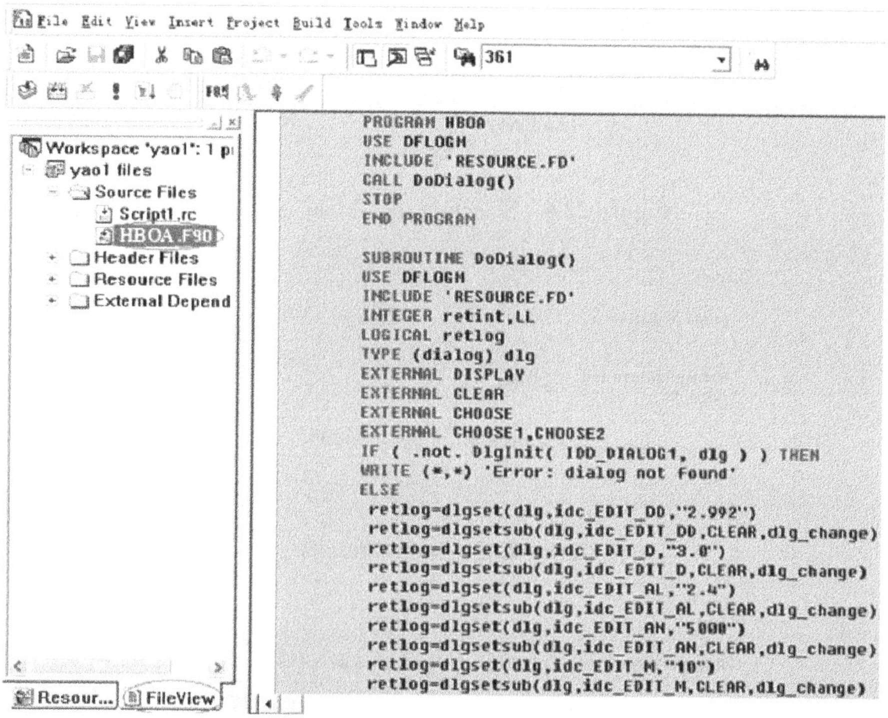

Figure 22.4 Fortran running interface.

dynamic viscosity (i.e. mm^2/s) and the unit of the density is g/cm^3. The parameters in the two columns can be modified according to the user's need.

The third column "Optimization option" is used to set the target function. The "Load" option is to select the maximum load capacity as the target while the "Flux" option is the minimum flux leakage.

The fourth column "Input" is used to select the target function. If the "Load" option is selected in the third column, the target function is only the load when the eccentricity ratio is given. If the "Flux" option is selected in the third column, the leakage flux can be analyzed when the eccentricity ratio or the load is given. If the eccentricity ratio is given, click "Eccentricity ratio" and input its value, as shown in Figure 22.5. If the load is given, click the "Load W" and input its value as shown in Figure 22.6.

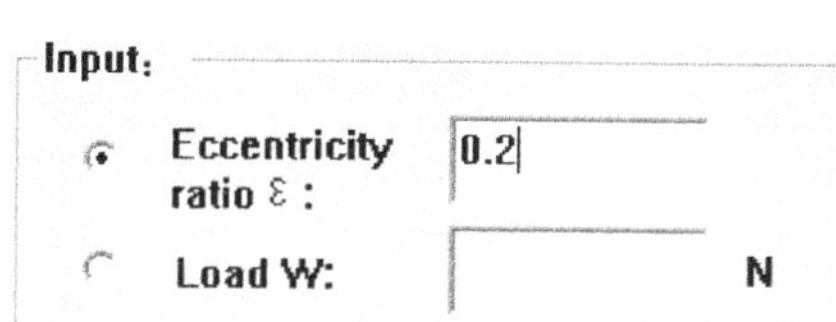

Figure 22.5 Selection eccentricity ratio or load.

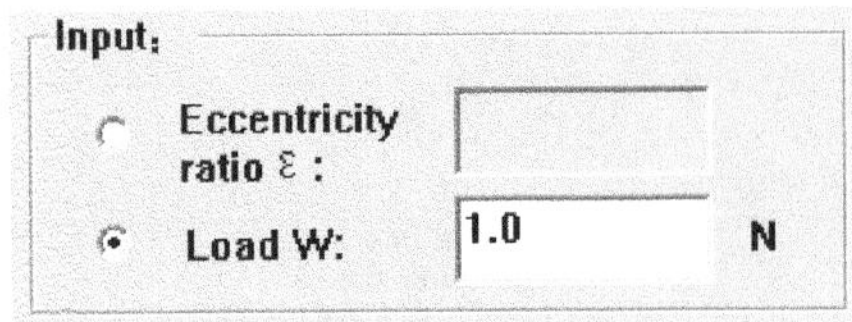

Figure 22.6 Selection the "Load" state.

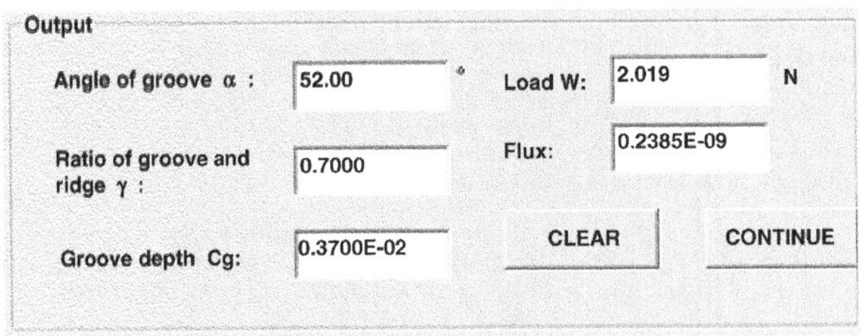

Figure 22.7 Output results.

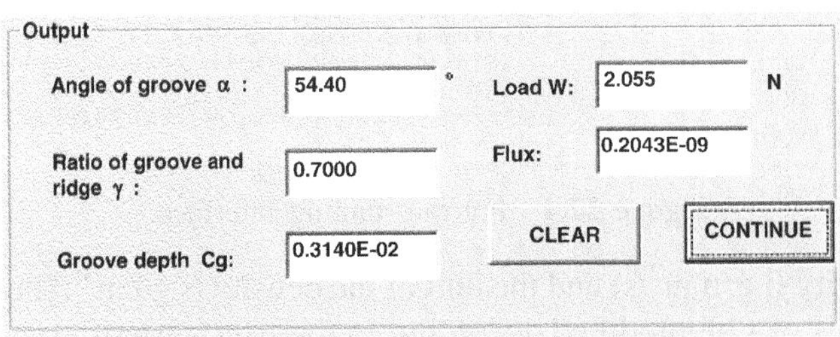

Figure 22.8 Second optimization results.

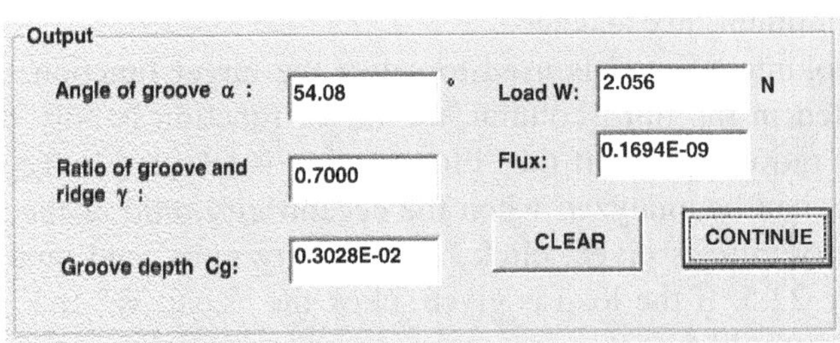

Figure 22.9 Third optimization results.

After the completion of the five-column parameters setting, click the "CACULATE" button as shown in Figure 22.3 and then enter the optimization process. In the program, the coordinates rotation method and the interval decreasing method are used to optimize the parameters, and the finite difference method and the iterative method are used to simulate the pressure.

The "Output" column will output the optimization results of the several parameters, including flux and load. A group of the first optimization results are shown in Figure 22.7 when the eccentricity ratio is equal to 0.2 and the "Load" option is selected.

When the accuracy of the optimization results is not satisfied, click the "CONTINUE" button to continue the optimization until the accuracy is satisfied. The "CONTINUE" button cannot be activated at initially, but it will activate after completion of the first optimization, as shown in Figure 22.7. Figures 22.8 and 22.9 give the second and third optimization results.

If the user needs to modify the parameters and optimize again, press the "CLEAR" button to clear the output data or modify the parameters directly.

In addition to the several characteristics above, two text files will be generated in the current directory: result.txt and oresult.txt. The result.txt contains the data of the numerical change during the process and the oresult.txt gives the final optimization results.

23

Calculation Program for Gas Lubrication of Hard Disk/Head in Ultra Thin Film

23.1 Basic Equations of Gas Lubricating Film of Hard Disk/Head

23.1.1 Basic Equations

The generalized steady Reynolds equation in a dimensionless form for actual sliders, as shown in Figure 23.1, is expressed as

$$\frac{\partial}{\partial X}\left(PH^3Q\frac{\partial P}{\partial X}\right)+A^2\frac{\partial}{\partial Y}\left(PH^3Q\frac{\partial P}{\partial Y}\right)=\Lambda_x\frac{\partial(PH)}{\partial X}+A\Lambda_y\frac{\partial(PH)}{\partial Y} \tag{23.1}$$

where P is dimensionless pressure ($P=p/p_a$), p is pressure, p_a is ambient pressure, H is dimensionless film thickness ($H=h/h_0$), h is film thickness, h_0 is reference thickness, X is a dimensionless coordinate in the x direction ($X=x/L$), Y is a dimensionless coordinate in the y direction ($Y=y/B$), L is slider length, B is slider width, A is the ratio of slider length to slider width ($A=L/B$), Λ_x and Λ_y are bearing numbers in the X and Y directions, respectively ($\Lambda_x=6\ \mu UL/p_a/h_0^2$; $\Lambda_y=6\ \mu VL/p_a/h_0^2$), μ is the viscocity of the gas, U and V are disk velocity components in the X and Y directions, respectively, Q is the dimensionless flow rate ($Q=\dfrac{Q_P}{Q_{Con}}$), Q_P is the pressure flow rate and Q_{Con} is the continuum pressure flow rate.

Numerical Calculation of Lubrication: Methods and Programs, First Edition. Ping Huang.

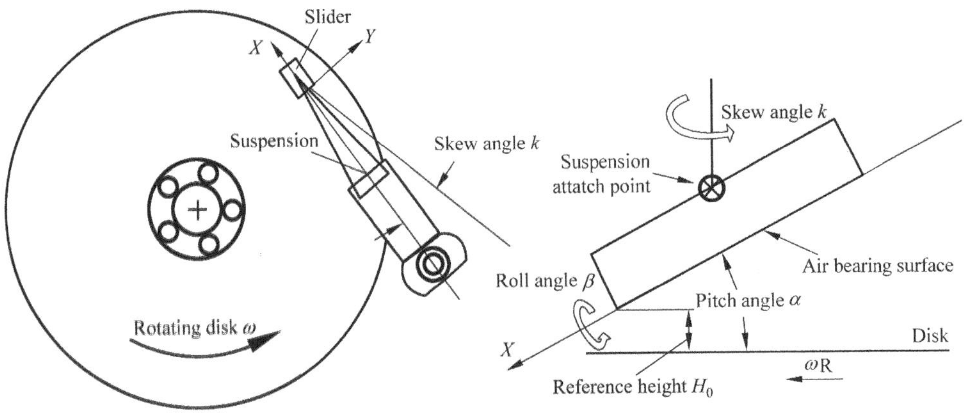

Figure 23.1 Sketch of disk/head system.

23.1.2 Gas Film Thickness

The film thickness can be given by

$$h(X,Y) = h_0(X,Y) + L(X_N - X)\sin(\alpha_p) + B(Y_M - Y)\sin\alpha_r + h_{\min}, \tag{23.2}$$

where $h_0(X,Y)$ is the original height decided by ABS design and $h_{\min}$ is the minimum height, whose coordinates are X_N and Y_M.

In Equation 23.2, the second and the third terms on the right side of the formula are the increment due to the slider's pitch and roll angles.

In the present program, the slider ABS is composed of a 114 nm shallow step and a 1.77 μm cavity, which are directly assigned to the corresponding nodes according to the ABS shape. The program can be revised to suit different ABS designs.

23.1.3 Poiseuille Flow Rate

Various models of the Poiseuille flow rate are proposed with the development of gas lubrication for magnetic heads. The nondimensional Poiseuille flow rate Q normalized by the continuum pressure flow rate (Equation 23.3), can be expressed as a piecewise function of the inverse Knudsen number (Equation 23.4) based on the linearized Boltzmann model. In Equation 23.4 D denotes the inverse Knudsen number.

$$Q = \frac{Q_P}{Q_{Con}} \tag{23.3}$$

$$Q = 1 + \frac{6.0972}{D} + \frac{6.3918}{D^2} - \frac{12.8124}{D^3} \quad (D \geq 5)$$

$$Q = 0.83112 + \frac{7.50522}{D} + \frac{0.93918}{D^2} - \frac{0.05814}{D^3} \quad (0.15 \leq D < 5) \tag{23.4}$$

$$Q = -13.7514 + \frac{12.64038}{D} + \frac{0.09918}{D^2} - \frac{0.0004164}{D^2} \quad (0.01 \leq D < 0.15)$$

where $D = D_0 PH$ and $D_0 = \frac{p_0 h_0}{\mu\sqrt{2RT_0}}$.

We present a continued model (Equation 23.5) which not only approximates Equation 23.4 but also covers the region of $D < 0.01$. The new equation of Poiseuille flow rate is used in the following program.

$$Q_p = \frac{D}{6} + 1.0162 + 0.40134\ln(1 + 1.2477/D) \tag{23.5}$$

23.2 Discrete Equation and Special Treatments

For the numerical calculation of ultra-thin gas film lubrication, there are two problems: high bearing numbers and abrupt changes between the ABS steps, besides the model of Poiseuille flow rate. These will influence the convergence and accuracy if not treated properly in the course of numerical calculation.

23.2.1 Iterative Scheme Considering High Bearing Numbers

In the conventional difference finite method to solve lubrication problems, the Poiseuille flow term is always taken as the main iterative part. Since Λ_x and Λ_y vary inversely with h_0^2 in Equation 23.1, the Couette flow terms will be much larger than the Poiseuille flow terms, considering the reference height h_0 is only several nanometers. This may cause a numerical solution to oscillate because a very small change in the Couette flow terms will cause the solution to vary dramatically. A new iterative algorithm of finite difference method must be developed to solve the present problems. Since the bearing numbers are very high and the Couette flow terms include the variable pressure (unlike the conventional difference finite method), the pressure included in the Couette flow term now serves as the main iterative part. This will effectively solve the problem of numerical oscillation at high bearing numbers.

Using the upwind scheme on the Couette flow terms, Equation 23.1 can be discretized as follows.

$$2\Lambda_x\left(P_{i,j}H_{i,j}-P_{i-1,j}H_{i-1,j}\right)/\Delta X_i+2A\Lambda_y\left(P_{i,j}H_{i,j}-P_{i,j-1}H_{i,j-1}\right)/\Delta Y_j$$
$$=\left[(QH^3)_{i+1/2,j}(P^2_{i+1,j}-P^2_{i,j})-(QH^3)_{i-1/2,j}(P^2_{i,j}-P^2_{i-1,j})\right]/\Delta X_i^2 \tag{23.6}$$
$$+A^2\left[(QH^3)_{i,j+1/2}(P^2_{i,j+1}-P^2_{i,j})-(QH^3)_{i,j-1/2}(P^2_{i,j}-P^2_{i,j-1})\right]/\Delta Y_j^2$$

According to the above analysis, the iterative scheme is deduced as

$$\hat{P}_{i,j}=\frac{\left\{\begin{array}{l}\hat{P}_{i-1,j}\Lambda_x H_{i-1,j}/\Delta X_i+0.5\left[(\hat{Q}H^3)_{i+1/2,j}\tilde{P}^2_{i+1,j}+(\hat{Q}H^3)_{i-1/2,j}\hat{P}^2_{i-1,j}\right]/\Delta X_i^2\\+\hat{P}_{i,j-1}A\Lambda_y H_{i,j-1}/\Delta Y_j+0.5A^2\left[(\hat{Q}H^3)_{i,j+1/2}\tilde{P}^2_{i,j+1}+(\hat{Q}H^3)_{i,j-1/2}\hat{P}^2_{i,j-1}\right]/\Delta Y_j^2\end{array}\right\}}{\left\{\begin{array}{l}\Lambda_x H_{i,j}/\Delta X_i+A\Lambda_y H_{i,j}/\Delta Y_j+0.5\tilde{P}_{i,j}\left[(\hat{Q}H^3)_{i+1/2,j}+(\hat{Q}H^3)_{i-1/2,j}\right]/\Delta X_i^2\\+0.5A^2\tilde{P}_{i,j}\left[(\hat{Q}H^3)_{i,j}+(\hat{Q}H^3)_{i,j-1/2}\right]/\Delta Y_j^2\end{array}\right\}} \tag{23.7}$$

where the above variables with $\sim$ denote old variables from the last iteration; those with $\wedge$ denote new values from the present iteration.

23.2.2 Abrupt Changes between Steps on ABS

Equation 23.7 comprises some interface fluxes ($q=H^3Q$) between two adjacent nodes. It should be pointed out that simple averages for the interface fluxes $q_{i+1/2,j}$, $q_{i-1/2,j}$, $q_{i,j+1/2}$ and $q_{i,j-1/2}$ may lead to rather incorrect implications in the present case because the film thickness H has abrupt changes between the steps of the magnetic head. A much better alternative is to use the harmonic average, such as Equation 23.8, when the interface is placed midway between two nodes.

$$q_{i+1/2,j}=\frac{2q_{i+1,j}q_{i,j}}{q_{i+1,j}+q_{i,j}} \tag{23.8}$$

This harmonic average has a nice property that, with abrupt changes in $q_{i,j}$ to $q_{i+1,j}$, it still produces physically correct results. A similar method can be used to obtain $q_{i-1/2,j}$, $q_{i,j+1/2}$ and $q_{i,j-1/2}$.

23.3 Calculation Program

23.3.1 Calculation Diagram

The calculation diagram for gas lubrication of hard disk/head in ultrathin film is shown in Figure 23.2.

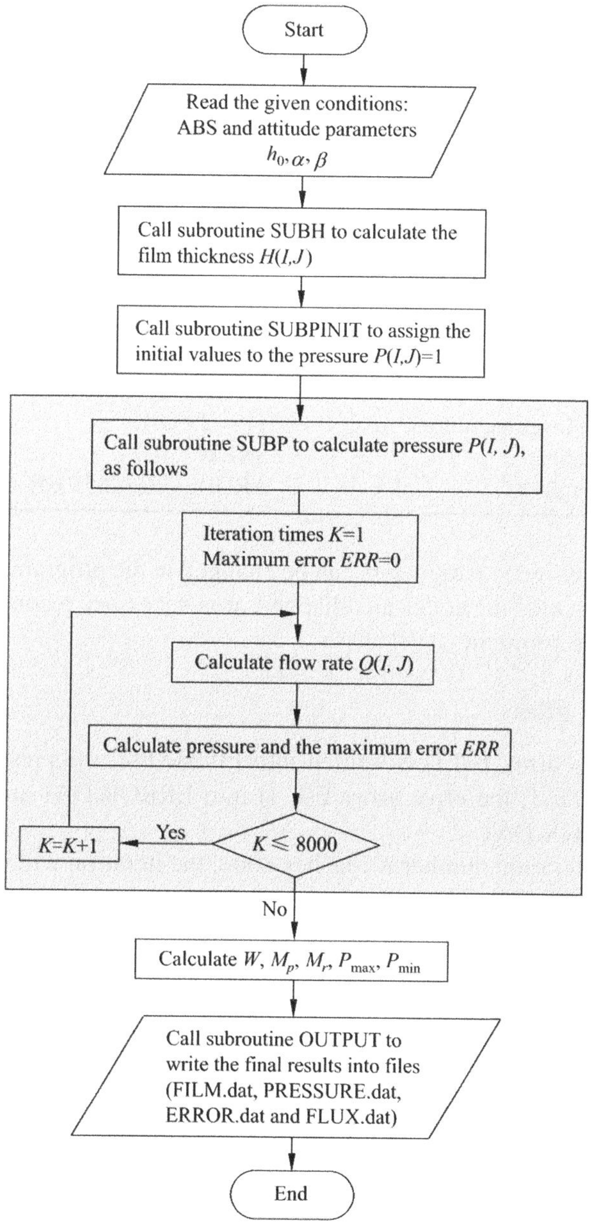

Figure 23.2 Calculation diagram for gas lubrication of hard disk/head in ultrathin film.

23.3.2 *Calculation Program*

23.3.2.1 Pre-assigned Parameters

Slider length	AL = 1.25E–3 m
Slider width	B = 1.0E–3 m
Minimum spacing	HM = 1.0 E–8 m
Velocity	U = 25.0 m s^{-1};
Gas viscosity	EDA = 1.806E–5 Pa s
Ambient pressure	PA = 1.0136E5 Pa
Mean free path	ALA = 63.5E–9 m
Pitch angle	ALFA = 0.95024E–4 rad
Gas constant	R = 287.03
Absolute temperature	T0 = 293.0 K
Skew angle	SKEW = 0.0
Roll angle	AROLL = 6.1465303E–6 rad

The above pre-assigned parameters can be changed in the program according to the actual conditions. Once the codes are changed, it is necessary recompile and rebuild before running the program.

23.3.2.2 Output Files

The film thickness array H(I,J) is written into FILM.DAT, the pressure array P(I,J) into PRESSURE.DAT, the error array F(I, J) into ERROR.DAT and the flux array QW(I, J) into FLUX.DAT.

Further, if the iteration number *K* reaches 8000, the iteration will end. If the errors displayed on the monitor are still big, or not randomly variable, the number of iterations needs to be increased to achieve a more accuracy solution.

```
PROGRAM MAGNETICHEAD
      IMPLICIT REAL*8 (A-H,O-Z)
      DIMENSION P(101,101),H(101,101),X(101),Y(101),F(101,101),
QW(101,101)
      DATA AL,B,HM,BETA1,U,EDA,PA/1.25E-3,1.0E-3,1.0E-8,0.01,25.0,
1.8060E-5,1.0135E5/
      DATA ALA,ALFA,R,T0,SKEW,PI,AROLL/63.5E-9,0.95024E-
4,287.03,293.0,0.0,3.14159265,6.1465303E-6/
      OPEN(8,FILE='FILM.DAT',STATUS='UNKNOWN')
      OPEN(9,FILE='PRESSURE.DAT',STATUS='UNKNOWN')
      OPEN(10,FILE='ERROR.DAT',STATUS='UNKNOWN')
      OPEN(11,FILE='FLUX.DAT',STATUS='UNKNOWN')
      N=101
      M=N
```

```
      N1=N-1
      M1=N1
      N2=N1/2+1
      M2=M1/2+1
      UX=U*COS(SKEW*PI/180.)
      UY=U*SIN(SKEW*PI/180.)
      HC=ALA/HM
      ALENDA=6.0*EDA*UX*AL/(HM**2*PA)
      ALENDAY=6.0*EDA*UY*AL/(HM**2*PA)
      DELTA=AL*SIN(ALFA)/HM
      Q=B/AL
      D0=PA*HM/EDA/SQRT(2.0*R*T0)
      CALL SUBH(N,M,M2,N1,M1,DX,DY,AL,ALFA,B,AROLL,HM,DELTA1,X,Y,H)
      CALL SUBPINIT(N,M,N1,M1,HMIN,H,P,F,QW)
      CALL SUBP(N,M,N1,M1,DX,DY,Q,ALENDA,ALENDAY,D0,DR,PA,AL,B,BETA1,
     PI,AD,HM,ALFA,U,H,P,F,QW)
      CALL OUTPUT(N,M,X,Y,H,P,F,QW)
      STOP
      END
      SUBROUTINE SUBH(N,M,M2,N1,M1,DX,DY,AL,ALFA,B,AROLL,HM,DELTA1,X,Y,H)
      IMPLICIT REAL*8 (A-H,O-Z)
      DIMENSION X(N),Y(M),H(N,M)
      DO I=1,N
      DO J=1,M
      H(I,J)=(1770.0E-9+HM)/HM
      ENDDO
      ENDDO
      DO I=1,0.1*N
      DO J=1,M
      H(I,J)=(114.0E-9+HM)/HM
      ENDDO
      ENDDO
      DO I=0.1*N+1,0.2*N
      DO J=1,M
      H(I,J)=1.0
      ENDDO
      ENDDO
      DO I=0.2*N+1,0.8*N
      DO J=1,0.1*M
      H(I,J)=1.0
      H(I,M-J)=1.0
      ENDDO
      ENDDO
      DO I=0.77*N+1,0.97*N
      DO J=0.4*M,M/2+1
```

```
H(I,J)=1.0
H(I,M-J)=1.0
ENDDO
ENDDO
DO I=0.85*N+1,0.95*N
DO J=0.2*M,0.3*M
H(I,J)=1.0
H(I,M-J)=1.0
ENDDO
ENDDO
X(1)=0.0
DX=1./N1
DO I=2,N
X(I)=X(I-1)+DX
ENDDO
Y(1)=-0.5
DY=1./M1
DO J=2,M
Y(J)=Y(J-1)+DY
ENDDO
X(N)=1.0
Y(M2)=0.0
HMIN=10.0
DO I=1,N
DO J=1,M
DELTA1=(AL*(X(N)-X(I))*SIN(ALFA)+B*(Y(M)-Y(J))*SIN(AROLL))/HM
H(I,J)=H(I,J)+DELTA1
IF(H(I,J).LT.HMIN)THEN
IMIN=I
JMIN=J
HMIN=H(I,J)
ENDIF
ENDDO
ENDDO
RETURN
END
SUBROUTINE SUBPINIT(N,M,N1,M1,HMIN,H,P,F,QW)
IMPLICIT REAL*8 (A-H,O-Z)
DIMENSION H(N,M),P(N,M),F(N,M),QW(N,M)
DO I=1,N
DO J=1,M
P(I,J)=1.0
F(I,J)=0.0
QW(I,J)=1.0
ENDDO
```

```
      ENDDO
      DO I=2,N1
      DO J=2,M1
      P(I,J)=P(I-1,J)*H(I-1,J)/H(I,J)
      ENDDO
      ENDDO
      RETURN
      END
      SUBROUTINE SUBP(N,M,N1,M1,DX,DY,Q,ALENDA,ALENDAY,D0,DR,PA,
AL,B,BETA1,PI,AD,HM,ALFA,U,H,P,F,QW)
      IMPLICIT REAL*8 (A-H,O-Z)
      DIMENSION H(N,M),P(N,M),F(N,M),QW(N,M)
      K=0
      DX1=1./DX
      DY1=1./(Q*DY)
      DX2=DX1*DX1
      DY2=DY1*DY1
      CX=2.*ALENDA*DX1
      CY=2.*ALENDAY*DY1
10    DO I=1,N
      DO J=1,M
      DR=D0*P(I,J)*H(I,J)
   QC=DR/6.0
      QP=QC+1.0162+0.40134*DLOG(1.0+1.2477/DR)
      QW(I,J)=QP/QC
      ENDDO
      ENDDO
      ERR=0.0
      DO J=2,M1
      DO I=2,N1
      C0=CX*H(I,J)
      C5=CX*P(I-1,J)*H(I-1,J)
      CY0=CY*H(I,J)
      CY5=CY*P(I,J-1)*H(I,J-1)
      TMP1 = QW(I+1,J)*H(I+1,J)*H(I+1,J)*H(I+1,J)
      TMP2 = QW(I,J)*H(I,J)*H(I,J)*H(I,J)
      QHP1 = 2.0*TMP1*TMP2/(TMP1+TMP2)
      TMP1 = QW(I-1,J)*H(I-1,J)*H(I-1,J)*H(I-1,J)
      QHM1 = 2.0*TMP1*TMP2/(TMP1+TMP2)
      C1=P(I,J)*(QHP1+ QHM1)*DX2
      C3=(P(I+1,J)*P(I+1,J)*QHP1+P(I-1,J)*P(I-1,J)*QHM1)*DX2
      TMP1 = QW(I,J+1)*H(I,J+1)*H(I,J+1)*H(I,J+1)
      QHP1 = 2.0*TMP1*TMP2/(TMP1+TMP2)
      TMP1 = QW(I,J-1)*H(I,J-1)*H(I,J-1)*H(I,J-1)
      QHM1 = 2.0*TMP1*TMP2/(TMP1+TMP2)
```

```
      C2=P(I,J)*(QHP1+QHM1)*DY2
      C4=(P(I,J+1)*P(I,J+1)*QHP1+P(I,J-1)*P(I,J-1)*QHM1)*DY2
      TMP=(C5+C3+C4+CY5)/(C0+C1+C2+CY0)
      P(I,J)=P(I,J)+BETA1*(TMP-P(I,J))
      F(I,J)=P(I,J)*(C0+C1+C2+CY0)-C3-C4-C5-CY5
      IF(ABS(F(I,J)).GT.ERR)ERR=ABS(F(I,J))
      ENDDO
      ENDDO
      K=K+1
      WRITE(*,*)'K=, ERR=',K,ERR
      IF(ERR.LT.1.E-5)GOTO 20
      IF(K.GT.500)BETA1=0.1
      IF(K.GT.2000)BETA1=0.5
      IF(K.GT.3000)BETA1=0.75
      IF(K.GT.5000)BETA1=0.95
      IF(K.LE.8000) GOTO 10
20    SM=0.0
      PMAX=0.0
      PMIN=0.0
      AD=2.0/SQRT(PI)
      DO I=1,N
      DO J=1,M
      P(I,J)=P(I,J)-1.
      SM=SM+P(I,J)
      IF(P(I,J).GT.PMAX)THEN
      PMAX=P(I,J)
      IMAX=I
      JMAX=J
      ENDIF
      IF(P(I,J).LT.PMIN)THEN
      PMIN=P(I,J)
      IMIN=I
      JMIN=J
      ENDIF
      QW(I,J)=AD*D0*(1.+P(I,J))*H(I,J)
      H(I,J)=HM*H(I,J)
      ENDDO
      ENDDO
      SM=SM*PA*AL*B/(N1*M1)
      WRITE(*,*)'ALFA=',ALFA,'HM=',HM,'U=',U,'LOAD=',
   SM,'PMAX=',PMAX,'PMIN=',PMIN
      RETURN
      END
      SUBROUTINE OUTPUT(N,M,X,Y,H,P,F,QW)
      IMPLICIT REAL*8 (A-H,O-Z)
```

```
      DIMENSION X(N),Y(M),H(N,M),P(N,M),F(N,M),QW(N,M)
      WRITE(8,30)(Y(J),J=1,M)
30    FORMAT(1X,101(E12.6,1X))
      WRITE(8,40)(X(I),(H(I,J),J=1,M),I=1,N)
40    FORMAT(102(E12.6,1X))
      WRITE(9,30)(Y(J),J=1,M)
      WRITE(9,40)(X(I),(P(I,J),J=1,M),I=1,N)
      WRITE(10,30)(Y(J),J=1,M)
      WRITE(10,40)(X(I),(F(I,J),J=1,M),I=1,N)
      WRITE(11,30)(Y(J),J=1,M)
      WRITE(11,40)(X(I),(QW(I,J),J=1,M),I=1,N)
      RETURN
      END
```

23.3.3 Calculation Results

Under the given working conditions, the pressure distribution and the corresponding film thickness can be acquired from the program, as shown in Figure 23.3.

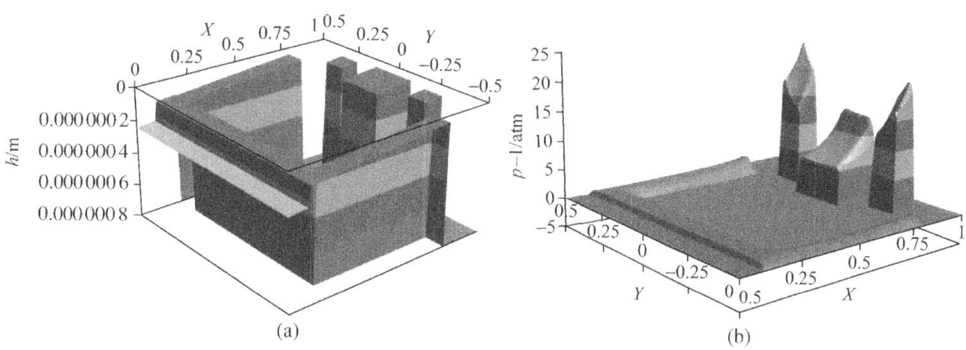

Figure 23.3 (a) Film thickness and (b) pressure distribution.

24

Calculation Program of Flight Attitude of Magnetic Head

24.1 Search Strategy for Flight Attitude

The numerical scheme developed in Chapter 23 is intended to solve the generalized lubrication equation for the ABS pressure under given flying attitudes (static flying height h_0, pitch angle σ, the roll angle β). However, during the air bearing design of magnetic heads, designers are more interested in determining the equilibrium flying attitudes under a given suspension preload and load location. This is an inverse problem to Chapter 23.

The flying attitudes can be found through iterating the lubrication equation. After a set of initial parameters for flying attitudes are given, the forces (load and moments) of the film pressure on the slider are calculated by using the method in Chapter 23. Then, we compare the results to the given forces and revise the flying attitude according to the deviation of the loads until the differences of the forces meet the accurate requirement so that the equilibrium position can be found.

Because h_0, σ and β have nonlinear influences on load and moments at the same time, finding the equilibrium position efficiently is a tough problem with three degrees of freedom.

The increments of force and attitude should satisfy Equation 24.1. The flying attitudes can be acquired by solving the three-dimensional linear equations.

Numerical Calculation of Lubrication: Methods and Programs, First Edition. Ping Huang.

$$\begin{bmatrix} dW \\ dM_P \\ dM_R \end{bmatrix} = \begin{bmatrix} \dfrac{\partial W}{\partial h_0} & \dfrac{\partial W}{\partial \alpha} & \dfrac{\partial W}{\partial \beta} \\ \dfrac{\partial M_P}{\partial h_0} & \dfrac{\partial M_P}{\partial \alpha} & \dfrac{\partial M_P}{\partial \beta} \\ \dfrac{\partial M_R}{\partial h_0} & \dfrac{\partial M_R}{\partial \alpha} & \dfrac{\partial M_R}{\partial \beta} \end{bmatrix} \begin{bmatrix} dh_0 \\ d\alpha \\ d\beta \end{bmatrix} \tag{24.1}$$

The matrix of coefficients on the right side of Equation 24.1is called the stiffness matrix, which can be approximately calculated later. When we give an increment to one of the flying attitudes, for example h_0, the variations of forces can be calculated by using the method in Chapter 23. In fact, the stiffness values are only the mean values of the current flying attitudes, depending on the increments of each attitude parameter. With the iteration method, we can have the slider approach the equilibrium attitudes so that the increments of the attitude parameters reduce to meet the required accuracy.

The column-pivoting elimination method is used to solve Equation 24.1, for it is robust in various situations.

It should be pointed out that the above stiffness matrix is based on the reference point of the slider, not on the pivot point of the suspension, because the increments of roll angle and pitch angle are centered over the reference point. Obviously, the stiffness matrix on the pivot point is more important, through which the stability of the slider can be judged. In fact, the two matrixes only differ from the elements in the middle column because only the increment of pitch angle $\Delta\alpha$ in the above two cases can lead to different film thicknesses if the pivot point is always on the x axis. When the slider rotates by a pitch increment $\Delta\alpha$ round the pivot point, there are two effects on the film thickness: an increase in the pitch angle and a decrease in the reference point height. The latter can be approximately calculated as

$$\Delta h_0 = (L - x_D)\Delta\alpha, \tag{24.2}$$

where x_D is the pivot point in the x coordinate.

By increasing the pitch angle and decreasing the reference point height, an attitude position can be obtained. Then, the new elements $\left(\dfrac{\partial W}{\partial \alpha}, \dfrac{\partial M_P}{\partial \alpha}, \dfrac{\partial M_R}{\partial \alpha}\right)$ can be solved.

24.2 Calculation Program

24.2.1 *Program Introduction*

24.2.1.1 Pre-assigned Parameters

Ambient pressure	PA = 1.0136E5 Pa
Gas viscosity	EDA = 1.806E–5 Pa s
Gas constant	R = 287.03
Absolute temperature	T0 = 293.0 K
Slider length	AL = 1.235 mm
Slider width	B = 0.7 mm
Reference point in the x coordinate	XHR = 1.21 mm
Reference point in the y coordinate	YHR = 0
Pivot point in the x coordinate	XC = 0.6175 mm
Pivot point in the y coordinate	YC = 0
Load of suspension	W0 = 1.5 g
Pitch moment of suspension	AM0 = 0.867 μNm
Roll moment of suspension	BM0 = 0
Speed of spinning disk	RPM = 4800 rpm
Track radius	CR = 17.561 mm
Skew angle	SKEW = −14.923°

The pre-assigned parameters can be modified according to the actual conditions, but we must recompile and relink the program before executing it.

24.2.1.2 Format of ABS File from Designers

The file named fort.20 includes the information about the heights from each node on the ABS to the reference plane. In other words, a series of data about dimensionless coordinates X, Y and the height at that position can be found in the file. Its format must adapt to the following codes.

```
      READ(20,3)X(1),(Y(J),J=1,161)
      DO I=1,161
      READ(20,3)X(I),(H0(I,J),J=1,161)
      ENDDO
3     FORMAT(162(E12.6,1X))
```

24.2.1.3 Output Files and Parameters

The pressure array P(I, J) is written into PRESSURE.DAT, the film thickness array H(I, J) into FILM.DAT and the following parameters into RESULT.DAT.

Maximum pressure	Pmax	atm
Minimum pressure	Pmin	atm
Load	W	g
Position of load	X0, Y0	mm
Moment of pitch	PM	μNm
Moment of roll	RM	μNm
Flying height of the reference point	H	nm
Pitch angle	ALFA	μrad
Roll angle	BETA	μrad

Elements of stiffness matrix:

WH	g·nm–1
WP	$g\cdot(\mu rad)^{-1}$
WR	$g\cdot(\mu rad)^{-1}$
PH	$\mu Nm\cdot nm^{-1}$
PP	$Nm\cdot rad^{-1}$
PR	$Nm\cdot rad^{-1}$
RH	$\mu Nm\cdot nm^{-1}$
RP	$Nm\cdot rad^{-1}$
RR	$Nm\cdot rad^{-1}$

Some important parameters for the working conditions are listed in Result.dat besides the above data for the results, which will be more convenient when comparing the flying characteristics at different parameters.

24.2.1.4 Real-time Running Window

Some data are shown on a monitor in real time when the program is being run, through which users may know the process of the iteration and convergence.

24.2.2 Calculation Diagram

The calculation diagram of the flight attitude of a magnetic head is shown in Figure 24.1.

24.2.3 Calculation Program

```
      PROGRAM FH
      IMPLICIT REAL*8 (A-H,O-Z)
      DIMENSION P(161,161),H(161,161),X(161),Y(161),PP0(161,161),
     HH0(161,161)
      DIMENSION H0(161,161),QPH3(161,161),HHH(161,161),A(3,4)
      REAL*8 L21,L31,L32,M22,M23,M24
      DATA BETA1/0.01/,R/287.03/,T0/293.0/,PI/3.14159265/,EDA/1.806E-5/
     ,PA/1.0136E5/,ALA/63.5E-9/
      DATA B/0.7/,AL/1.235/,CR/17.561/,RPM/0.48E4/,SKEW/-14.923/,XHR/
     1.21/,YHR/0./,H00/6.0/,ALFA0/50.0/,BETA/0./
      DATA XC/0.6175/,YC/0./,W0/1.5/,AM0/0.867/,BM0/0./
      KG=0
```

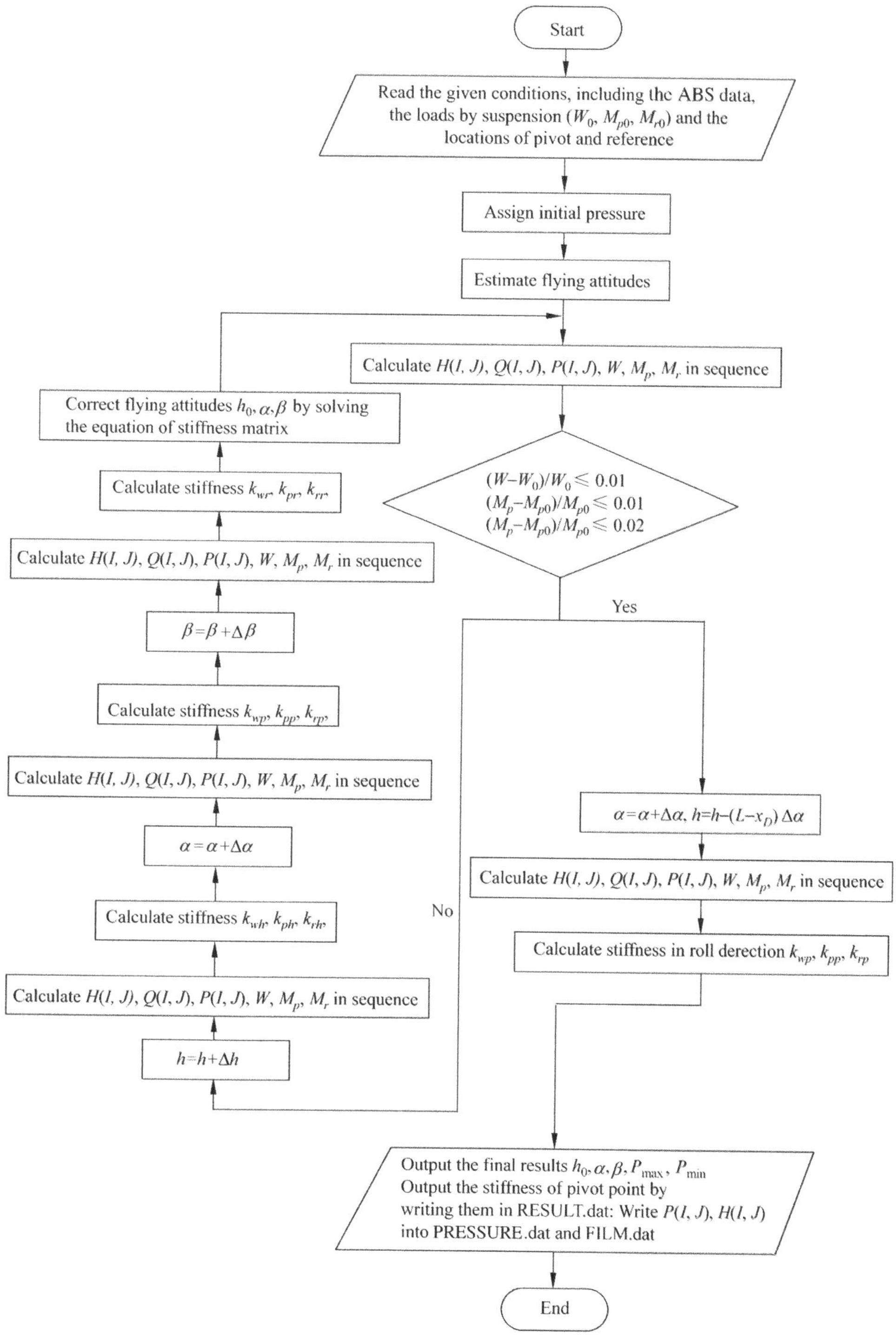

Figure 24.1 Calculation diagram of flight attitude of magnetic head.

```
      KL=0
      KG1=0
      KG2=0
      AL=AL*1.0E-3
      B=B*1.0E-3
      ALFA0=ALFA0*1.0E-6
      BETA=BETA*1.0E-6
      H00=H00*1.0E-9
      XCD=XC*1.0E-3/AL
      YCD=YC*1.0e-3/B
      U=2*PI*CR*RPM/60.E3
      N1=161
      M1=N1
      N=N1-1
      N2=N/2+1
      M=M1-1
      M2=M/2+1
      UX=U*COS(SKEW*PI/180.)
      UY=U*SIN(SKEW*PI/180.)
      Q=B/AL
      AD=2.0/SQRT(3.14159265)
      DX=1./N
      DY=DX
      DX1=1./DX
      DY1=1./(Q*DY)
      DX2=DX1*DX1
      DY2=DY1*DY1
      SPA=PA/(EDA*SQRT(2.0*R*T0))
      ALENDA0=6.0*EDA*AL/PA
      CW=PA*AL*B/(N*M)
      CM=PA*AL**2*B/(N*M)
      CMR=PA*B**2*AL/(N*M)
      READ(20,3)X(1),(Y(J),J=1,161)
      DO I=1,161
      READ(20,3)X(I),(H0(I,J),J=1,161)
      ENDDO
  3   FORMAT(162(E12.6,1X))
      IF(SKEW.LT.0)THEN
      UY=ABS(UY)
      BM0=-1.*BM0
      DO I=1,N1
      DO J=1,M1
      QPH3(I,J)=H0(I,162-J)
      ENDDO
      ENDDO
      DO I=1,N1
      DO J=1,M1
      H0(I,J)=QPH3(I,J)
      ENDDO
```

```
      ENDDO
      ENDIF
      X(1)=0.0
      DO I=2,N1
      X(I)=X(I-1)+DX
      ENDDO
      Y(1)=-0.5
      DO J=2,M1
      Y(J)=Y(J-1)+DY
      ENDDO
      Y(M2)=0.0
      X0=0.0
      AM=0.0
      H1=H00
      ALFA=ALFA0
      SDH=3.0E-9
      SDALFA=20.E-6
      SDBETA=5.E-6
      XHR=XHR*1.0e-3/AL
      YHR=YHR*1.0e-3/B
      DO I=1,N1
      IF(ABS(XHR/DX-I).LT.0.5)IXHR=I
      ENDDO
      DO J=1,M1
      IF(ABS(YHR-Y(J)).LT.(DY/2))JYHR=J
      ENDDO
1     IF(KG2==0)THEN
      K=0
      BETA1=0.01
      ENDIF
      IF(KG2==1)THEN
      K=1000
      BETA1=0.75
      ENDIF
      H2=H1*H1
      ALENDAX=ALENDA0*UX/H2
      ALENDAY=ALENDA0*UY/H2
      SALFA=AL*SIN(ALFA)
      SBETA=B*SIN(BETA)
      D0=SPA*H1
      HMIN=10.0
      DO I=1,N1
      DO J=1,M1
      H(I,J)=((H0(I,J)-H0(IXHR,JYHR))/COS(ALFA)/COS(BETA)+(X(IXHR)-X(I))
     *SALFA+(Y(J)-Y(JYHR))*SBETA)/H1+1.0
      IF(H(I,J).LT.0.0)THEN
      H1=1.5*H1
      ALFA=8.E-5
```

```
    WRITE(*,*)'H1 was forced to multiply by 1.5 automatically To avoid
negative film thickness dots'
    GOTO 1
    ENDIF
    HHH(I,J)=H(I,J)*H(I,J)*H(I,J)
    IF(KG1.EQ.0)THEN
    P(I,J)=1.0
    ENDIF
    ENDDO
    ENDDO
    IF(KG1.EQ.0)THEN
    KG1=1
    DO I=2,N
    DO J=2,M
    P(I,J)=P(I-1,J)*H(I-1,J)/H(I,J)
    ENDDO
    ENDDO
    ENDIF
10  DO I=1,N1
    DO J=1,M1
    DR=D0*P(I,J)*H(I,J)
    QC=DR/6.0
    QP= 1.720805*DR**(-0.25953)+3.173598*SQRT(3.14159265)/2.+QC
    QPH3(I,J)=P(I,J)*QP*HHH(I,J)/QC
    ENDDO
    ENDDO
    ERR=0.0
    DO J=2,M
    J1=J-1
    J2=J+1
    DO I=2,N
    I1=I-1
    I2=I+1
    HX1=(H(I,J)+H(I1,J))/2.
    HX2=(H(I,J)+H(I2,J))/2.
    HY1=(H(I,J)+H(I,J1))/2.
    HY2=(H(I,J)+H(I,J2))/2.
    FW=ALENDAX*HX1
    FE=ALENDAX*HX2
    FN=ALENDAY*HY2/Q
    FS=ALENDAY*HY1/Q
    TMP=QPH3(I,J)
    TMP1=QPH3(I2,J)
    QHP1=2.*DX1*TMP*TMP1/(TMP+TMP1)
    PL1=FE/QHP1
    CE=QHP1*MAX(0.,(1.0-0.1*PL1)**5)
    TMP1=QPH3(I1,J)
    QHM1=2.*DX1*TMP*TMP1/(TMP+TMP1)
    PL1=FW/QHM1
```

```
      CWW=QHM1*MAX(0.,(1.0-0.1*PL1)**5)
      TMP1=QPH3(I,J2)
      QHP1=2.*TMP*TMP1/(TMP+TMP1)/(DY*Q**2)
      PL1=FN/QHP1
      CN=QHP1*MAX(0.,(1.0-0.1*PL1)**5)
      TMP1=QPH3(I,J1)
      QHM1=2.*TMP*TMP1/(TMP+TMP1)/(DY*Q**2)
      PL1=FS/QHM1
      CS=QHM1*MAX(0.,(1.0-0.1*PL1)**5)
      TMP1=CE*P(I2,J)+(CWW+FW)*P(I1,J)+CN*P(I,J2)+(CS+FS)*P(I,J1)
      TMP2=CE+CWW+CN+CS+FE+FN
      P(I,J)=P(I,J)+BETA1*(TMP1/TMP2-P(I,J))
      F=P(I,J)*TMP2-TMP1
      IF(ABS(F).GT.ERR)ERR=ABS(F)
      ENDDO
      ENDDO
      K=K+1
      IF(ERR.LT.10.)GOTO 20
      IF(K.GT.300)BETA1=0.1
      IF(K.GT.600)BETA1=0.5
      IF(K.GT.1000)BETA1=0.75
      IF(K.GT.1500)BETA1=0.95
      IF(K.GT.2000)BETA1=1.1
      IF(K.GT.2500)BETA1=1.2
      IF(K.GT.3000)BETA1=1.25
      IF(K.GT.15000)BETA1=0.95
      WRITE(*,*)'K=',K,'   ERR=',ERR
      IF(ERR.GT.20.AND.K.LE.18000)GOTO 10
20    W=0.0
      AM=0.0
      BM=0.0
      DO I=1,N1
      DO J=1,M1
      W=W+P(I,J)-1.0
      AM=AM+(P(I,J)-1.0)*(X(I)-XCD)
      BM=BM+(P(I,J)-1.0)*(Y(J)-YCD)
      ENDDO
      ENDDO
      WRITE(*,*) K,ERR
      W=W*CW
      AM=AM*CM
      BM=BM*CMR
      IF(KL.EQ.1) THEN
      KL=0
      WP=(W-W00)*1.E3/9.80665
      PP=(AM-AM00)*1.0E6
      RP=(BM-BM00)*1.0E6
      W=W00
      AM=AM00
```

```
    BM=BM00
    ALFA=ALFA-1.0E-6
    H1=H1+(AL-XC*1.0E-3)*1.0E-6
    DO I=1,161
    DO J=1,161
    P(I,J)=PP0(I,J)
    H(I,J)=HH0(I,J)
    ENDDO
    ENDDO
    WRITE(*,*)
    WRITE(*,*)'stiffness_Matrix:'

    WRITE(*,'(A15,3(A12,1X))')'        \        ',' H(nm) ','PITCH
(urad)','ROLL(urad)'

    WRITE(*,21) 'W(g)         ',WH,WP,WR
    WRITE(*,21) 'P-TORQUE(uNm)',PH,PP,PR
    WRITE(*,21) 'R-TORQUE(uNm)',RH,RP,RR
    WRITE(*,*)
    GOTO 29
    ENDIF
    IF(KG2.EQ.0) THEN
    DW=W0-W*1.E3/9.80665
    DM=AM0-AM*1.E6
    DMR=BM0-BM*1.E6
    WRITE(*,21)'H1_ALFA_BETA:',H1,ALFA,BETA
    WRITE(*,21)'DW_DM_DMR:  ',DW,DM,DMR
21  FORMAT(A15,1X,3(ES13.6,3X))
    IF(ABS(DW)/W0.LT.0.01.AND.ABS(DM)/AM0.LT.0.01.AND.ABS(DMR).
LT.0.02) THEN
    DO I=1,161
    DO J=1,161
    PP0(I,J)=P(I,J)
    HH0(I,J)=H(I,J)
    ENDDO
    ENDDO
    W00=W
    AM00=AM
    BM00=BM
    ALFA=ALFA+1.0E-6
    H1=H1-(AL-XC*1.0E-3)*1.0E-6
    KL=1
    GOTO 1
    ENDIF
    IF(ABS(DW).LT.0.5.AND.ABS(DM).LT.1.0.AND.ABS(DMR).LT.0.5)THEN
    SDH=1.0E-9
    SDALFA=5.0E-6
    SDBETA=2.0E-6
    ENDIF
    IF(ABS(DW).LT.0.1.AND.ABS(DM).LT.0.1.AND.ABS(DMR).LT.0.1)THEN
```

```
      SDH=1.E-9
      SDALFA=5.0E-6
      SDBETA=5.0E-6
      ENDIF
      WRITE(*,21)'SDH_AFA_BTA',SDH,SDALFA,SDBETA
      W00=W
      AM00=AM
      BM00=BM
      KG2=1
      ENDIF
      IF(KG.EQ.0)THEN
      KG=1
      H1=H1+SDH
      GOTO 1
      ENDIF
      IF(KG.EQ.1)THEN
      KG=2
      WH=(W-W00)*1.E3/9.80665/(SDH*1.E9)
      PH=(AM-AM00)*1.E-3/SDH
      RH=(BM-BM00)*1.E-3/SDH
      ALFA=ALFA+SDALFA
      H1=H1-SDH
      GOTO 1
      ENDIF
      IF(KG.EQ.2)THEN
      KG=3
      WP=(W-W00)*1.E-3/9.80665/SDALFA
      PP=(AM-AM00)/SDALFA
      RP=(BM-BM00)/SDALFA
      BETA=BETA+SDBETA
      ALFA=ALFA-SDALFA
      GOTO 1
      ENDIF
      IF(KG.EQ.3)THEN
      WR=(W-W00)*1.E-3/9.80665/SDBETA
      PR=(AM-AM00)/SDBETA
      RR=(BM-BM00)/SDBETA
      ENDIF
      WRITE(*,*)'stiffness_Matrix:'

      WRITE(*,'(A15,3(A12,1X))')'     \      ',' H(nm) ','PITCH
(urad)','ROLL(urad)'

      WRITE(*,21) 'W(g)     ',WH,WP,WR
      WRITE(*,21) 'P-TORQUE(uNm)',PH,PP,PR
      WRITE(*,21) 'R-TORQUE(uNm)',RH,RP,RR
      BETA=BETA-SDBETA
      A(1,1)=MAX(WH,PH,RH)
      IF(ABS(A(1,1)-WH).LT.1.0E-6)THEN
      A(1,2)=WP
```

```
A(1,3)=WR
A(1,4)=DW
L21=PH/WH
A(2,1)=0.0
A(2,2)=PP-L21*WP
A(2,3)=PR-L21*WR
A(2,4)=DM-L21*DW
L31=RH/WH
A(3,1)=0.0
A(3,2)=RP-L31*WP
A(3,3)=RR-L31*WR
A(3,4)=DMR-L31*DW
ELSE IF(ABS(A(1,1)-PH).LT.1.0E-6)THEN
A(1,1)=PH
A(1,2)=PP
A(1,3)=PR
A(1,4)=DM
L21=WH/PH
A(2,1)=0.0
A(2,2)=WP-L21*PP
A(2,3)=WR-L21*PR
A(2,4)=DW-L21*DM
L31=RH/PH
A(3,1)=0.0
A(3,2)=RP-L31*PP
A(3,3)=RR-L31*PR
A(3,4)=DMR-L31*DM
ELSE
A(1,1)=RH
A(1,2)=RP
A(1,3)=RR
A(1,4)=DMR
L21=PH/RH
A(2,1)=0.0
A(2,2)=PP-L21*RP
A(2,3)=PR-L21*RR
A(2,4)=DM-L21*DMR
L31=WH/RH
A(3,1)=0.0
A(3,2)=WP-L31*RP
A(3,3)=WR-L31*RR
A(3,4)=DW-L31*DMR
ENDIF
W22=MAX(A(2,2),A(3,2))
IF(ABS(W22-A(2,2)).LT.1.0E-6)THEN
L32=A(3,2)/A(2,2)
A(3,2)=0.0
A(3,3)=A(3,3)-L32*A(2,3)
A(3,4)=A(3,4)-L32*A(2,4)
ELSE
```

```
    L32=A(2,2)/A(3,2)
    M22=A(3,2)
    M23=A(3,3)
    M24=A(3,4)
    A(3,2)=0.0
    A(3,3)=A(2,3)-L32*A(3,3)
    A(3,4)=A(2,4)-L32*A(3,4)
    A(2,2)=M22
    A(2,3)=M23
    A(2,4)=M24
    ENDIF
    DBETA=A(3,4)/A(3,3)
    DALFA=(A(2,4)-A(2,3)*DBETA)/A(2,2)
    DH=(A(1,4)-A(1,2)*DALFA-A(1,3)*DBETA)/A(1,1)
    ERRDW=ABS(WH*DH+WP*DALFA+WR*DBETA-DW)
    ERRDM=ABS(PH*DH+PP*DALFA+PR*DBETA-DM)
    ERRDMR=ABS(RH*DH+RP*DALFA+RR*DBETA-DMR)
    WRITE(*,21)' Equation_Err:   ',ERRDW,ERRDM,ERRDMR
    WRITE(*,21)'Revise_Atti_P',DH,DALFA,DBETA
    KG=0
    KG2=0
    H1=H1+DH*1.0E-9
    IF(H1.LE.0)THEN
    H1=1.E-8
    WRITE(*,*)'H1 was revised from a negative value to 10nm'
    ENDIF
    ALFA=ALFA+DALFA*1.0E-6
    IF(ALFA.LE.0)THEN
    ALFA=8.E-5
    WRITE(*,*)'ALFA was revised from a negative value to 80urad'
    ENDIF
    BETA=BETA+DBETA*1.0E-6
    WRITE(*,*)'**************'
    GOTO 1
29  X0=AM/W+XCD*AL
    Y0=BM/W+YCD*B
    PMAX=0.0
    PMIN=0.0
    IF(SKEW.LT.0)THEN
    BM=-1.*BM
    Y0=-1.*Y0
    BETA=-1.*BETA
    RH=-1.*RH
    RP=-1.*RP
    WR=-1.*WR
    PR=-1.*PR
    DO I=1,N1
    DO J=1,M1
    QPH3(I,J)=H(I,162-J)
    HHH(I,J)=P(I,162-J)
```

```
      ENDDO
      ENDDO
      DO I=1,N1
      DO J=1,M1
      H(I,J)=QPH3(I,J)
      P(I,J)=HHH(I,J)
      ENDDO
      ENDDO
      ENDIF
      DO I=1,N1
      DO J=1,M1
      P(I,J)=P(I,J)-1.0
      IF(P(I,J).GT.PMAX)PMAX=P(I,J)
      IF(P(I,J).LT.PMIN)PMIN=P(I,J)
      H(I,J)=H1*H(I,J)
      ENDDO
      ENDDO
      OPEN(8,FILE='FILM.DAT',STATUS='UNKNOWN')
      OPEN(9,FILE='PRESSURE.DAT',STATUS='UNKNOWN')
      WRITE(8,30)Y(1),(Y(J),J=1,M1)
30    FORMAT(162(E12.6,1X))
      WRITE(8,40)(X(I),(H(I,J),J=1,M1),I=1,N1)
40    FORMAT(162(E12.6,1X))
      WRITE(9,30)Y(1),(Y(J),J=1,M1)
      WRITE(9,40)(X(I),(P(I,J),J=1,M1),I=1,N1)
50    FORMAT(161(E12.6,1X))
      H1=H1*1.0E9
      ALFA=ALFA*1.0E6
      BETA=BETA*1.0E6
      W=W*1.0E3/9.80665
      AM=AM*1.0E6
      BM=BM*1.0E6
      X0=X0*1.0E3
      Y0=Y0*1.0E3
      XHR=XHR*1.0e3*AL
      YHR=YHR*1.0e3*B
51    FORMAT(30(G12.6,1X))
      OPEN(10,FILE='RESULT.DAT',STATUS='UNKNOWN')
      WRITE(10,*)RPM,CR,SKEW,U,XC,YC,W0,AM0,BM0,PMAX,PMIN,W,AM,BM,X0,
Y0,H1,XHR,YHR,ALFA,BETA,WH,WP,WR,PH,PP,PR,RH,RP,RR
      STOP
      END
```

24.2.4 Calculation Results

It may take several hours at a PC computer (current general configuration) to find the equilibrium position finally because of the number of iterations. A window screenshot is shown as Figure 24.2, which gives the information for the calculation. The film thickness and pressure distribution are shown in Figure 24.3. The pressure

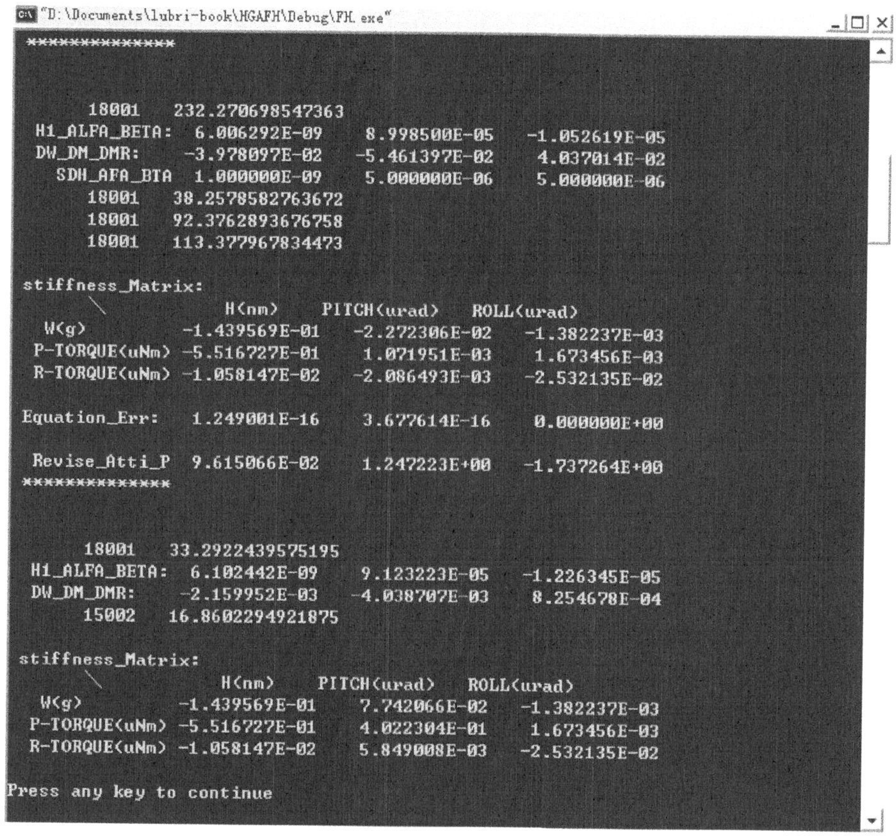

Figure 24.2 Result data from screenshot.

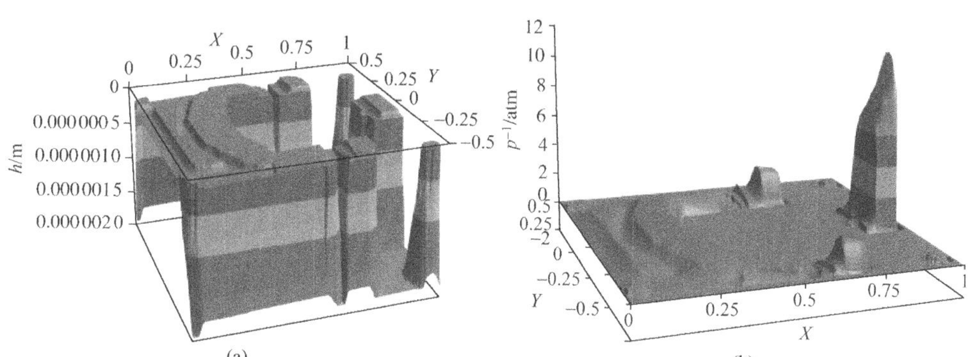

Figure 24.3 Results of (a) film thickness and (b) pressure distribution.

distribution shows that there is an asymmetrical pattern at the trail pad due to the asymmetrical structure of the ABS and the skew angle (−14.923°).

The preassigned values are reference height $h_0 = 6$ nm, pitch angle $\alpha = 50 \times 10^{-6}$ rad and roll angle $\beta = 0$, while the equilibrium position is $h_0 = 6.10$ nm, $\alpha = 91.23 \times 10^{-6}$ rad and $\beta = 12.26 \times 10^{-6}$ rad, respectively. The forces are load $w = 1.500$ g, pitch moment $M_\alpha = 0.8655$ μNm and roll moment $M_\beta = 6.687 \times 10^{-5}$ μNm. They are very close, even equal to the forces imposed by the suspension, which are 1.5 g, 0.867 μNm and 0, respectively. This proves the success of searching for the equilibrium flying attitude.

References

Fukui, S. and Kaneko, R. (1988) Analysis of ultrathin gas film lubrication based on linearized Boltzmann equation: first report – derivation of a generalized lubrication equation including thermal creep flow. *ASME J. Tribol.*, **11**, 253–262.

Fukui, S. (1990) A database interpolation of Poiseuille flow rates for high Knudsen number lubrication problems. *ASME J. Tribol.*, **112**, 78–83.

Wen, S. and Huang, P. (2011) *Principles of Tribology*, Wiley and Tsinghua, Beijing.

Yang, P. (1998) *Numerical Fluid Lubrication Analysis*, National Defense Industrial Press, Beijing.

Numerical Calculation of Lubrication: Methods and Programs, First Edition. Ping Huang.

Index

Numerical Calculation of Lubrication: Methods and Programs, First Edition. Ping Huang.

Printed and bound by CPI Group (UK) Ltd, Croydon, CR0 4YY
16/09/2026
14963390-0001